Problems and Solutions for Undergraduate Real Analysis

by Kit-Wing Yu, PhD

kitwing@hotmail.com

Copyright © 2020 by Kit-Wing Yu. All rights reserved. No part of this publication may be reproduced, stored in a retrieval system, or transmitted, in any form or by any means, electronic, mechanical, photocopying, recording, or otherwise, without the prior written permission of the author.

ISBN: 978-988-74155-2-7 (eBook)
ISBN: 978-988-74155-3-4 (Paperback)

About the author

Dr. Kit-Wing Yu received his B.Sc. (1st Hons), M.Phil. and Ph.D. degrees in Math. at the HKUST, PGDE (Mathematics) at the CUHK. After his graduation, he has joined United Christian College to serve as a mathematics teacher for at least seventeen years. He has also taken the responsibility of the mathematics panel since 2002. Furthermore, he was appointed as a part-time tutor (2002 – 2005) and then a part-time course coordinator (2006 – 2010) of the Department of Mathematics at the OUHK.

Apart from teaching, Dr. Yu has been appointed to be a marker of the HKAL Pure Mathematics and HKDSE Mathematics (Core Part) for over thirteen years. Between 2012 and 2014, Dr. Yu was invited to be a Judge Member by the World Olympic Mathematics Competition (China). In the area of academic publication, he is the author of the following six books

- *A Complete Solution Guide to Complex Analysis.*
- *Problems and Solutions for Undergraduate Real Analysis II.*
- *A Complete Solution Guide to Real and Complex Analysis I.*
- *Problems and Solutions for Undergraduate Real Analysis I.*
- *Mock Tests for the ACT Mathematics.*
- *A Complete Solution Guide to Principles of Mathematical Analysis.*

Besides, he has published over twelve research papers in international mathematical journals, including some well-known journals such as J. Reine Angew. Math., Proc. Roy. Soc. Edinburgh Sect. A and Kodai Math. J.. His research interests are inequalities, special functions and Nevanlinna's value distribution theory.

Preface

The present book *Problems and Solutions for Undergraduate Real Analysis* is the combined volume of author's two books [30] and [31]. By offering 456 exercises with different levels of difficulty, this book gives a brief exposition of the foundations of first-year undergraduate real analysis. Furthermore, we believe that students and instructors may find that the book can also be served as a source for some advanced courses or as a reference.

The wide variety of problems, which are of varying difficulty, include the following topics: (1) Elementary Set Algebra, (2) The Real Number System, (3) Countable and Uncountable Sets, (4) Elementary Topology on Metric Spaces, (5) Sequences in Metric Spaces, (6) Series of Numbers, (7) Limits and Continuity of Functions, (8) Differentiation, (9) The Riemann-Stieltjes Integral, (10) Sequences and Series of Functions, (11) Improper Integrals, (12) Lebesgue Measure, (13) Lebesgue Measurable Functions, (14) Lebesgue Integration, (15) Differential Calculus of Functions of Several Variables and (16) Integral Calculus of Functions of Several Variables. Furthermore, the main features of this book are listed as follows:

- The book contains 456 problems of undergraduate real analysis, which cover the topics mentioned above, with *detailed* and *complete* solutions. In fact, the solutions show every detail, every step and every theorem that I applied.

- Each chapter starts with a brief and concise note of introducing the notations, terminologies, basic mathematical concepts or important/famous/frequently used theorems (without proofs) relevant to the topic. As a consequence, students can use these notes as a quick review before midterms or examinations.

- Problems are classified as three levels of difficulty. The classification is as follows:

Symbol	Level of difficulty	Meaning
⋆	Introductory	These problems are basic and every student must be familiar with them.
⋆ ⋆	Intermediate	The depth and the complexity of the problems increase. Students who targets for higher grades must study them.
⋆ ⋆ ⋆	Advanced	These problems are very difficult and they may need some specific skills.

- Different colors are used frequently in order to highlight or explain problems, examples, remarks, main points/formulas involved, or show the steps of manipulation in some complicated proofs. (ebook only)

- An appendix about mathematical logic is included. It tells students what concepts of logic (e.g. techniques of proofs) are necessary in advanced mathematics. If you are familiar with these, you may skip it. Otherwise, you are strongly recommended to spend time to read at least §A.3 to §A.5.

Finally, if you find such any typos or mistakes, please send your valuable comments or opinions to

kitwing@hotmail.com

so that I will post the updated errata on my website

https://sites.google.com/view/yukitwing/

from time to time.

Kit-Wing Yu
February 2020

List of Figures

10.1 An example of pointwise convergence. 188
10.2 An example of uniform convergence. 188

15.1 The Inverse Function Theorem. 331

16.1 The subinterval in $\mathbb{R}^2$. 354
16.2 The outer and the inner Jordan measures. 357
16.3 The mapping $\phi : E \to D$. 375
16.4 The mapping $\phi : E \to R$. 376

List of Tables

4.1	Properties of the interior and the closure of E in X.	29
A.1	The truth table of $p \wedge q$.	384
A.2	The truth table of $p \vee q$.	384
A.3	The truth table of $\sim p$.	385
A.4	The truth table of $p \rightarrow q$.	385
A.5	The truth table of $p \rightarrow q$.	385
A.6	The truth table of $p \rightarrow q$ and $\sim q \rightarrow \sim p$.	386
A.7	The tautology $(p \vee q) \vee (\sim p)$	386
A.8	The contradiction $(p \wedge q) \wedge (\sim p)$	386

Contents

Preface	v
List of Figures	vii
List of Tables	ix

1 Elementary Set Algebra — 1
  1.1 Fundamental Concepts 1
  1.2 Sets, Functions and Relations 4
  1.3 Mathematical Induction 6

2 The Real Number System — 9
  2.1 Fundamental Concepts 9
  2.2 Rational and Irrational Numbers 10
  2.3 Absolute Values 12
  2.4 The Completeness Axiom 13

3 Countable and Uncountable Sets — 19
  3.1 Fundamental Concepts 19
  3.2 Problems on Countable and Uncountable Sets 20

4 Elementary Topology on Metric Spaces — 27
  4.1 Fundamental Concepts 27
  4.2 Open Sets and Closed Sets 31
  4.3 Compact Sets 38
  4.4 The Heine-Borel Theorem 44
  4.5 Connected Sets 45

5 Sequences in Metric Spaces — 49
  5.1 Fundamental Concepts 49

5.2	Convergence of Sequences	53
5.3	Upper and Lower Limits	59
5.4	Cauchy Sequences and Complete Metric Spaces	65
5.5	Recurrence Relations	70

6 Series of Numbers   75

6.1	Fundamental Concepts	75
6.2	Convergence of Series of Nonnegative Terms	79
6.3	Alternating Series and Absolute Convergence	87
6.4	The Series $\Sigma_{n=1}^{\infty} a_n b_n$ and Multiplication of Series	90
6.5	Power Series	94

7 Limits and Continuity of Functions   97

7.1	Fundamental Concepts	97
7.2	Limits of Functions	103
7.3	Continuity and Uniform Continuity of Functions	108
7.4	The Extreme Value Theorem and the Intermediate Value Theorem	116
7.5	Discontinuity of Functions	120
7.6	Monotonic Functions	122

8 Differentiation   127

8.1	Fundamental Concepts	127
8.2	Properties of Derivatives	132
8.3	The Mean Value Theorem for Derivatives	138
8.4	L'Hôspital's Rule	146
8.5	Higher Order Derivatives and Taylor's Theorem	149
8.6	Convexity and Derivatives	153

9 The Riemann-Stieltjes Integral   157

9.1	Fundamental Concepts	157
9.2	Integrability of Real Functions	162
9.3	Applications of Integration Theorems	170
9.4	The Mean Value Theorems for Integrals	182

10 Sequences and Series of Functions   187

10.1	Fundamental Concepts	187
10.2	Uniform Convergence for Sequences of Functions	192
10.3	Uniform Convergence for Series of Functions	201

| Contents | xiii |

    10.4 Equicontinuous Families of Functions 209
    10.5 Approximation by Polynomials 214

11 Improper Integrals   219
    11.1 Fundamental Concepts 219
    11.2 Evaluations of Improper Integrals 223
    11.3 Convergence of Improper Integrals 228
    11.4 Miscellaneous Problems on Improper Integrals 236

12 Lebesgue Measure   243
    12.1 Fundamental Concepts 243
    12.2 Lebesgue Outer Measure 247
    12.3 Lebesgue Measurable Sets 251
    12.4 Necessary and Sufficient Conditions for Measurable Sets 264

13 Lebesgue Measurable Functions   269
    13.1 Fundamental Concepts 269
    13.2 Lebesgue Measurable Functions 271
    13.3 Applications of Littlewood's Three Principles 282

14 Lebesgue Integration   291
    14.1 Fundamental Concepts 291
    14.2 Properties of Integrable Functions 295
    14.3 Applications of Fatou's Lemma 309
    14.4 Applications of Convergence Theorems 315

15 Differential Calculus of Functions of Several Variables   327
    15.1 Fundamental Concepts 327
    15.2 Differentiation of Functions of Several Variables 333
    15.3 The Mean Value Theorem for Differentiable Functions 341
    15.4 The Inverse Function Theorem and the Implicit Function Theorem . . 343
    15.5 Higher Order Derivatives 347

16 Integral Calculus of Functions of Several Variables   353
    16.1 Fundamental Concepts 353
    16.2 Jordan Measurable Sets 360
    16.3 Integration on $\mathbb{R}^n$ 364
    16.4 Applications of the Mean Value Theorem 372
    16.5 Applications of the Change of Variables Theorem 375

Contents

Appendix — 382

A Language of Mathematics — 383
- A.1 Fundamental Concepts 383
- A.2 Statements and Logical Connectives 384
- A.3 Quantifiers and their Basic Properties 386
- A.4 Necessity and Sufficiency 387
- A.5 Techniques of Proofs 388

Index — 391

Bibliography — 397

CHAPTER 1

Elementary Set Algebra

1.1 Fundamental Concepts

In this section, we briefly record some basic properties of sets, functions, equivalence relations and order relations. For detailed introduction to these topics, the reader can read [16, §1 - §4, pp. 4 - 36], [27, Chap. 3] and [33, §1.2 - §1.4].

1.1.1 Sets

Capital letters $E, F, \ldots$ are usually used to represent **sets** and lowercase letters $x, y, \ldots$ refer to elements of sets. The notation $x \in E$ means that x belongs to the set E. Similarly, the notation $x \notin E$ means that x does not belong to E.

Several sets are common in analysis. They are $\mathbb{N}, \mathbb{Z}, \mathbb{Q}$ and $\mathbb{R}$ which are called positive integers, integers, rational numbers and real numbers respectively. Furthermore, the set $\mathbb{R} \setminus \mathbb{Q}$ is the set of all irrational numbers. If a "+"sign (resp. "−" sign) is put in the superscript of each of the above set (except $\mathbb{N}$), then the new set takes only the positive (resp. negative) part of the base set. For instance, $\mathbb{R}^+$ means the set of all positive real numbers.

1.1.2 Basic Operations with Sets

Two sets E and F are said to be **equal**, namely $E = F$, if they consist of precisely the same elements. A set E is called a **subset** of F, in symbols $E \subseteq F$, if every element of E is an element of F. If $E \subseteq F$ but $E \neq F$, then E is called a **proper subset** of F. The set without any elements is called the **empty set** which is denoted by $\varnothing$.

Suppose that E and F are subsets of a set S. Then we define their **union**, their **intersection** and their **difference** as follows:

- **Union:** $E \cup F = \{x \in S \,|\, x \in E \text{ or } x \in F\}$.

- **Intersection:** $E \cap F = \{x \in S \,|\, x \in E \text{ and } x \in F\}$.

- **Difference:** $E \setminus F = \{x \in S \,|\, x \in E \text{ and } x \notin F\}$.

In particular, we call
$$E^c = S \setminus E$$
the **complement** of E. The following theorem states some basic operations of union, intersection and difference of sets:

Theorem 1.1. *Suppose that A, B and C are sets. Then we have*

- $(A \cup B) \cap C = (A \cap C) \cup (B \cap C)$;

- $(A \cap B) \cup C = (A \cup C) \cap (B \cup C)$;

- $(A \cup B) \setminus C = (A \setminus C) \cup (B \setminus C)$;

- $(A \cap B) \setminus C = (A \setminus C) \cap (B \setminus C)$.

1.1.3 Functions

Given two sets X and Y, we define their **Cartesian product** $X \times Y$ to be the set of all **ordered pairs** (x, y), where $x \in X$ and $y \in Y$. Formally, we have

$$X \times Y = \{(x, y) \,|\, x \in X \text{ and } y \in Y\}.$$

By a **function** (or **mapping**) f from X to Y, in symbols $f : X \to Y$, we mean that a specific "rule" that assigns to each element x of the set X a unique element y in Y. More precisely,

Definition 1.2. *[16, p. 15] A **rule of assignment** is a subset r of the Cartesian product $X \times Y$ such that each element of X appears as the first coordinate of at most one ordered pair in r.*

Now the element y is called the **value** of f at x and it is expressed as $y = f(x)$. The set X is called the **domain** of f and the set

$$f(X) = \{f(x) \in Y \,|\, x \in X\}$$

is called the **range** of f. A function $f : X \to Y$ is called **injective** (or **one-to-one**) if $x_1 \neq x_2$ implies that $f(x_1) \neq f(x_2)$. Besides, f is said to be **surjective** (or **onto**) if $f(X) = Y$. In particular, a function f is **bijective** if it is both injective and surjective.

If $E \subseteq Y$, then we define
$$f^{-1}(E) = \{x \in X \,|\, f(x) \in E\}.$$

We call $f^{-1}(E)$ the **inverse image** of E under f. In particular, if $E = \{y\}$, then we have

$$f^{-1}(y) = \{x \in X \,|\, f(x) = y\}.$$

Definition 1.3. *Given functions $f : X \to Y$ and $g : Y \to X$. Their **composition** $g \circ f$ is the function $g \circ f : X \to Z$ defined by*

$$(g \circ f)(x) = g(f(x))$$

for each $x \in X$.

1.1.4 Equivalence Relations

Definition 1.4. *Given a set X. By a **relation** on a set X, we mean a subset $\mathcal{R}$ of the Cartesian product $X \times X$. If $(x, y) \in \mathcal{R}$, then x is said to be in the relation $\mathcal{R}$ with y. We use the notation $x\mathcal{R}y$ to mean this situation.*

One of the most interesting and important relations is the **equivalence relation**. In fact, a relation $\mathcal{R}$ on a set X is said to be an **equivalence relation** if it satisfies the following three conditions:

- **(1) Reflexivity:** $x\mathcal{R}x$ for every $x \in X$.
- **(2) Symmetry:** If $x\mathcal{R}y$, then $y\mathcal{R}x$.
- **(3) Transitivity:** If $x\mathcal{R}y$ and $y\mathcal{R}z$, then $x\mathcal{R}z$.

To make things simple, we will use the notation "$\sim$" to replace the letter "$\mathcal{R}$" in an equivalence relation.

Given an equivalence relation $\sim$ on a set X. Pick $x \in X$. We define a subset $E \subseteq X$, called the **equivalence class** determined by x, by

$$E_x = \{y \in X \mid y \sim x\}.$$

A basic fact about equivalence classes is the following result:

Theorem 1.5. *Two equivalence classes E_x and E_y are either disjoint or coincide.*

By this concept, we know that the family of equivalence classes form a **partition** of the set X.

1.1.5 Order Relations

Besides equivalence relations, **order relations** are another important types of relations. In fact, a relation "$<$" (means "less than") is called an order relation if it satisfies the following properties:

- **(1) Comparability:** If $x \neq y$, then either $x < y$ or $y < x$.
- **(2) Nonreflexivity:** If $x < y$, then $x \neq y$.
- **(3) Transitivity:** If $x < y$ and $y < z$, then $x < z$.

With the order relation "$<$", concepts of inequality, interval, (upper and lower) bound, supremum and infimum of real numbers can be developed.

1.2 Sets, Functions and Relations

Problem 1.1

(⋆) Suppose that E and F are subsets of a set S. Let E^c be the complement of E in S. Prove De Morgan's laws: $(E \cap F)^c = E^c \cup F^c$ and $(E \cup F)^c = E^c \cap F^c$.

Proof. Suppose that $x \in (E \cap F)^c$. Then it must be the case that $x \notin E$ or $x \notin F$ which mean $x \in E^c$ and $x \in F^c$, i.e.,

$$(E \cap F)^c \subseteq E^c \cup F^c. \tag{1.1}$$

Conversely, if $x \in E^c \cup F^c$, then we have $x \notin E$ or $x \notin F$. Thus we have $x \notin E \cap F$, i.e., $x \in (E \cap F)^c$ and then

$$E^c \cup F^c \subseteq (E \cap F)^c. \tag{1.2}$$

Hence the first identity follows from the set relations (1.1) and (1.2).

Since $(E^c)^c = E$, the second identity follows immediate from the first identity if we replace E and F by E^c and F^c respectively. This finishes the proof of the problem. ∎

Problem 1.2

(⋆) Suppose that E and F are subsets of a set S. Prove that $E \setminus F = E \cap F^c$.

Proof. Suppose that $x \in E \setminus F$. Then it is equivalent to the statement $x \in E$ and $x \notin F$ or $x \in E$ and $x \in F^c$. Finally, the last statement is equivalent to $x \in E \cap F^c$. This proves

$$E \setminus F \subseteq E \cap F^c. \tag{1.3}$$

Conversely, let $x \in E \cap F^c$. Then we have $x \in E$ and $x \in F^c$, so $x \in E$ and $x \notin F$. Therefore, it means that $x \in E \setminus F$, i.e.,

$$E \cap F^c \subseteq E \setminus F. \tag{1.4}$$

Now the identity follows from the set relations (1.3) and (1.4). This completes the proof of the problem. ∎

Problem 1.3

(⋆) Let E and F be subsets of S. Prove that $(E \setminus F) \cup F = E$ if and only if $F \subseteq E$.

Proof. By Problem 1.2, we know from the fact $S = F \cup F^c$ that

$$(E \setminus F) \cup F = (E \cap F^c) \cup F = (E \cup F) \cap (F^c \cup F) = (E \cup F) \cap S = E \cup F.$$

Thus $(E \setminus F) \cup F = E$ if and only if $E = E \cup F$ if and only if $F \subseteq E$. This completes the proof of the problem. ∎

1.2. Sets, Functions and Relations

Problem 1.4

(★) Let I, X and Y be sets. Let $f : X \to Y$ be a function and $E_i \subseteq X$ for $i \in I$. Show that

$$f\Big(\bigcup_{i \in I} E_i\Big) = \bigcup_{i \in I} f(E_i) \quad \text{and} \quad f\Big(\bigcap_{i \in I} E_i\Big) \subseteq \bigcap_{i \in I} f(E_i). \tag{1.5}$$

Show, by an example, that the inclusion of the set relation in (1.5) can be proper.

Proof. Let $y \in f\Big(\bigcup_{i \in I} E_i\Big)$. Then there exists $x \in \bigcup_{i \in I} E_i$ such that $f(x) = y$. In other words, there exists $x \in E_i$ such that $f(x) = y$ *for some* $i \in I$ which imply that $y \in f(E_i)$ for some $i \in I$, i.e., $y \in \bigcup_{i \in I} f(E_i)$ or

$$f\Big(\bigcup_{i \in I} E_i\Big) \subseteq \bigcup_{i \in I} f(E_i). \tag{1.6}$$

Conversely, if $y \in \bigcup_{i \in I} f(E_i)$, then $y \in f(E_i)$ *for some* $i \in I$ so that $y = f(x)$ for some $x \in E_i \subseteq \bigcup_{i \in I} E_i$. Thus we have $y \in f\Big(\bigcup_{i \in I} E_i\Big)$ which means

$$\bigcup_{i \in I} f(E_i) \subseteq f\Big(\bigcup_{i \in I} E_i\Big). \tag{1.7}$$

Now the first assertion follows from the set relations (1.6) and (1.7).

For the second assertion, since we always have $\bigcap_{i \in I} E_i \subseteq E_i$ *for every* $i \in I$, we must have $f\Big(\bigcap_{i \in I} E_i\Big) \subseteq f(E_i)$ for every $i \in I$ and this proves that

$$f\Big(\bigcap_{i \in I} E_i\Big) \subseteq \bigcap_{i \in I} f(E_i).$$

To show that the inclusion can be proper, let us consider $E_1 = \{1\}, E_2 = \{2\}, X = \{1, 2\}$ and $Y = \mathbb{R}$. Suppose that $f : X \to \mathbb{R}$ is given by $f(x) = 1$ for $x \in X$. Since $f(E_1) = f(E_2) = \{1\}$, we have $f(E_1) \cap f(E_2) = \{1\}$. However, we know that $E_1 \cap E_2 = \varnothing$ which means that $f(E_1 \cap E_2) = \varnothing$. In this case, we have[a]

$$f(E_1 \cap E_2) \subset f(E_1) \cap f(E_2).$$

This ends the proof of the problem. ∎

Problem 1.5

(★) Let X and Y be sets. Let $f : X \to Y$ be a function and $E \subseteq X$. Prove that

$$f^{-1}(E^c) = [f^{-1}(E)]^c. \tag{1.8}$$

---

[a] Here we use the fact that the empty set is a subset of every set, see [23, Chap. 2, Exercise 1, p. 43].

Proof. Let $x \in f^{-1}(E^c)$. Then $f(x) \in E^c$ or $f(x) \notin E^c$. In other words, we have $x \notin f^{-1}(E)$ and so $x \in [f^{-1}(E)]^c$, i.e.,
$$f^{-1}(E^c) \subseteq [f^{-1}(E)]^c. \tag{1.9}$$
Conversely, if $x \in [f^{-1}(E)]^c$, then $x \notin f^{-1}(E)$ so that $f(x) \notin E$. Therefore, we have $f(x) \in E^c$ and this implies that $x \in f^{-1}(E^c)$, i.e.,
$$[f^{-1}(E)]^c \subseteq f^{-1}(E^c). \tag{1.10}$$
Combining the set relations (1.9) and (1.10), we have the desired result (1.8), completing the proof of the problem. ∎

Problem 1.6

(⋆) Suppose that $f : X \to Y$, $g : Y \to Z$ and $E \subseteq Z$. Prove that
$$(g \circ f)^{-1}(E) = f^{-1}(g^{-1}(E)).$$

Proof. We note that $x \in (g \circ f)^{-1}(E)$ if and only if $g(f(x)) \in E$ if and only if $f(x) \in g^{-1}(E)$ which is obviously equivalent to $x \in f^{-1}(g^{-1}(E))$. Hence the identity holds and we complete the proof of the problem. ∎

Problem 1.7

(⋆) Let p be a prime. For every $x, y \in \mathbb{Z}$, define the relation $x \equiv_p y$ to be such that $x - y$ is divisible by p. Prove that $\equiv_p$ is an equivalence relation.

Proof. For every $x \in \mathbb{Z}$, since $x - x = 0$ is divisible by p, we have $x \equiv_p x$. If $x - y$ is divisible by p, then so is $y - x$ because $y - x = -(x - y)$. Thus we have $y \equiv_p x$ if $x \equiv_p y$. Finally, if $x - y = pM$ and $z - y = pN$ for some integers M and N, then we have
$$z - x = (z - y) + (y - x) = pN + pM = p(M + N).$$
In other words, we have $z \equiv_p x$ whenever $x \equiv_p y$ and $y \equiv_p z$. In conclusion, $\equiv_p$ is an equivalence relation and this completes the proof of the problem. ∎

1.3 Mathematical Induction

Problem 1.8

(⋆) Prove that
$$\frac{1}{2} \times \frac{3}{4} \times \cdots \times \frac{2n-1}{2n} < \frac{1}{\sqrt{2n+1}}$$
for every positive integer n.

1.3. Mathematical Induction

Proof. When $n = 1$, since $\frac{1}{2} < \frac{1}{\sqrt{3}}$, the inequality holds in this case. Assume that

$$\frac{1}{2} \times \frac{3}{4} \times \cdots \times \frac{2k-1}{2k} < \frac{1}{\sqrt{2k+1}}$$

for some positive integer k. When $n = k+1$, we have

$$\frac{1}{2} \times \frac{3}{4} \times \cdots \times \frac{2k-1}{2k} \times \frac{2k+1}{2k+2} < \frac{1}{\sqrt{2k+1}} \times \frac{2k+1}{2k+2} = \frac{\sqrt{2k+1}}{2k+2}. \tag{1.11}$$

Since $4k^2 + 8k + 3 < 4k^2 + 8k + 4$, we have $(2k+1)(2k+3) < (2k+2)^2$ which implies that

$$\frac{\sqrt{2k+1}}{2k+2} < \frac{1}{\sqrt{2k+3}}. \tag{1.12}$$

Combining the inequalities (1.11) and (1.12), we obtain

$$\frac{1}{2} \times \frac{3}{4} \times \cdots \times \frac{2k-1}{2k} \times \frac{2k+1}{2k+2} < \frac{1}{\sqrt{2k+3}}.$$

Hence the inequality is true for $n = k+1$ when it is true for $n = k$. By induction, we know that the inequality holds for all positive integers n. We end the proof of the problem. ∎

Problem 1.9

(⋆) *Use induction to prove Bernoulli's inequality*

$$(1+x_1)(1+x_2)\cdots(1+x_n) \geq 1 + x_1 + \cdots + x_n,$$

where $x_1, x_2, \ldots, x_n$ have the same sign and $x_1, x_2, \ldots, x_n \geq -1$.

Proof. It is obvious that the statement is true for $n = 1$. Assume that

$$(1+x_1)(1+x_2)\cdots(1+x_k) \geq 1 + x_1 + \cdots + x_k \tag{1.13}$$

for some positive integer k. When $n = k+1$, since $1 + x_{k+1} > 0$, we deduce from the assumption (1.13) that

$$(1+x_1)(1+x_2)\cdots(1+x_k)(1+x_{k+1})$$
$$\geq (1 + x_1 + \cdots + x_k)(1 + x_{k+1})$$
$$= 1 + x_1 + \cdots + x_k + x_{k+1} + (x_1 x_{k+1} + \cdots + x_k x_{k+1}). \tag{1.14}$$

Since $x_1, x_2, \ldots, x_{k+1}$ have the same sign, we have $x_i x_{k+1} \geq 0$ for all $i = 1, \ldots, k$. Therefore, we follow from this that $x_1 x_{k+1} + \cdots + x_k x_{k+1} \geq 0$ and then we yield from the inequality (1.14) that

$$(1+x_1)(1+x_2)\cdots(1+x_k)(1+x_{k+1}) \geq 1 + x_1 + \cdots + x_k + x_{k+1}.$$

Hence the statement is true for $n = k+1$ when it is true for $n = k$. By induction, we know that the inequality holds for all positive integers n. We complete the proof of the problem. ∎

> **Problem 1.10**
>
> (★)(★) The Fibonacci numbers $\{F_n\}$ are defined by $F_1 = F_2 = 1$ and $F_{n+2} = F_{n+1} + F_n$, where $n \in \mathbb{N}$. Verify that
>
> $$F_n = \frac{1}{2^n\sqrt{5}}[(1+\sqrt{5})^n - (1-\sqrt{5})^n] \qquad (1.15)$$
>
> for every $n \in \mathbb{N}$.

Proof. It is easy to show that the formula (1.15) holds for $n = 1$ and $n = 2$. Assume that

$$F_k = \frac{1}{2^k\sqrt{5}}[(1+\sqrt{5})^k - (1-\sqrt{5})^k] \quad \text{and} \quad F_{k+1} = \frac{1}{2^{k+1}\sqrt{5}}[(1+\sqrt{5})^{k+1} - (1-\sqrt{5})^{k+1}]$$

for some positive integer k. For $n = k+2$, we obtain from the definition that

$$\begin{aligned}
F_{k+2} &= F_{k+1} + F_k \\
&= \frac{1}{2^k\sqrt{5}}[(1+\sqrt{5})^k - (1-\sqrt{5})^k] + \frac{1}{2^{k+1}\sqrt{5}}[(1+\sqrt{5})^{k+1} - (1-\sqrt{5})^{k+1}] \\
&= \frac{1}{2^{k+1}\sqrt{5}}[2(1+\sqrt{5})^k - 2(1-\sqrt{5})^k + (1+\sqrt{5})^{k+1} - (1-\sqrt{5})^{k+1}] \\
&= \frac{1}{2^{k+1}\sqrt{5}}[(1+\sqrt{5})^k(3+\sqrt{5}) - (1-\sqrt{5})^k(3-\sqrt{5})] \\
&= \frac{1}{2^{k+1}\sqrt{5}}\left[(1+\sqrt{5})^k \times \frac{(1+\sqrt{5})^2}{2} - (1-\sqrt{5})^k \times \frac{(1-\sqrt{5})^2}{2}\right] \\
&= \frac{1}{2^{k+2}\sqrt{5}}[(1+\sqrt{5})^{k+2} - (1-\sqrt{5})^{k+2}].
\end{aligned}$$

Hence the statement is true for $n = k+2$ when it is true for $n = k+1$ and $n = k$. By induction, we know that the formula (1.15) holds for all positive integers n. We complete the proof of the problem. ∎

CHAPTER 2

The Real Number System

2.1 Fundamental Concepts

The set of real numbers include all the rational numbers (e.g. the integer -4 and the fraction $\frac{1}{2}$) and all the irrational numbers (e.g. $\sqrt{3}$). Besides, a real number can also be thought of as a point lying on an infinitely long line called the **real number line**. In this section, fundamental properties regarding real numbers are reviewed and the main references here are [5, §2.2 - §2.4], [6, Chap. 1] [23, §1.12 - §1.22] and [27, Chap. 4 and 8].

2.1.1 Rational Numbers, Irrational Numbers and Real Numbers

The set of all rational numbers $\mathbb{Q}$ is defined by
$$\mathbb{Q} = \left\{ x \in \mathbb{R} \,\middle|\, x = \tfrac{p}{q}, \text{ where } p, q \in \mathbb{Z} \text{ and } q \neq 0 \right\}.$$
Recall that the set of all irrational numbers is denoted by $\mathbb{R} \setminus \mathbb{Q}$.

2.1.2 Absolute Values

Given $a \in \mathbb{R}$, the **absolute value** of a is defined by
$$|a| = \begin{cases} a, & \text{if } a \geq 0; \\ -a, & \text{if } a < 0. \end{cases}$$
It has the following useful properties:

- For every $a \in \mathbb{R}$, we have $|a| \geq 0$.
- $|a| = 0$ if and only if $a = 0$.
- For every $a, b \in \mathbb{R}$, we have $|a \times b| = |a| \times |b|$.
- $|a + b| \leq |a| + |b|$.
- $|a - b|$ can be treated as the **distance** between the two real numbers a and b.

2.1.3 The Completeness Axiom

What axioms "characterize" the set $\mathbb{R}$? In fact, there are three axioms which characterize $\mathbb{R}$. They are the "field axioms", the "order axioms" and the "completeness axiom". For details of the first two axioms, please refer to [23, §1.12 - §1.18, pp. 5 - 8]. The last axiom (completeness axiom) is a bit nontrivial:

The Completeness Axiom. *[3, §1.10] Every nonempty subset of $\mathbb{R}$ that is bounded above has a least upper bound (or a supremum) in $\mathbb{R}$.*

There are two useful results about $\mathbb{R}$ which can be deduced from the Completeness Axiom. The first one is known as the **Archimedean Property**:[a]

Theorem 2.1 (The Archimedean Property). *If $x, y \in \mathbb{R}$ and $x > 0$, then there exists a $n \in \mathbb{N}$ such that $nx > y$.*

Another important result is about the *density* of $\mathbb{Q}$ in $\mathbb{R}$.

Theorem 2.2 (Density of Rationals). *If $x, y \in \mathbb{R}$ and $x < y$, then there exists a $r \in \mathbb{Q}$ such that $x < r < y$, i.e.,*
$$\mathbb{Q} \cap (x, y) \neq \varnothing.$$

2.2 Rational and Irrational Numbers

Problem 2.1

(★) *Show that $\sqrt{7}$ is irrational.*

Proof. Assume that $\sqrt{7}$ was rational. We write $\sqrt{7} = \frac{m}{n}$, where m and n are integers, $n \neq 0$ and they have no common divisor other than 1. Then we have
$$7n^2 = m^2. \tag{2.1}$$

- **Case (1):** n **is even.** Then $7n^2$ is also even. By the equation (2.1) and this fact, we have m^2 and thus m is even. However, this contradicts to the fact that m and n have no common divisor other than 1.

- **Case (2):** n **is odd.** In this case, $7n^2$ is odd and the equation (2.1) shows that m^2 and then m is also odd. Let $m = 2p + 1$ and $n = 2q + 1$ for some integers p and q. Now the equation (2.1) becomes
$$28q^2 + 28q + 7 = 4p^2 + 4p + 1. \tag{2.2}$$

After simplification, the equation (2.2) reduces to
$$14q^2 + 14q + 3 = 2p^2 + 2p. \tag{2.3}$$

We note that $14p^2 + 14p + 3$ is always odd, but $2p^2 + 2p$ is even. Therefore, the equation (2.3) cannot hold.

---

[a] For proofs of Theorems 2.1 and 2.2, please read [23, Theorem 1.20].

2.2. Rational and Irrational Numbers

Hence we conclude from these that $\sqrt{7}$ is irrational, completing the proof of the problem. ∎

Problem 2.2

(⋆) Prove that $\sqrt{7}+\sqrt{2}$ is irrational.

Proof. Assume that $\sqrt{7}+\sqrt{2}$ was rational. By the fact that
$$(\sqrt{7}+\sqrt{2})(\sqrt{7}-\sqrt{2}) = 5,$$
the number $\sqrt{7}-\sqrt{2}$ is also rational. Thus the number $\sqrt{7}$ is also rational which contradicts Problem 2.1. Hence $\sqrt{7}+\sqrt{2}$ is irrational and this completes the proof of the problem. ∎

Problem 2.3

(⋆)(⋆) Suppose that n is a positive integer and $\theta = \sqrt{n} \in \mathbb{R}^+ \setminus \mathbb{Q}$. Prove that there exists a positive constant A such that
$$|p\theta - q| > \frac{A}{p} \qquad (2.4)$$
for all integers p and q such that $p > 0$.

Proof. Let p and q be integers such that $p > 0$. Note that
$$(p\theta - q)(-p\theta - q) = -p^2\theta^2 + q^2 = -p^2 n + q^2$$
is an integer. Furthermore, $-p^2 n + q^2 \neq 0$; otherwise, $\theta = \sqrt{n} = \frac{q}{p}$, a contradiction. Therefore, it must be the case that
$$|p\theta - q| \times |p\theta + q| \geq 1 \qquad (2.5)$$
Let $A = \min(\theta, \frac{1}{4\theta})$. This number is well-defined because $\theta \neq 0$. If $|p\theta - q| < p\theta$, then it follows from the triangle inequality that
$$|p\theta + q| = |-2p\theta + (p\theta - q)| \leq |2p\theta| + |p\theta - q| < 3p\theta. \qquad (2.6)$$
Therefore, we know from the inequalities (2.5) and (2.6) that
$$|p\theta - q| \geq \frac{1}{|p\theta + q|} > \frac{1}{3p\theta} > \frac{A}{p}.$$
If $|p\theta - q| \geq p\theta$, then recall that $p \in \mathbb{N}$ so that
$$|p\theta - q| \geq p\theta \geq Ap > \frac{A}{p}.$$
Hence we have established the required inequality (2.4) and this completes the proof of the problem. ∎

> **Problem 2.4** (Dirichlet's Approximation Theorem)
>
> (⋆)(⋆) *Suppose that θ is irrational and N is a positive integer. Prove that there exist integers h, k with $0 < k \leq N$ such that*
> $$|k\theta - h| < \frac{1}{N}. \tag{2.7}$$

Proof. Since θ is irrational, we have $i\theta \notin \mathbb{Z}$ for every nonzero integer i. By this fact, for $i = 0, 1, \ldots, N$, the $N+1$ *distinct* numbers
$$t_i = i\theta - [i\theta]$$
belong to $[0, 1)$, where $[x]$ is the greatest integer less than or equal to x. By the Pigeonhole Principle,[b] there exist *at least* one interval $[\frac{j}{N}, \frac{j+1}{N})$ containing the numbers t_r and t_s, where $j = 0, 1, \ldots, N-1$ and $r, s \in \{0, 1, \ldots, N\}$. Without loss of generality, we may assume that $r > s$ and
$$\frac{j}{N} \leq t_r - t_s < \frac{j+1}{N} \tag{2.8}$$
for some $j = 0, 1, \ldots, N-1$. By the definition of t_i, we have
$$t_r - t_s = (r-s)\theta - ([r\theta] - [s\theta]) = k\theta - h, \tag{2.9}$$
where $h = [r\theta] - [s\theta]$ and $k = r - s$. It is clear that both h and k are integers with $0 < k \leq N$. By these and putting the identity (2.9) into the inequality (2.8), we obtain the result (2.7). This completes the proof of the problem. ∎

2.3 Absolute Values

> **Problem 2.5**
>
> (⋆) *For every $x, y \in \mathbb{R}$, prove that $||x| - |y|| \leq |x - y|$.*

Proof. By the triangle inequality, we have $|x| = |x - y + y| \leq |x - y| + |y|$ so that
$$|x| - |y| \leq |x - y| \tag{2.10}$$
Exchanging the roles of x and y in the inequality (2.10), we have $|y| - |x| \leq |x - y|$ and this implies that
$$|x| - |y| \geq -|x - y|. \tag{2.11}$$
Combining the inequalities (2.10) and (2.11), we obtain the desired inequality. This ends the proof of the problem. ∎

> **Problem 2.6**
>
> (⋆) *Solve the inequality $|x - 3| + |x + 3| \leq 14$.*

---

[b]If n items are put into m containers, where $n > m$, then *at least* one container contains more than one item.

2.4. The Completeness Axiom

Proof. Taking square to both sides of the inequality, we have

$$x^2 - 6x + 9 + x^2 + 6x + 9 + 2|x^2 - 9| \leq 196$$
$$x^2 + 9 + |x^2 - 9| \leq 98. \qquad (2.12)$$

If $x^2 \leq 9$, then the inequality (2.12) becomes $18 \leq 98$ which is always true. Thus it means that the inequality holds when

$$|x| \leq 3. \qquad (2.13)$$

If $x^2 > 9$, then we deduce from the inequality (2.12) that $x^2 \leq 49$. In this case, we have

$$3 < |x| \leq 7. \qquad (2.14)$$

By the results (2.13) and (2.14), we know that the inequality holds for $-7 \leq x \leq 7$. We complete the proof of the problem. ∎

Problem 2.7

(★) *Suppose that a and b are real numbers. Prove that*

$$\frac{|a+b|}{1+|a+b|} \leq \frac{|a|}{1+|a|} + \frac{|b|}{1+|b|}.$$

Proof. Since $0 \leq |a+b| \leq |a| + |b|$, we have $1 \leq 1 + |a+b| \leq 1 + |a| + |b|$ and then

$$\begin{aligned}
\frac{|a+b|}{1+|a+b|} &= \frac{1+|a+b|-1}{1+|a+b|} \\
&= 1 - \frac{1}{1+|a+b|} \\
&\leq 1 - \frac{1}{1+|a|+|b|} \\
&= \frac{|a|+|b|}{1+|a|+|b|} \\
&\leq \frac{|a|}{1+|a|} + \frac{|b|}{1+|b|},
\end{aligned}$$

completing the proof of the problem. ∎

2.4 The Completeness Axiom

Problem 2.8

(★)(★) *Prove that $\mathbb{N}$ has no upper bound.*

Proof. Assume that $\mathbb{N}$ had an upper bound. By the Completeness Axiom, we know that $\sup \mathbb{N}$ exists in $\mathbb{R}$. By definition, $\sup \mathbb{N} - 1$ is not an upper bound of $\mathbb{N}$, so there exists $n \in \mathbb{N}$ such that

$$\sup \mathbb{N} - 1 < n. \tag{2.15}$$

Since $n + 1 \in \mathbb{N}$, we also have

$$n + 1 \le \sup \mathbb{N}. \tag{2.16}$$

It follows from the inequalities (2.15) and (2.16) that

$$\sup \mathbb{N} < n + 1 \le \sup \mathbb{N}$$

which is a contradiction. Hence $\mathbb{N}$ has no upper bound and this ends the proof of the problem. ∎

Problem 2.9

(★)(★) Let $a, b \in \mathbb{R}$ and $a < b$. Prove that there is an irrational number θ such that $a < \theta < b$.

Proof. Since $a < b$, Theorem 2.2 (Density of Rationals) implies that there exists a $r \in \mathbb{Q}$ such that $a < r < b$. Since $b - r > 0$ and $\sqrt{2} > 0$, Theorem 2.1 (The Archimedean Property) ensures that one can find a positive integer n such that $n(b - r) > \sqrt{2} > 0$. Note that $\frac{\sqrt{2}}{n}$ is irrational and so the number

$$\theta = r + \frac{\sqrt{2}}{n}$$

is also irrational. It is easy to see that

$$a < r < \theta = r + \frac{\sqrt{2}}{n} < b.$$

Hence we complete the proof of the problem. ∎

Remark 2.1

Problem 2.9 can be rephrased as the density of irrational numbers. There is another proof which uses the fact that the interval (a, b) is uncountable, see Problem 3.1.

Problem 2.10

(★)(★) Let $x \ge 0$ and n be a positive integer. Show that there exists a unique $y \ge 0$ such that

$$y^n = x. \tag{2.17}$$

Proof. If $x = 0$, then we take $y = 0$. Thus we may suppose that $x > 0$. Furthermore, the case $n = 1$ is trivial, so we may assume that $n \ge 2$.

We first prove the uniqueness part of this problem. If $y_1 > 0$ and $y_2 > 0$ satisfy the equation (2.17), then we have

$$0 = y_1^n - y_2^n = (y_1 - y_2)(y_1^{n-1} + y_1^{n-2} y_2 + \cdots + y_1 y_2^{n-2} + y_2^{n-1}), \tag{2.18}$$

2.4. The Completeness Axiom

where n is a positive integer. Since y_1 and y_2 are positive, $y_1^{n-1} + y_1^{n-2}y_2 + \cdots + y_1 y_2^{n-2} + y_2^{n-1}$ must be positive. Therefore, we deduce from the equation (2.18) that $y_1 = y_2$. This proves the uniqueness.

Now we show the existence of the y as follows: Suppose that
$$E = \{y \geq 0 \,|\, y^n \leq x\}$$
which is a subset of $\mathbb{R}$. Since $x > 0$, Theorem 2.1 (The Archimedean Property) guarantees the existence of a positive integer m such that $mx > 1$. Since $m \in \mathbb{N}$, we have
$$0 < \left(\frac{1}{m}\right)^n < \frac{1}{m} < x$$
for every $n \geq 2$. This shows that $\frac{1}{m} \in E$. Next, if p is a positive integer such that $p \geq x$, then we have $y^n \leq x \leq p$ for every $y \in E$. Thus we have shown that E is a nonempty subset of $\mathbb{R}$ that is bounded above. Hence the Completeness Axiom implies that $\alpha = \sup E$ exists in $\mathbb{R}$. This number must be positive because E contains the positive number $\frac{1}{m}$.

Assume that $\alpha^n < x$. Now for each $p \in \mathbb{N}$, we have
$$\left(\alpha + \frac{1}{p}\right)^n = \alpha^n + \sum_{k=1}^{n} C_k^n \alpha^{n-k} p^{-k} \leq \alpha^n + p^{-1} \sum_{k=1}^{n} C_k^n \alpha^{n-k}. \tag{2.19}$$

Denote $A = \sum_{k=1}^{n} C_k^n \alpha^{n-k}$. Then it follows from Theorem 2.1 (The Archimedean Property) that there is a positive integer q such that $q(x - \alpha^n) > A$, i.e.,
$$x - \alpha^n > \frac{A}{q}.$$
We substitute $p = q$ in the inequality (2.19) to obtain
$$\left(\alpha + \frac{1}{q}\right)^n \leq \alpha^n + \frac{A}{q} < \alpha^n + x - \alpha^n = x.$$
By definition, $\alpha + \frac{1}{q} \in E$ so that $\alpha + \frac{1}{q} \leq \alpha$ which is a contradiction. Thus the case $\alpha^n < x$ is impossible. The impossibility of the other case $\alpha^n > x$ can be shown similarly. Hence we conclude that $\alpha^n = x$ and this completes the proof of the problem. ∎

Problem 2.11

(★) Let E be a nonempty subset of $\mathbb{R}$ having least upper bounds α and β. Prove that $\alpha = \beta$.

Proof. Since α and β are least upper bounds of E, they are upper bounds of E. By definition, both inequalities
$$\alpha \leq \beta \quad \text{and} \quad \beta \leq \alpha$$
hold simultaneously. Hence we definitely have $\alpha = \beta$, completing the proof of our problem. ∎

Problem 2.12

(★)(★) Let E be a nonempty subset of $\mathbb{R}$ and $y = \sup E$. Let

$$-E = \{-x \in \mathbb{R} \,|\, x \in E\}.$$

Prove that $-y$ is the greatest lower bound of $-E$.

Proof. Since $x \leq y$ for all $x \in E$, we have $-y \leq -x$ for all $x \in E$. Therefore, $-y$ is a lower bound of $-E$. Assume that there was $w \in \mathbb{R}$ such that

$$-y < w \leq z \tag{2.20}$$

for all $z \in -E$. Since $z = -x$ for some $x \in E$, we deduce from the inequalities (2.20) that $-y < w \leq -x$ or equivalently,

$$y > -w \geq x \tag{2.21}$$

for all $x \in E$. However, the inequalities (2.21) say that $y \neq \sup E$, a contradiction. Hence we must have $-y = \inf(-E)$, completing the proof of the problem. ∎

Problem 2.13

(★)(★) Suppose that E and F are two nonempty subsets of positive real numbers and they are bounded above. Let
$$EF = \{x \times y \,|\, x \in E \text{ and } y \in F\}.$$
Prove that $\sup(EF) = \sup E \times \sup F$.

Proof. Since E and F are nonempty and bounded above, EF is also nonempty and bounded above. By the Completeness Axiom, $\sup E$, $\sup F$ and $\sup(EF)$ exist in $\mathbb{R}^+$. It is clear that $\sup E \times \sup F$ is an upper bound of EF, i.e.,

$$\sup(EF) \leq \sup E \times \sup F. \tag{2.22}$$

Now we are going to show that the equality in (2.22) holds. Given that $\epsilon > 0$. We claim that there exist $x_0 \in E$ and $y_0 \in F$ such that

$$\sup E \times \sup F - \epsilon < x_0 y_0.$$

To this end, since $\sup E >$ and $\sup F > 0$, we can find a large positive integer N such that

$$\sup E - \frac{\epsilon}{N} > 0, \quad \sup F - \frac{\epsilon}{N} > 0 \quad \text{and} \quad \sup E + \sup F < N. \tag{2.23}$$

By the definition of supremum, we must have $\sup E - \frac{\epsilon}{N} < x_0$ and $\sup F - \frac{\epsilon}{N} < y_0$ *for some* $x_0 \in E$ and $y_0 \in F$, so they imply that

$$\sup E \times \sup F - \frac{\epsilon}{N}\left(\sup E + \sup F - \frac{\epsilon}{N}\right) = \left(\sup E - \frac{\epsilon}{N}\right)\left(\sup F - \frac{\epsilon}{N}\right) < x_0 y_0. \tag{2.24}$$

2.4. The Completeness Axiom 17

Now we are able to derive from the inequalities (2.23) and (2.24) that

$$\sup E \times \sup F - \epsilon < \sup E \times \sup F - \frac{\epsilon}{N}\left(\sup E + \sup F\right)$$
$$< \sup E \times \sup F - \frac{\epsilon}{N}\left(\sup E + \sup F - \frac{\epsilon}{N}\right)$$
$$< x_0 y_0$$

for some $x_0 y_0 \in EF$. Hence this proves our claim and it means that $\sup E \times \sup F$ is in fact the least upper bound of EF, i.e., $\sup(EF) = \sup E \times \sup F$ and this completes the proof of the problem. ∎

Problem 2.14

(★) Let p and q be two primes. Let $E = \{p^{-s} + q^{-t} \mid s, t \in \mathbb{N}\}$. What are $\sup E$ and $\inf E$?

Proof. Since $0 < \frac{1}{p^s} \le \frac{1}{p}$ and $0 < \frac{1}{q^t} \le \frac{1}{q}$ for every $s, t \in \mathbb{N}$, we have

$$0 < x \le \frac{1}{p} + \frac{1}{q}$$

for every $x \in E$. Since $\frac{1}{p} + \frac{1}{q} \in E$, it is the maximum of E. In other words, we have

$$\sup E = \max E = \frac{1}{p} + \frac{1}{q}.$$

Let $\epsilon > 0$. We know from Theorem 2.2 that there exists a $s_0 \in \mathbb{N}$ such that $0 < \frac{1}{p^{s_0}} < \frac{\epsilon}{2}$. Similarly, by Theorem 2.2 again, there exists a $t_0 \in \mathbb{N}$ such that $0 < \frac{1}{q^{t_0}} < \frac{\epsilon}{2}$. Therefore, it is clear that they give

$$0 < \frac{1}{p^{s_0}} + \frac{1}{q^{t_0}} < \epsilon$$

which implies that ϵ is not a lower upper of E. Hence we have $\inf E = 0$ and we finish the proof of the problem. ∎

Problem 2.15

(★)(★) Let

$$E = \left\{\frac{p}{q} \,\middle|\, p, q \in \mathbb{N} \text{ and } 0 < p < q\right\}.$$

Prove that E does not contain a maximum element or a minimum element. Find $\sup E$ and $\inf E$.

Proof. Since $q > p > 0$, we have $pq + q > pq + p > 0$ and then

$$\frac{p+1}{q+1} > \frac{p}{q}.$$

Thus E does not contain the maximum element. Similarly, since $0 < \frac{p}{q} < 1$, we must have

$$0 < \frac{p^2}{q^2} < \frac{p}{q}$$

which implies that E does not contain the minimum element.

To find $\sup E$ and $\inf E$, we first note that
$$0 < \frac{p}{q} < 1$$
for every $\frac{p}{q} \in E$. Therefore, 1 and 0 are an upper bound and a lower bound of E. If $\sup E \neq 1$, then we can find a rational $\frac{m}{n}$ such that
$$\frac{p}{q} < \frac{m}{n} < 1 \tag{2.25}$$
for every $\frac{p}{q} \in E$. Here m and n must satisfy $0 < m < n$. In other words, we have $\frac{m}{n} \in E$. By this and the inequality (2.25), E contains the maximum element $\frac{m}{n}$, a contradiction. Hence we establish the result that
$$\sup E = 1.$$

The case for $\inf E = 0$ is similar and so we omit the details here. We have completed the proof of the problem. ■

CHAPTER 3

Countable and Uncountable Sets

3.1 Fundamental Concepts

When we are given a set containing a *finite* number of objects, then we can simply count the number of objects and determine the **cardinality** of that set. Can we do a similar thing when the set has *infinitely* many objects? The answer to this question is "yes" if we define the concept of **countability** of such an infinite set. In fact, countability of a set A is to "measure how many" elements the set A contains. The main references for this chapter are [3, §2.12 & §2.15], [23, §2.8, §2.12 & §2.13] and [33, §2.4].

3.1.1 Definitions of Countable and Uncountable Sets

Definition 3.1. *Given two sets A and B. If there exists a function $f : A \to B$ which is one-to-one and onto, then we say that A and B **equivalent**.*

It can be shown easily that this is indeed an equivalent relation, see §1.1.3. Now we are ready to apply this equivalent relation to define the countability of a set E. Since there are two sets in Definition 3.1, we need to choose a set to compare with A. The most natural one is the set of all positive integers $\mathbb{N}$.

Definition 3.2 (Countability). *A set A is said to be **countable** if it is equivalent to $\mathbb{N}$. In other words, there exists a function $f : \mathbb{N} \to A$ which is one-to-one and onto. If A is not finite or countable, then it is **uncountable**.*

Definition 3.3. *We say that A is **at most countable** if A is either finite or countable.*

3.1.2 Properties of Countable Sets

Here we list some important and useful results about the countability of a set A.

Theorem 3.4. *Every infinite subset of a countable set A must be countable*

Theorem 3.5. *Suppose that* $\{A_1, A_2, \ldots\}$ *is a countable collection of countable sets. Then the set*

$$A = \bigcup_{k=1}^{\infty} A_k \tag{3.1}$$

is countable.

Theorem 3.6. *Suppose that* $\{A_1, A_2, \ldots, A_n\}$ *is a finite collection of countable sets. Then the set*

$$A = A_1 \times A_2 \times \cdots \times A_n$$

is countable.

As an immediate application of Theorem 3.6, we see that the sets $\mathbb{Z}$ and $\mathbb{Q}$ are countable. However, we note that Theorem 3.6 cannot hold for *infinite* collection of countable (or even finite) sets, see Problem 3.10 for a counterexample.

3.2 Problems on Countable and Uncountable Sets

> **Problem 3.1**
>
> (⋆) Reprove Problem 2.9 by using the fact that (a, b) is uncountable.

Proof. Since $\mathbb{Q} \cap (a, b) \subset \mathbb{Q}$, Theorem 3.4 implies that it is countable. Since (a, b) is given uncountable, the set

$$(a, b) \setminus [\mathbb{Q} \cap (a, b)] \tag{3.2}$$

must be uncountable. Since elements in the set (3.2) are *all* irrational numbers in (a, b), there exists an irrational number θ such that $a < \theta < b$. This completes the proof of the problem. ∎

> **Problem 3.2**
>
> (⋆) Suppose that $(0, 1)$ is uncountable. Is $\mathbb{R}$ uncountable?

Proof. The answer is affirmative. Assume that $\mathbb{R}$ was countable. Since $(0, 1) \subset \mathbb{R}$, $(0, 1)$ is also countable by Theorem 3.4, a contradiction to the hypothesis. Hence we have $\mathbb{R}$ is uncountable. We end the proof of the problem. ∎

> **Problem 3.3 (Cantor's Theorem)**
>
> (⋆)(⋆) Suppose that A is a nonempty set. Denote 2^A to be the **power set** of A (the set of all subsets of A). Show that there does not exist a surjective function $f : A \to 2^A$.

3.2. Problems on Countable and Uncountable Sets

Proof. Assume that there was a surjective function $f : A \to 2^A$. We consider the subset $B \subseteq A$ defined by
$$B = \{x \in A \mid x \notin f(x)\}. \tag{3.3}$$
It is clear that $B \subset A$ so that $B \in 2^A$. Since f is sufjective, there exists a unique $x_0 \in A$ such that
$$f(x_0) = B.$$
Does $x_0 \in B$? On the one hand, if $x_0 \in B$, then the definition (3.3) gives $x_0 \notin f(x_0) = B$, a contradiction. On the other hand, if $x_0 \notin B$, then the definition (3.3) again implies that $x_0 \in f(x_0) = B$ which is a contradiction again. Hence these contradictions show that the existence of such element x_0 and then the existence of such surjective function f is impossible, finishing the proof of the problem. ∎

Problem 3.4

(⋆) Prove that $2^{\mathbb{N}}$ is uncountable.

Proof. Assume that $2^{\mathbb{N}}$ was countable. Then there exists an one-to-one and onto function
$$f : \mathbb{N} \to 2^{\mathbb{N}}$$
which certainly contradicts the result of Problem 3.3. Hence $2^{\mathbb{N}}$ is uncountable and this completes the proof of the problem. ∎

Problem 3.5

(⋆)(⋆) Prove Theorem 3.5.

Proof. Let A be defined as in the equation (3.1). For each k, since A_k is countable, we can list its elements as
$$A_k = \{a_{k1}, a_{k2}, \ldots\}.$$
Consider the following array:

$$
\begin{array}{llll}
A_1: & a_{11} & a_{12} & a_{13} & \cdots \\
A_2: & a_{21} & a_{22} & a_{23} & \cdots \\
A_3: & a_{31} & a_{32} & a_{33} & \cdots \\
\cdots & & & &
\end{array}
$$

As indicated by the colored elements, we see that they can be arranged in a sequence
$$a_{11}, a_{21}, a_{12}, a_{31}, a_{22}, \ldots, a_{13}, \ldots \tag{3.4}$$

If x is a common element of two distinct sets A_i and A_j, then it appears more than once in the sequence (3.4). Therefore there exists a subset $B \subseteq \mathbb{N}$ and a function $f : B \to A$ such that f is one-to-one and onto. By Definition 3.3, A is at most countable. Since $A_1 \subseteq A$ and A_1 is countable, A is an infinite set and we see from this that A is also countable. This ends the proof of the problem. ∎

Problem 3.6

(★) Prove that the set $E = \{p + q\sqrt{3} \,|\, p, q \in \mathbb{Q}\}$ is countable.

Proof. Since $\mathbb{Q}$ is countable, $\mathbb{Q} \times \mathbb{Q}$ is also countable by Theorem 3.6. Define $f : \mathbb{Q} \times \mathbb{Q} \to E$ by

$$f(p, q) = p + q\sqrt{3}.$$

We see easily that f is surjective, i.e., $E = f(\mathbb{Q} \times \mathbb{Q})$. Note that this function is also injective because if $(p_1, q_1), (p_2, q_2)$ are two distinct elements in E such that $f(p_1, q_1) = f(p_2, q_2)$. Then we have

$$p_1 + q_1\sqrt{3} = p_2 + q_2\sqrt{3}$$
$$\sqrt{3} = \frac{p_1 - p_2}{q_2 - q_1}$$

which implies that $\sqrt{3}$ is irrational, a contradiction. Hence E is also countable by Definition 3.2 and this completes the proof of the problem. ∎

Problem 3.7

(★) Prove that the set of straight lines passing through the origin in $\mathbb{R}^2$ is uncountable.

Proof. Any line L passing through the origin in $\mathbb{R}^2$ must be in the form

$$y = mx \quad \text{or} \quad x = 0,$$

where $m \in \mathbb{R}$. Let this set be A. Since A has a subset B which is equivalent to $\mathbb{R}$, it follows from Problem 3.2 that B is uncountable. Assume that A was countable. Then every infinite subset of A must be countable by Theorem 3.4. In particular, B is countable which is a contradiction. Hence A must be uncountable and we finish the proof of the problem. ∎

Problem 3.8

(★) Denote A to the set of circles in $\mathbb{R}^2$ having rational radii and centres with rational coordinates. Verify that A is countable.

Proof. If a circle has rational radius r_1 and rational centre (r_2, r_3), where $r_1, r_2, r_3 \in \mathbb{Q}$, then it is *uniquely determined* by the triple (r_1, r_2, r_3). Therefore, A is a subset of $\mathbb{Q} \times \mathbb{Q} \times \mathbb{Q}$. By Theorem 3.6, $\mathbb{Q} \times \mathbb{Q} \times \mathbb{Q}$ is countable. Since $(n, n, n) \in A$ for every $n \in \mathbb{N}$, A must be infinite and by Theorem 3.4, it is countable. This completes the proof of the problem. ∎

Problem 3.9

(★)(★) Suppose that $F \subset E$, where E is uncountable and F is countable. Show that $E \setminus F$ is uncountable. Find $|E \setminus F|$, where $|E|$ means the cardinality of the set E.

3.2. Problems on Countable and Uncountable Sets

Proof. Note that $E = (E \setminus F) \cup F$. Thus if both $E \setminus F$ and F are countable, then Theorem 3.4 implies that E must be countable which is impossible. Hence this proves our first assertion that $E \setminus F$ is uncountable.

For the second assertion, we claim that

$$|E \setminus F| = |E|. \tag{3.5}$$

Since F is a proper subset of E, we have $E \setminus F \neq \varnothing$ and then we can find a $x_1 \in E \setminus F$. Now we consider the set $E \setminus (F \cup \{x_1\})$. Since F is countable, $F \cup \{x_1\}$ is also countable by Theorem 3.5 and we follow from the first assertion that $E \setminus (F \cup \{x_1\})$ is still uncountable. In particular, we must have $E \setminus (F \cup \{x_1\}) \neq \varnothing$. Therefore, there exists a $x_2 \in E \setminus (F \cup \{x_1\})$. Now this process can be done inductively and we can find a subset $X = \{x_1, x_2, \ldots\} \subset E \setminus F$, where all $x_i \neq x_j$ if $i \neq j$ and the set $E \setminus (X \cup F)$ is uncountable.

It is clear that both X and $X \cup F$ are countable, so we can find a bijective function $f : X \cup F \to X$ by Definition 3.2. We define $g : E \to E \setminus F$ by

$$g(x) = \begin{cases} x, & \text{if } x \in E \setminus (X \cup F); \\ f(x), & \text{if } x \in X \cup F. \end{cases} \tag{3.6}$$

If we can show that g is bijective, then we obtain the result (3.5). To this end, suppose that $g(x) = g(y)$. We claim that it is impossible to have either

"$x \in E \setminus (X \cup F)$ and $y \in X \cup F$" or "$y \in E \setminus (X \cup F)$ and $x \in X \cup F$".

In fact, if $x \in E \setminus (X \cup F)$ and $y \in X \cup F$, then we have $g(x) = x \notin X \cup F$ but $g(y) = f(y) \in X$, so we cannot have

$$g(x) = g(y)$$

and therefore the situation "$x \in E \setminus (X \cup F)$ and $y \in X \cup F$" is impossible. Similarly, the situation "$y \in E \setminus (X \cup F)$ and $x \in X \cup F$" can also be shown impossible. Now there are two remaining cases which are

- **Case (1):** $x, y \in E \setminus (X \cup F)$. In this case, we have $g(x) = x$ and $g(y) = y$ so that $x = y$.

- **Case (2):** $x, y \in X \cup F$. In this case, we have $g(x) = f(x)$ and $g(y) = f(y)$ so that $f(x) = f(y)$ which implies that $x = y$.

In conclusion, we have shown that g is injective.

Next we check that g is surjective. Let $y \in E \setminus F$. If $y \in X$, then we take $x = f^{-1}(y) \in X \cup F$ so that the definition (3.6) implies

$$g(x) = f(x) = y.$$

If $y \notin X$, then $y \in E \setminus (X \cup F)$. Thus we take $x = y$ and the definition (3.6) implies

$$g(x) = x = y.$$

In conclusion, we have shown that g is surjective. Hence, this proves our desired result (3.5), completing the proof of the problem. ∎

Problem 3.10

$(\star)(\star)$ For each $k \in \mathbb{N}$, define $A_k = \{0,1\}$. Prove that the set

$$A = \prod_{k=1}^{\infty} A_k = \{0,1\} \times \{0,1\} \times \cdots$$

is uncountable.

Proof. Assume that A was countable. By Definition 3.2, a bijective function $f : \mathbb{N} \to A$ exists. For every $n \in \mathbb{N}$, denote $f(n)_n \in \{0,1\}$ to be the n-th "coordinate" of $f(n)$, i.e.,

$$f(n) = (*, *, \ldots, *, \underbrace{f(n)_n}_{n\text{-th coordinate}}, *, \ldots).$$

Next, we define $a = (a_1, a_2, \ldots, a_n, \ldots)$ by

$$a_n = 1 - f(n)_n$$

for each $n \in \mathbb{N}$. It is clear that we always have $a \in A$. However, $f(n) \neq a$ for every $n \in \mathbb{N}$ because the n-digits of $f(n)$ and a are different. This fact contradicts the surjective property of f. Hence A must be uncountable and this finishes the proof of the problem. ∎

Problem 3.11

$(\star)(\star)$ A real number is said to be **algebraic** if it is a root of an equation

$$a_n x^n + \cdots + a_1 x + a_0 = 0$$

with integer coefficients, where $a_n \neq 0$. Show that the set of all algebraic numbers, denoted by $\mathcal{A}$, is countable.

Proof. For each $n = 0, 1, 2, \ldots$, define

$$A_n = \{a_n x^n + \cdots + a_1 x + a_0 \,|\, a_n \in \mathbb{Z} \setminus \{0\}, a_0, a_1, \ldots, a_{n-1} \in \mathbb{Z}\}.$$

Next, we define $f : A_n \to (\mathbb{Z} \setminus \{0\}) \times \mathbb{Z}^n$ by

$$f(a_n x^n + \cdots + a_1 x + a_0) = (a_n, \ldots, a_1, a_0)$$

which is obviously bejective. Since $\mathbb{Z} \setminus \{0\}$ and $\mathbb{Z}$ are countable, Theorem 3.6 implies that $(\mathbb{Z} \setminus \{0\}) \times \mathbb{Z}^n$ is also countable. By Definition 3.2, each A_n is countable. By Theorem 3.5, the set

$$A = \bigcup_{n=0}^{\infty} A_n$$

is countable. Since $\mathcal{A}$ is equivalent to a subset of A and each integer is algebraic, $\mathcal{A}$ must be infinite and thus it is countable by Theorem 3.4. We complete the proof of the problem. ∎

3.2. Problems on Countable and Uncountable Sets

> **Problem 3.12**
>
> (★) Let $\mathcal{A}$ be the set of all algebraic real numbers. Show that $\mathbb{R} \setminus \mathcal{A}$ is uncountable.

Proof. By Problem 3.11, $\mathcal{A}$ is countable. If $\mathcal{A} = \mathbb{R}$, then $\mathbb{R}$ is also countable which contradicts Problem 3.2. Thus we must have $\mathcal{A} \subset \mathbb{R}$ and Problem 3.9 shows that $\mathbb{R} \setminus \mathcal{A}$ is uncountable. This completes the proof of the problem. ∎

> **Problem 3.13**
>
> (★) Construct a bijection between the sets $(0, 1)$ and $(0, 1]$.

Proof. Define $A = \{\frac{1}{2}, \frac{1}{3}, \ldots\} \subseteq (0, 1)$ and $B = \{1, \frac{1}{2}, \frac{1}{3}, \ldots\} \subseteq (0, 1]$. Define $f : A \to B$ by

$$f\left(\frac{1}{n}\right) = \frac{1}{n-1}$$

which is clearly a bijection. Next, we define $F : (0, 1) \to (0, 1]$ by

$$F(x) = \begin{cases} f(x), & \text{if } x \in A; \\ x, & \text{if } x \in (0, 1) \setminus A. \end{cases}$$

Now it is easy to check that F is in fact a bijection, completing the proof of the problem. ∎

CHAPTER 4

Elementary Topology on Metric Spaces

4.1 Fundamental Concepts

In the real number line $\mathbb{R}$, everyone knows that we can find the **distance** between two points $a, b \in \mathbb{R}$ by simply evaluating the absolute value $|a - b|$ (see §2.1.2). This situation can be generalized without any difficulties. In fact, most of the properties of the real number $\mathbb{R}$ equipped with the absolute value $|\cdot|$ can be treated as special cases of the more general setting of a **metric space** with a **distance function** (or a **metric**). Therefore, instead of studying topological properties of the real number system $\mathbb{R}$, we choose to review some topological results of a metric space. For details of the theory, please read [3, Chap. 3] or [23, §2.15 - §2.47, pp. 30 - 43].

4.1.1 Metric Spaces

Suppose that X is a set. Now elements of X are called **points**.

Definition 4.1. *A metric space is a pair (X, d) of a set X and a function $d : X \times X \to \mathbb{R}$ satisfying the following conditions for any points $p, q, r \in X$:*

(1) $d(p,p) = 0$ and $d(p,q) > 0$ if $p \neq q$;

(2) $d(p,q) = d(q,p)$;

(3) $d(p,q) \leq d(p,r) + d(r,q)$.

Any function satisfying the conditions in Definition 4.1 is called a **distance function** or a **metric**. It is well-known that the set $\mathbb{R}^n$ is a metric space, where n is a positive integer. In the following discussion and problems, we assume that the notation X means a metric space with a metric d.

4.1.2 Open Sets and Closed Sets

Definition 4.2. *A **neighborhood** of a point $p \in X$ with **radius** $r > 0$, denoted by $N_r(p)$, is defined by*
$$N_r(p) = \{q \in X \mid d(p,q) < r\}.$$

Definition 4.3. Let E be a set. A point $p \in E$ is called an **interior point** of E if $N_r(p) \subseteq E$ for some $r > 0$. We call E an **open set** in X if every point of E is an interior point of E.

Theorem 4.4. Let $p \in X$. Every neighborhood $N_r(p)$ is open in X.

Definition 4.5. Let E be a set. A point $p \in X$ is called a **limit point** of E if

$$N_r(p) \cap (E \setminus \{p\}) \neq \varnothing$$

for every $r > 0$. Denote the set of all limit points of E to be E'. We call E a **closed set** in X if $E' \subseteq E$.

Theorem 4.6. A set E is open in X if and only if E^c is closed in X.

Theorem 4.7.

(a) For any families $\{E_\alpha\}$ of open sets and $\{F_\alpha\}$ of closed sets, $\bigcup E_\alpha$ is open in X and $\bigcap F_\alpha$ is closed in X.

(b) For any finite families $\{E_1, E_2, \ldots, E_n\}$ of open sets and $\{F_1, F_2, \ldots, F_n\}$ of closed sets, $\bigcap_{k=1}^n E_k$ is open in X and $\bigcup_{k=1}^n F_k$ is closed in X.

4.1.3 Interiors and Closures

Definition 4.8. Let E be a subset of X. The set

$$\overline{E} = E \cup E'$$

is called the **closure** of E and the collection of all interior points of E is denoted by E° and it is called the **interior** of E.

Theorem 4.9. Given that $E \subseteq X$. We have

(a) $\overline{E}$ is closed in X.

(b) $E = \overline{E}$ if and only if E is closed.

(c) if $E \subseteq F$ and F is closed in X, then $\overline{E} \subseteq F$.

By Theorem 4.9(a) and (c), we see that $\overline{E}$ is the **smallest** closed subset of X containing E. There is a similar result for the interior of E. In fact, we see from Definition 4.8 that an interior point of E lies in E so that we must have $E^\circ \subseteq E$. Then it can be shown that E° is the **largest** open subset of X inside E, see Problem 4.11.

Interiors and closures of sets have many properties in very "similar" forms. The following table summarizes and compares such properties of the interior and the closure of a set E in the metric space X.

4.1. Fundamental Concepts

	Interior of E	Closure of E
Subset relation	$E \subset X$ implies $E^\circ \subseteq X^\circ$	$E \subset X$ implies $\overline{E} \subseteq \overline{X}$
Finite union	$E^\circ \cup F^\circ \subseteq (E \cup F)^\circ$	$\overline{E \cup F} = \overline{E} \cup \overline{F}$
Finite intersection	$(E \cap F)^\circ = E^\circ \cap F^\circ$	$\overline{E \cap F} \subseteq \overline{E} \cap \overline{F}$
Arbitrary union	$\bigcup_\alpha E_\alpha^\circ \subseteq \left(\bigcup_\alpha E_\alpha\right)^\circ$	$\bigcup_\alpha \overline{E_\alpha} \subseteq \overline{\bigcup_\alpha E_\alpha}$
Arbitrary intersection	$\left(\bigcap_\alpha E_\alpha\right)^\circ \subseteq \bigcap_\alpha E_\alpha^\circ$	$\overline{\bigcap_\alpha E_\alpha} \subseteq \bigcap_\alpha \overline{E_\alpha}$
Finite product relation	$(E \times F)^\circ = E^\circ \times F^\circ$	$\overline{E \times F} = \overline{E} \times \overline{F}$

Table 4.1: Properties of the interior and the closure of E in X.

Proofs of the finite union and intersection of interiors can be seen in Problem 4.12 and a proof of the finite intersection of closures can be found in Problem 4.13.

4.1.4 Sets in Metric Subspaces

Let Y be a subset of a metric space X. Then it is natural to think that Y is also a metric space with the same metric as X because all points in Y satisfy Definition 4.1. Therefore, we have to pay particular attention when we talk about open (resp. closed) sets in X and open (resp. closed) sets in Y.

Let $E \subseteq Y \subseteq X$, where both X and Y are metric spaces. It may happen that E is open in Y, but *not* open in X. An example can be found in [23, Example 2.21(g), p. 33]. Fortunately, there is a test for E to be open or closed in X:

Theorem 4.10. *Let $E \subseteq Y \subseteq X$. Then E is open (resp. closed) in Y if and only if*
$$E = Y \cap F$$
for some open (resp. closed) subset F of X.

Is there any similar result for interiors and closures of subsets of X and of Y? The answer to this question is "yes" to closures, but "no" to interiors. In fact, we have the following result:

Theorem 4.11. *Given that $E \subseteq Y \subseteq X$. Denote E_X° and $\overline{E}_X$ to be the interior and the closure of E in X respectively. Similarly, E_Y° and $\overline{E}_Y$ are the interior and the closure of E in Y respectively. Then we have*

(a) $\overline{E}_Y = \overline{E}_X \cap Y$;

(b) $E_Y^\circ \supseteq E_X^\circ \cap Y$.

We remark that the inclusion in (b) can be proper, see Problem 4.15 for an example.

4.1.5 Compact Sets in Metric Spaces

Let K be a subset of X. A collection $\{V_\alpha\}$ of subsets of X is said to **cover** K if
$$K \subseteq \bigcup_\alpha V_\alpha.$$

By an **open cover** of K, we mean that each V_α is open in X.

Definition 4.12 (Compact Sets). *A subset K of X is said to be* **compact** *if for every open cover $\{V_\alpha\}$ of K, there are finitely many indices $\alpha_1, \ldots, \alpha_n$ such that*

$$K \subseteq V_{\alpha_1} \cup V_{\alpha_2} \cup \cdots \cup V_{\alpha_n}.$$

In particular, if the metric space X is itself compact, then we call it a **compact metric space**. There are many nice properties for compact sets. One of them is that it **does not** depend on the space in which K lies. More explicitly, this says that if $K \subseteq Y \subseteq X$, then K is compact in X if and only if K is compact in Y. See [23, Theorem 2.33, p. 37] for a proof of it. Other important and useful properties of compact sets are given as follows:

Theorem 4.13. *Let K be a compact subset of a metric space X.*

(a) *K is closed in X.*

(b) *If E is closed in K, then E is compact. In particular, $E \cap K$ is compact for every closed subset E of X.*

Another important property of compact sets is linked to the so-called **Cantor's Intersection Theorem** which concentrates on intersections of **decreasing nested sequences of nonempty compact sets**. In fact, the result is stated as follows:

Theorem 4.14. *Suppose that $\{K_\alpha\}$ is a family of compact subsets of X. If the intersection of every finite subcollection of $\{K_\alpha\}$ is nonempty, then we have*

$$\bigcap_\alpha K_\alpha \neq \varnothing.$$

In particular, if each K_n is nonempty and $K_{n+1} \subseteq K_n$ for every $n \in \mathbb{N}$, then we have

$$\bigcap_{n=1}^\infty K_n \neq \varnothing.$$

4.1.6 The Heine-Borel Theorem

A set E of a metric space X is said to be **bounded** if there exists a positive number M and a (fixed) point $q \in X$ such that
$$d(p, q) \leq M$$
for all $p \in E$.

In the previous subsection, we consider compact sets in an arbitrary metric space. If we take the metric space to be the Euclidean space $\mathbb{R}^n$, then there is a powerful and nice result for testing the compactness of a set. This is stated as follows:[a]

The Heine-Borel Theorem. *A subset $E \subseteq \mathbb{R}^n$ is compact if and only if E is closed and bounded in $\mathbb{R}^n$.*

---

[a] For a proof of it, please refer to [23, Theorem 2.41, p. 40].

4.1.7 Connected Sets

Definition 4.15. *Let X be a metric space. A **separation** of X is a pair E, F of disjoint nonempty open sets in X such that $X = U \cup V$. We call X **connected** if there is no separation of X.*

If the metric space X is $\mathbb{R}$, then its connected subsets have a nice structure:

Theorem 4.16. *The set $E \subseteq \mathbb{R}$ is connected if and only if for any $x, y \in E$ with $x < z < y$, we have $z \in E$.*

4.2 Open Sets and Closed Sets

Problem 4.1

(★) Let E be an open subset of X and $p \in E$. Prove that $E \setminus \{p\}$ is open.

Proof. Given $x \in E \setminus \{p\}$. Since $x \neq p$, we have $d(x, p) > 0$. Since E is open, we know that $N_r(x) \subseteq E$ for some $r > 0$. Take $r' = \min(r, \frac{d(x,p)}{2}) > 0$. If $p \in N_{r'}(x)$, then the definition gives

$$d(x, p) < r' \leq \frac{d(x, p)}{2},$$

a contradiction. Thus we have $N_{r'}(x) \subseteq E \setminus \{p\}$. Hence $E \setminus \{p\}$ is open and we finish the proof of the problem. ∎

Problem 4.2

(★) Let E be a closed set of X and p be a limit point of E. Is the set $E \setminus \{p\}$ closed?

Proof. Since p is a limit point of E, every neighborhood $N_r(p)$ contains a point $q \neq p$ such that $q \in E$. Thus q must be an element of $E \setminus \{p\}$ and then the point p is also a limit point of $E \setminus \{p\}$. However, it is clear that $E \setminus \{p\}$ does not contain p and hence Definition 4.5 shows that it is not closed. This completes the proof of the problem. ∎

Problem 4.3

(★) Suppose that $x, y \in X$ and $x \neq y$. Prove that there exist neighborhoods $N_r(x)$ and $N_R(y)$ of x and y respectively such that

$$N_r(x) \cap N_R(y) = \varnothing.$$

Proof. Since $x \neq y$, we have $d(x, y) > 0$. Let $\delta = \frac{d(x,y)}{4}$. Consider

$$N_\delta(x) = \{p \in X \mid d(x, p) < \delta\} \quad \text{and} \quad N_\delta(y) = \{p \in X \mid d(y, p) < \delta\}.$$

We claim that
$$N_\delta(x) \cap N_\delta(y) = \varnothing. \tag{4.1}$$
Otherwise, there exists a $p_0 \in N_\delta(x) \cap N_\delta(y)$. Since $p_0 \in N_\delta(x)$ and $p_0 \in N_\delta(y)$, we have
$$d(x, p_0) < \delta \quad \text{and} \quad d(y, p_0) < \delta. \tag{4.2}$$
By the triangle inequality and the inequalities (4.2), we obtain
$$d(x,y) \le d(x, p_0) + d(y, p_0) < \delta + \delta = 2\delta = \frac{d(x,y)}{2}$$
which is a contradiction. Hence we get the claim (4.1). This ends the proof of the problem. ∎

Problem 4.4

(★) Suppose that E is open and F is closed. Prove that $E \setminus F$ is open and $F \setminus E$ is closed.

Proof. Recall from Problem 1.2 that $E \setminus F = E \cap F^c$. Since F is closed, F^c is open by Theorem 4.6. By Theorem 4.7(b), $E \cap F^c$ is open. Similarly, we have $F \setminus E = F \cap E^c$ and Theorem 4.7(b) implies that $F \cap E^c$ is closed. Hence we have completed the proof of the problem. ∎

Problem 4.5

(★) Let E and F be subsets of $\mathbb{R}$. Define
$$E + F = \{x + y \mid x \in E \text{ and } y \in F\}.$$
Prove that $E + F$ is open in X if either E or F is open in X.

Proof. We just prove the case when E is open because the other case is similar. For each $y \in \mathbb{R}$, since E is open in X, $E + \{y\}$ is also open in X.[b] By definition, we have
$$E + F = \bigcup_{y \in F} (E + \{y\}).$$
By Theorem 4.7(a), $E + F$ is open in X which completes the proof of the problem. ∎

Problem 4.6

(★)(★) Prove that E is open in X if and only if E is an union of neighborhoods.

Proof. For every $p \in E$, since E is open, there exists $r_p > 0$ such that
$$N_{r_p}(p) \subseteq E. \tag{4.3}$$

---

[b]You can imagine that the set $E + \{y\}$ is a "translation" of the set E, so every neighborhood of E is translated y units to become a neighborhood of $E + \{y\}$.

4.2. Open Sets and Closed Sets

It is clear that
$$\{p\} \subseteq N_{r_p}(p) \tag{4.4}$$
for every $p \in E$. Combining the set relations (4.3) and (4.4), we have the relations
$$E = \bigcup_{p \in E} \{p\} \subseteq \bigcup_{p \in E} N_{r_p}(p) \subseteq \bigcup_{p \in E} E = E.$$
In other words, we have
$$E = \bigcup_{p \in E} N_{r_p}(p)$$
which proves the sufficient part of the problem.

Next, for the necessary part of the problem, we suppose that
$$E = \bigcup_{\alpha} N_{\alpha},$$
where each N_α is a neighborhood of some point of X. By Theorem 4.4, each N_α is open in X and we deduce from Theorem 4.7(a) that E is also open in X. Hence we complete the proof of the problem. ∎

Problem 4.7

(★)(★) *Let E be closed in X. Prove that it is a countable intersection of open sets in X.*

Proof. For each $n \in \mathbb{N}$ and $p \in E$, it is clear that $N_{\frac{1}{n}}(p)$ is open in X by Theorem 4.4. Therefore, Theorem 4.7(a) ensures that
$$F_n = \bigcup_{p \in E} N_{\frac{1}{n}}(p)$$
is also open in X. Now we note from the definition that $E \subseteq F_n$ for each $n \in \mathbb{N}$, so
$$E \subseteq \bigcap_{n=1}^{\infty} F_n. \tag{4.5}$$

Let $q \in \left(\bigcap_{n=1}^{\infty} F_n\right) \setminus E$. Then we have $q \in E^c$. Since E is closed in X, E^c is open in X by Theorem 4.6. Thus there is a $r > 0$ such that $N_r(q) \subseteq E^c$. By Theorem 2.1 (the Archimedean Property), one can find $m \in \mathbb{N}$ such that $mr > 1$, i.e., $\frac{1}{m} < r$. Recall that $q \in F_n$ for all $n \in \mathbb{N}$, so we fix $n = m$ to get $q \in F_m$ and thus $q \in N_{\frac{1}{m}}(p)$ *for some* $p \in E$. By definition, we have
$$d(p, q) < \frac{1}{m} < r$$
so that $p \in N_{\frac{1}{m}}(q) \subseteq N_r(q) \subseteq E^c$, but this implies that $E \cap E^c \neq \varnothing$, a contradiction. Hence we have the equality in the set relation (4.5), completing the proof of the problem. ∎

Problem 4.8

(★) Suppose that $\mathcal{V} = \{V_\alpha\}$ is a family of nonempty disjoint open subsets of $\mathbb{R}$. If $\mathcal{V}$ is infinite, prove that $\mathcal{V}$ is countable.

Proof. Let $V_\alpha \in \mathcal{V}$. Pick an arbitrary point $p_\alpha \in V_\alpha$. Since V_α is open in $\mathbb{R}$, there exists a $\epsilon_\alpha > 0$ such that
$$(p_\alpha - \epsilon_\alpha, p_\alpha + \epsilon_\alpha) \subseteq V_\alpha.$$
By Theorem 2.2 (Density of Rationals), the interval $(p_\alpha - \epsilon_\alpha, p_\alpha + \epsilon_\alpha)$ contains a rational number r_α. Take $V_\beta \in \mathcal{V}$, where $\beta \neq \alpha$. Repeat the above analysis, we can get a rational number r_β in V_β. Now $r_\alpha \neq r_\beta$ because $V_\alpha \cap V_\beta = \varnothing$. Thus what we have shown is that we take a rational number r_α from each V_α as a representative and different open subsets gives us different rational representatives. Hence the collection $\{r_\alpha\}$ is a subset of $\mathbb{Q}$. Since $\mathcal{V}$ is infinite, the set $\{r_\alpha\}$ must be infinite too. By Theorem 3.4, $\{r_\alpha\}$ must be countable, finishing our proof of the problem. ∎

Problem 4.9

(★)(★) What sets are both open and closed in $\mathbb{R}$?

Proof. Suppose that E is both open and closed in $\mathbb{R}$. By Theorem 4.6, E^c is also open and closed in $\mathbb{R}$. Assume that
$$E \neq \varnothing \quad \text{and} \quad E^c \neq \varnothing.$$
Then we pick $x \in E$ and $y \in E^c$. Without loss of generality, we may assume that $x < y$. Consider the set
$$F = \{a \in \mathbb{R} \mid [x, a] \subseteq E\}.$$
Since E is open in $\mathbb{R}$, we have $(x - \epsilon, x + \epsilon) \subseteq E$ for some $\epsilon > 0$. This means $x + \frac{\epsilon}{2} \in F$ and F is nonempty. If F is not bounded above, then we have $[x, n] \subseteq E$ for infinitely many positive integers n. Take $n \geq y$ so that
$$[x, y] \subseteq [x, n] \subseteq E$$
which gives a contradiction that $y \in E$. Thus F must be bounded above. By the Completeness Axiom, we know that $\alpha = \sup F$ exists in $\mathbb{R}$.

If $\alpha \in E$, then since E is open in $\mathbb{R}$, $\alpha + \delta \in E$ for some $\delta > 0$. Thus $[x, \alpha + \delta] \subseteq E$ and then $\alpha + \delta \in F$. However, this implies the contradiction that
$$\alpha + \delta \leq \alpha.$$

Suppose that $\alpha \in E^c$. On the one hand, since E^c is open in $\mathbb{R}$, we have $\alpha - \delta' \in E^c$ for some $\delta' > 0$. Since $\alpha - \delta' < \alpha$, we follow from the definition of supremum that
$$\alpha - \delta' \leq a$$
for some $a \in F$. If $\alpha - \delta' \leq x$, then
$$[x, y] \subseteq [\alpha - \delta', y] \subseteq E^c$$

4.2. Open Sets and Closed Sets

which implies that $x \in E^c$, a contradiction. Therefore, we have $x < \alpha - \delta'$ and this fact yields

$$[x, \alpha - \delta'] \subseteq [x, a] \subseteq E. \tag{4.6}$$

On the other hand, the fact $\alpha - \delta' \in E^c$ implies that $\alpha - \delta' \notin E$ and so $[x, \alpha - \delta'] \nsubseteq E$ which contradicts the set relation (4.6). Hence we obtain either $E = \varnothing$ or $E^c = \varnothing$ and each case implies that $\varnothing$ and $\mathbb{R}$ are the *only* sets that are both open and closed in $\mathbb{R}$. This completes the proof of the problem. ∎

Problem 4.10

(★)(★) Let E be a subset of X and $x \in X$. Define $\rho_E(x) = \inf\{d(x, p) \,|\, p \in E\}$. Prove that $x \in \overline{E}$ if and only if $\rho_E(x) = 0$.

Proof. It is obvious that $\rho_E(x) \geq 0$. Suppose that $x \in \overline{E}$. Then $x \in E$ or $x \in E'$. If $x \in E$, then it is clear that $\rho_E(x) = 0$. If $x \in E'$, then it is a limit point of E. Thus *for every* $\epsilon > 0$, there is a point

$$p \in N_\epsilon(x) \cap (E \setminus \{x\})$$

In other words, we have $d(x, p) < \epsilon$ *for every* $\epsilon > 0$ and by definition, $\rho_E(x) = 0$.

Conversely, we suppose that $\rho_E(x) = 0$. If $x \in E$, then we are done. Now without loss of generality, we may assume that $x \notin E$. We claim that for every $\epsilon > 0$, there exists a $p \in E$ such that $d(x, p) < \epsilon$. Otherwise, there is a positive integer N such that

$$\frac{1}{N} \leq d(x, p)$$

for all $p \in E$ but this means that $\frac{1}{N}$ is a lower bound of the set $A_x = \{d(x, p) \,|\, p \in E\}$. Thus we have

$$\rho_E(x) \geq \frac{1}{N} > 0,$$

a contradiction. This proves our claim. In addition, it is easy to see that $p \neq x$ because $x \notin E$. In other words, what we have shown is that

$$N_\epsilon(x) \cap (E \setminus \{x\}) \neq \varnothing$$

for every $\epsilon > 0$. By Definition 4.8, we have $x \in E' \subseteq \overline{E}$, completing the proof of the problem. ∎

Problem 4.11

(★)(★) Suppose that E° is the interior of E.

(a) Prove that E° is an open set in X.

(b) Prove that E is open in X if and only if $E^\circ = E$.

(c) Prove that if $F \subseteq E$ and F is open in X, then $F \subseteq E^\circ$.

Proof.

(a) Let $x \in E^\circ$. By Definition 4.8, x is an interior point of E and then there exists a $\epsilon > 0$ such that $N_\epsilon(x) \subseteq E$. If $y \in N_\epsilon(x)$, then Theorem 4.4 implies that

$$N_{\epsilon'}(y) \subseteq N_\epsilon(x) \subseteq E$$

for some $\epsilon' > 0$. By Definition 4.8, $y \in E^\circ$. Since y is arbitrary, we have $N_\epsilon(x) \subseteq E^\circ$ and hence E° is open in X.

(b) If $E^\circ = E$, then part (a) implies that E is open in X. Suppose that E is open in X. Recall that we always have $E^\circ \subseteq E$. Since E is open in X, every $x \in E$ is an interior point of E by Definition 4.3. Thus $x \in E^\circ$, i.e., $E \subseteq E^\circ$. Hence we have $E^\circ = E$.

(c) Let $\mathcal{F}$ be the union of all open subsets $F \subseteq E$ of X. Since $E^\circ \subseteq E$, we have

$$E^\circ \subseteq \mathcal{F}. \tag{4.7}$$

By Theorem 4.7(a), $\mathcal{F}$ is open in X. Thus for every $x \in \mathcal{F}$, there exists a $\delta > 0$ such that

$$N_\delta(x) \subseteq \mathcal{F}. \tag{4.8}$$

Since $\mathcal{F} \subseteq E$, we obtain from the set relation (4.8) that

$$N_\delta(x) \subseteq E$$

and thus $x \in E^\circ$, i.e., $\mathcal{F} \subseteq E^\circ$. Combining this and the set relation (4.7), we have $E^\circ = \mathcal{F}$. This completes the proof of the problem. ∎

Problem 4.12

(⋆) (⋆) For any E and F, show that

$$E^\circ \cap F^\circ = (E \cap F)^\circ \quad \text{and} \quad E^\circ \cup F^\circ \subseteq (E \cup F)^\circ.$$

Find an example to show that the inclusion of the second assertion can be proper.

Proof. Since $E^\circ \subseteq E$ and $F^\circ \subseteq F$, we have

$$E^\circ \cap F^\circ \subseteq E \cap F \quad \text{and} \quad E^\circ \cup F^\circ \subseteq E \cup F. \tag{4.9}$$

By Problem 4.11(a), both E° and F° are open in X so we deduce from Theorem 4.7(b) that $E^\circ \cap F^\circ$ and $E^\circ \cup F^\circ$ are open in X. Now Problem 4.10(c) shows that $(E \cap F)^\circ$ and $(E \cup F)^\circ$ are the largest open subsets of X inside $E \cap F$ and $E \cup F$ respectively. This fact and the set relations (4.9) implies that

$$E^\circ \cap F^\circ \subseteq (E \cap F)^\circ \quad \text{and} \quad E^\circ \cup F^\circ \subseteq (E \cup F)^\circ. \tag{4.10}$$

Thus the second set relation in (4.10) is one of our expected results.

Since $(E \cap F)^\circ \subseteq E \cap F$, we acquire that

$$(E \cap F)^\circ \subseteq E \quad \text{and} \quad (E \cap F)^\circ \subseteq F. \tag{4.11}$$

4.2. Open Sets and Closed Sets

Since E° and F° are the largest open subsets of X inside E and F respectively, it follows from this and the set relations (4.11) that

$$(E \cap F)^\circ \subseteq E^\circ \subseteq E \quad \text{and} \quad (E \cap F)^\circ \subseteq F^\circ \subseteq F.$$

Therefore, we have
$$(E \cap F)^\circ \subseteq E^\circ \cap F^\circ. \tag{4.12}$$

Combining the first set relation in (4.10) and the set relation (4.12), we obtain the remaining expected result that

$$(E \cap F)^\circ = E^\circ \cap F^\circ.$$

Consider the sets $E = [0, 1]$ and $F = [1, 2]$. Then we have

$$E^\circ = (0, 1), \quad F^\circ = (1, 2) \quad \text{and} \quad (E \cup F)^\circ = (0, 2).$$

In this case, we have $E^\circ \cup F^\circ \subset (E \cup F)^\circ$, completing the proof of the problem. ∎

Problem 4.13

(★)(★) Let E and F be subsets of X.

(a) Show that $\overline{E \cap F} \subseteq \overline{E} \cap \overline{F}$.

(b) Show that $E \cap \overline{F} \subseteq \overline{E \cap F}$ if E is open.

(c) Find an example to show that the inclusion in parts (a) and (b) can be proper.

Proof.

(a) Note that
$$E \cap F \subseteq E \subseteq \overline{E} \tag{4.13}$$

and $\overline{E}$ is closed in X by Theorem 4.9(a). Since $\overline{E \cap F}$ is the smallest closed subset of X containing $E \cap F$, this and the set relation (4.13) implies that $\overline{E \cap F} \subseteq \overline{E}$. Similarly, $\overline{E \cap F} \subseteq \overline{F}$ also holds and so

$$\overline{E \cap F} \subseteq \overline{E} \cap \overline{F}.$$

(b) Suppose that E is open in X. Given $\epsilon > 0$. If $x \in E \cap \overline{F}$, then $x \in E$ and so *there exists* a $\delta > 0$ such that $N_\delta(x) \subseteq E$. Since $x \in \overline{F}$, x is a limit point of F. Take $\epsilon' = \min(\epsilon, \delta)$. By Definition 4.5, we have

$$N_{\epsilon'}(x) \cap (F \setminus \{x\}) \neq \varnothing.$$

Let $y \in N_{\epsilon'}(x) \cap (F \setminus \{x\})$. Since $N_{\epsilon'}(x) \subseteq N_\delta(x) \subseteq E$, we always have

$$y \in N_{\epsilon'}(x) \cap (F \setminus \{x\}) \subseteq E \cap (F \setminus \{x\}). \tag{4.14}$$

Furthermore, $y \in N_{\epsilon'}(x) \subseteq N_\epsilon(x)$. It follows from this and the set relation (4.14) that

$$N_\epsilon(x) \cap [(E \cap F) \setminus \{x\}] \neq \varnothing$$

for every $\epsilon > 0$. By Definition 4.5, $x \in \overline{E \cap F}$ and hence $E \cap \overline{F} \subseteq \overline{E \cap F}$.

(c) Take $E = (0,1)$ and $F = (1,2)$. Then it is easy to check that
$$\overline{E \cap F} = \overline{\varnothing} = \varnothing \quad \text{and} \quad \overline{E} \cap \overline{F} = \{1\},$$
so the inclusion in parts (a) and (b) can be proper.

This completes the proof of the problem. ∎

> **Problem 4.14**
>
> (⋆) Prove that $E^\circ = X \setminus (\overline{X \setminus E})$.

Proof. Now $x \in E^\circ$ if and only if *there exists* a $\epsilon > 0$ such that $N_\epsilon(x) \subseteq E$ (by Definition 4.3) if and only if $N_\epsilon(x) \cap (X \setminus E) = \varnothing$ (by difference of sets in §1.1.1) if and only if $x \notin \overline{X \setminus E}$ (by Definition 4.5) if and only if $x \in X \setminus (\overline{X \setminus E})$ (by difference of sets in §1.1.1). Hence we have the desired formula and we complete the proof of the problem. ∎

> **Problem 4.15**
>
> (⋆) Find an example to show that the inclusion in Theorem 4.11(b) can be proper.

Proof. Consider $X = \mathbb{R}$, $Y = \mathbb{Q}$ and $E = Y \cap (0,1)$. On the one hand, since $(0,1)$ is open in $\mathbb{R}$, E is open in Y by Theorem 4.10. By Problem 4.11(b), we know that
$$E_Y^\circ = E = Y \cap (0,1). \tag{4.15}$$
On the other hand, if $x \in Y_X^\circ$, then x is an interior point of Y and thus there exists a $\epsilon > 0$ such that $(x - \epsilon, x + \epsilon) \subseteq Y$. However, Problem 2.9 guarantees that the existence of an irrational number θ lying in $(x - \epsilon, x + \epsilon)$ and this implies that $\theta \in Y$, a contradiction. Hence we must have
$$Y_X^\circ = \varnothing. \tag{4.16}$$
Since $E \subseteq Y$ implies that $E^\circ \subseteq Y^\circ$,[c] we obtain from this and the fact (4.16) that
$$E_X^\circ \subseteq Y_X^\circ = \varnothing.$$
In other words, we have $E_X^\circ \cap Y = \varnothing$ which is clearly a proper subset of the set (4.15). This ends the proof of the problem. ∎

4.3 Compact Sets

> **Problem 4.16**
>
> (⋆)(⋆) Suppose that $K \subseteq U \cup V$, where U and V are disjoint open subsets of X and K is compact. Prove that $K \cap U$ is compact.

---

[c]See Table 4.1.

4.3. Compact Sets

Proof. Since $U \cap V = \emptyset$, we have $U \subseteq V^c$. Since $U \cup U^c = X$, we have

$$K \cap V^c = (K \cap V^c) \cap X$$
$$= (K \cap V^c) \cap (U \cup U^c)$$
$$= [(K \cap V^c) \cap U] \cup [(K \cap V^c) \cap U^c]. \qquad (4.17)$$

Let $x \in K$. Then $x \in U \cup V$ so that $x \notin U^c \cap V^c$. In other words, $K \cap (U^c \cap V^c) = \emptyset$. Thus we put this into (4.17) to get

$$K \cap V^c = (K \cap U) \cup \emptyset = K \cap U. \qquad (4.18)$$

Since V^c is closed in X, Theorem 4.13(b) implies that $K \cap V^c$ is compact. By the representation (4.18), we conclude that $K \cap U$ is also compact. This finishes the proof of the problem. ■

Problem 4.17

(★)(★) Suppose that $K_1, \ldots, K_n$ are compact sets. Prove that $K = K_1 \cup K_2 \cup \cdots \cup K_n$ is also compact.

Proof. Let $\{V_\alpha\}$ be an open cover of K. Since K_i is compact, there exists $V_{\alpha_{i_1}}, V_{\alpha_{i_2}}, \ldots, V_{\alpha_{i_m}}$ such that

$$K_i \subseteq V_{\alpha_{i_1}} \cup V_{\alpha_{i_2}} \cup \cdots \cup V_{\alpha_{i_m}} = \bigcup_{j=1}^{m} V_{\alpha_{i_j}}$$

which implies that

$$K = \bigcup_{i=1}^{n} K_i \subseteq \bigcup_{i=1}^{n} \bigcup_{j=1}^{m} V_{\alpha_{i_j}}$$

By Definition 4.12 (Compact Sets), K is compact and this completes the proof of the problem. ■

Problem 4.18

(★)(★) Suppose that $\{K_\alpha\}$ is a collection of compact sets. Prove that

$$K = \bigcap_\alpha K_\alpha$$

is also compact.

Proof. Since each K_α is compact, it is closed in X by Theorem 4.13(a). By Theorem 4.7(a), the set K must be closed in X. It is clear that

$$K = K \cap K_\alpha$$

for every α. Now an application of Theorem 4.13(b) shows that K is compact and we finish the proof of the problem. ■

Problem 4.19

(★)(★) Prove that the closed interval $[0, 1]$ is compact from the definition.

Proof. Let $\{V_\alpha\}$ be an open cover of $[0, 1]$. Consider the set

$$E = \{x \in [0, 1] \mid [0, x] \text{ can be covered by finitely many } V_\alpha\}.$$

It is clear that $E \neq \varnothing$ because $0 \in E$. Furthermore, it is also bounded above by 1. By the Completeness Axiom, we know that $\beta = \sup E$ exists in $\mathbb{R}$. Since $0 \in V_\alpha$ for some α and V_α is open in $\mathbb{R}$, we have $(-\eta, \eta) \subseteq V_\alpha$ for some $\eta > 0$. Then it implies that $[0, \frac{\eta}{2}] \subseteq V_\alpha$ and this means that $\frac{\eta}{2} \in E$. By this fact and the definition of supremum, we must have

$$\beta > 0. \tag{4.19}$$

Assume that $\beta < 1$. Since $\{V_\alpha\}$ covers $[0, 1]$, $\beta \in V_{\alpha_0}$ for some α_0. Since V_{α_0} is open in $\mathbb{R}$, there exists a $\delta > 0$ such that

$$\beta \in (\beta - \delta, \beta + \delta) \subseteq V_{\alpha_0}. \tag{4.20}$$

We claim that $[0, \beta]$ can be covered by finitely many V_α. By the fact (4.19) and Theorem 2.2 (Density of Rationals), we can choose δ so small that $\beta - \delta > 0$. Since $\beta - \frac{\delta}{2} < \beta$, there exists a $x \in E$ such that $\beta - \frac{\delta}{2} \leq x < \beta$. Since $[0, \beta - \frac{\delta}{2}] \subseteq [0, x]$, the interval $[0, \beta - \frac{\delta}{2}]$ is covered by finitely many V_α. Therefore, we follow from the these, the set relation (4.20) and the fact $[0, \beta] \subseteq [0, \beta - \frac{\delta}{2}] \cup (\beta - \delta, \beta]$ that $[0, \beta]$ is covered by finitely many V_α. This prove the claim, i.e.,

$$[0, \beta] \subseteq V_{\alpha_1} \cup V_{\alpha_2} \cup \cdots \cup V_{\alpha_m}. \tag{4.21}$$

Furthermore, it follows from the set relations (4.20) and (4.21) that

$$\left[0, \beta + \frac{\delta}{2}\right] = [0, \beta] \cup \left[\beta, \beta + \frac{\delta}{2}\right]$$
$$\subseteq V_{\alpha_1} \cup V_{\alpha_2} \cup \cdots \cup V_{\alpha_m} \cup (\beta - \delta, \beta + \delta)$$
$$\subseteq V_{\alpha_1} \cup V_{\alpha_2} \cup \cdots \cup V_{\alpha_m} \cup V_{\alpha_0}$$

which implies that $\beta + \frac{\delta}{2} \in E$, but this certainly contradicts to the fact that $\beta = \sup E$. Hence we must have $\beta = 1$ and then $[0, 1]$ is compact, completing the proof of the problem. ∎

Problem 4.20

(★)(★) Construct an open cover of $(0, 1)$ which has no finite subcover so that $(0, 1)$ is not compact.

Proof. For each $n = 2, 3, \ldots$, we consider the interval $V_n = (\frac{1}{n}, 1)$. If $x \in (0, 1)$, then it follows from Theorem 2.1 (the Archimedean property) that there exists a positive integer n such that $nx > 1$, i.e., $x \in V_n$. Furthermore, we have

$$(0, 1) \subseteq \bigcup_{i=2}^{\infty} V_n,$$

4.3. Compact Sets

i.e., $\{V_2, V_3, \ldots\}$ is an open cover of the segment $(0, 1)$.

Assume that $\{V_{n_1}, V_{n_2}, \ldots, V_{n_k}\}$ was a finite subcover of $(0,1)$, where $n_1, n_2, \ldots, n_k$ are positive integers and $2 \leq n_1 < n_2 < \cdots < n_k$. By definition, we have

$$V_{n_1} \subseteq V_{n_2} \subseteq \cdots \subseteq V_{n_k}$$

and so

$$(0,1) \subseteq \bigcup_{i=1}^{k} V_{n_i} \subseteq V_{n_k},$$

contradicting to the fact that $\frac{1}{2n_k} \in (0,1)$ but $\frac{1}{2n_k} \notin (\frac{1}{n_k}, 1)$. Hence $\{V_2, V_3, \ldots\}$ does not have a finite subcover for $(0, 1)$. This completes the proof of the problem. ∎

Remark 4.1

Problems 4.19 and 4.20 tell us the fact that arbitrary subsets of a compact set need not be compact.

Problem 4.21

(⋆)(⋆) *Suppose that $\{K_\alpha\}$ is a collection of compact subsets of X. Prove that if $\bigcap_\alpha K_\alpha = \varnothing$, then there is a choice of finitely many indices $\alpha_1, \ldots, \alpha_n$ such that*

$$K_{\alpha_1} \cap K_{\alpha_2} \cap \cdots \cap K_{\alpha_n} = \varnothing.$$

Proof. Let $K \in \{K_\alpha\}$. By Theorem 4.13(a), each K_α is closed in X so that K_α^c is open in X by Theorem 4.6. For every $x \in K$, since $\bigcap_\alpha K_\alpha = \varnothing$, $x \notin K_\alpha$ for some α and this is equivalent to saying that

$$x \in K_\alpha^c.$$

Therefore, $\{K_\alpha^c\}$ is an open cover of K. Since K is compact, a finite subcover $K_{\alpha_1}^c, K_{\alpha_2}^c, \ldots, K_{\alpha_n}^c$ of K exists, i.e.,

$$K \subseteq K_{\alpha_1}^c \cup K_{\alpha_2}^c \cup \ldots \cup K_{\alpha_n}^c. \tag{4.22}$$

Applying complements to both sides in (4.22) and using Problem 1.1, we get

$$K_{\alpha_1} \cap K_{\alpha_2} \cap \cdots \cap K_{\alpha_n} \subseteq K^c$$

which implies that

$$K_{\alpha_1} \cap K_{\alpha_2} \cap \cdots \cap K_{\alpha_n} \cap K = \varnothing.$$

This ends the proof of the problem. ∎

> **Problem 4.22**
>
> $(\star)(\star)$ Suppose that $K \neq \varnothing$ is a compact metric space and $\{E_n\}$ is a sequence of nonempty closed subsets of K such that $E_{n+1} \subseteq E_n$ for every $n \in \mathbb{N}$. Verify that
>
> $$E = \bigcap_{n=1}^{\infty} E_n \neq \varnothing.$$

Proof. Assume that $E = \varnothing$. We claim that

$$K \subseteq E^c.$$

Otherwise, there exists a $x \in K$ such that $x \notin E^c$. This means that $x \notin E_n^c$ for every $n \in \mathbb{N}$ or equivalently, $x \in E_n$ for every $n \in \mathbb{N}$. However, this implies that $x \in E$ so that $E \neq \varnothing$, a contradiction.

Recall that each E_n is closed in K, so we get from Theorem 4.6 that E_n^c is open in K and then $\{E_n^c\}$ forms an open covering of K becuase

$$E^c = \bigcup_{n=1}^{\infty} E_n^c.$$

By Definition 4.12 (Compact Sets), one can find a finitely many indices $n_1, \ldots, n_k$ such that

$$K \subseteq E_{n_1}^c \cup E_{n_2}^c \cup \cdots \cup E_{n_k}^c. \tag{4.23}$$

Let x be a point of K such that $x \in E_{n_1} \cap \cdots \cap E_{n_k}$. Then Problem 1.1 implies that

$$x \notin E_{n_1}^c \cup E_{n_2}^c \cup \cdots \cup E_{n_k}^c. \tag{4.24}$$

Combining the two facts (4.23) and (4.24), we must have $x \notin K$, but this contradicts to our hypothesis that $x \in K$. Therefore, we have

$$E_{n_1} \cap \cdots \cap E_{n_k} = \varnothing. \tag{4.25}$$

Without loss of generality, we assume that $n_1 \leq n_2 \leq \cdots \leq n_k$. Since $E_{n+1} \subseteq E_n$ for every $n \in \mathbb{N}$, we see that

$$E_{n_k} \subseteq E_{n_{k-1}} \subseteq \cdots \subseteq E_{n_2} \subseteq E_{n_1}.$$

Put this chain into the left-hand side of the expression (4.25), we obtain

$$E_{n_k} = \varnothing$$

which contradicts the hypothesis that each E_n is nonempty. Hence we must have $E \neq \varnothing$ and thus finishing the proof of the problem. ∎

4.3. Compact Sets

> **Problem 4.23**
>
> (⋆)(⋆)(⋆) Let $\mathcal{F}$ be a collection of subsets of X. We say $\mathcal{F}$ has the **finite intersection property** if for every finite intersection of sets from $\mathcal{F}$, $\{F_1, F_2, \ldots, F_n\} \subseteq \mathcal{F}$, we have
>
> $$\bigcap_{k=1}^{n} F_k \neq \varnothing.$$
>
> Prove that a metric space K is compact if and only if every collection of closed sets $\mathcal{F}$ with the finite intersection property has a nonempty intersection.

Proof. Let $\{F_\alpha\}$ be a collection of closed sets with the finite intersection property. Suppose that the metric space K is compact. If $\bigcap_\alpha F_\alpha = \varnothing$, then we have

$$K = \bigcup_\alpha F_\alpha^c. \tag{4.26}$$

Since each F_α is closed in K, we know from Theorem 4.6 that F_α^c is open in K. Therefore, it is easy to see from the expression (4.26) that $\{F_\alpha^c\}$ is an open cover of K. By the compactness of K, a choice of finitely many indices $\alpha_1, \ldots, \alpha_n$ exists so that

$$K = F_{\alpha_1}^c \cup \cdots \cup F_{\alpha_n}^c. \tag{4.27}$$

By Problem 1.1, this fact (4.27) implies that

$$F_{\alpha_1} \cap \cdots \cap F_{\alpha_n} = (F_{\alpha_1}^c \cup \cdots \cup F_{\alpha_n}^c)^c = K^c = \varnothing$$

which contradicts the basic assumption that $\{F_\alpha\}$ has the finite intersection property. Therefore, we conclude that

$$\bigcap_\alpha F_\alpha \neq \varnothing. \tag{4.28}$$

Conversely, let $\{V_\beta\}$ be an open covering of K. Note that each V_β^c must be closed in K by Theorem 4.6 and thus $\{V_\beta^c\}$ forms a collection of closed subsets of K. Assume that $K \neq \bigcup_{k=1}^{n} V_{\beta_k}$ for *all* finite indices $\beta_1, \ldots, \beta_n$. By this and Problem 1.1, we know that

$$\bigcap_{k=1}^{n} V_{\beta_k}^c = \Big(\bigcup_{k=1}^{n} V_{\beta_k}\Big)^c \neq \varnothing.$$

Therefore, it means that $\{V_\beta^c\}$ satisfies the finite intersection property. Hence we follow from the hypothesis that the condition (4.28) holds for $\{V_\beta^c\}$. Finally, Problem 1.1 again implies that

$$\varnothing \neq \bigcap_\beta V_\beta^c = \Big(\bigcup_\beta V_\beta\Big)^c$$

or equivalently,

$$\bigcup_\beta V_\beta \neq K$$

which definitely contradicts to the fact that $\{V_\beta\}$ is an open cover of K. Hence $\{V_\beta\}$ must have a finite subcover for K and then K is compact. We have completed the proof of the problem. ■

4.4 The Heine-Borel Theorem

Problem 4.24

(⋆) *Prove that the sets $\mathbb{Q}$, $\mathbb{R}$ and $E = \{\frac{1}{n} \mid n = 1, 2, \ldots\}$ are not compact.*

Proof. The first two sets are not compact because they are not bounded. Since 0 is a limit point of E but $0 \notin E$, E is not closed. By the Heine-Borel Theorem, E is not compact. We end the proof of the problem. ∎

Problem 4.25

(⋆) *Prove that $[0, 1]$ is compact but $(0, 1)$ is not by using the Heine-Borel Theorem.*

Proof. Since $[0, 1]$ is closed and bounded in $\mathbb{R}$, it is compact by the Heine-Borel Theorem. Since $(0, 1)$ is not closed in $\mathbb{R}$, it is not compact by the Heine-Borel Theorem. This ends the proof of the problem. ∎

Remark 4.2

If you compare Problem 4.25 with Problems 4.19 and 4.20, you will see how powerful the Heine-Borel Theorem is.

Problem 4.26

(⋆) *Prove that every finite subset of $\mathbb{R}$ is compact.*

Proof. Let $E = \{x_1, \ldots, x_n\}$ be a finite subset of $\mathbb{R}$. Without lost of generality, we may assume that $x_1 \leq x_2 \leq \cdots \leq x_n$. Then it is easy to see that E is bounded by $\max(|x_1|, |x_n|)$. For each x_k ($k = 1, 2, \ldots, n$), since the set

$$\mathbb{R} \setminus \{x_k\} = (-\infty, x_k) \cup (x_k, \infty)$$

is open in $\mathbb{R}$, the point set $\{x_k\}$ is closed in $\mathbb{R}$. By Theorem 4.7(b), E is also closed in $\mathbb{R}$. Hence it follows from the Heine-Borel Theorem that E is compact. This ends the proof of the problem. ∎

Problem 4.27

(⋆)(⋆) *Let K be a nonempty compact set in $\mathbb{R}$. Prove that $\sup K, \inf K \in K$.*

Proof. By the Heine-Borel Theorem, K must be bounded. Besides, K is nonempty. Thus, by the Completeness Axiom, we conclude that $\sup K$ and $\inf K$ exist in $\mathbb{R}$. Let $\beta = \sup K$ and $\alpha = \inf E$.

Assume that $\beta \notin K$. Since K is compact, it is closed in $\mathbb{R}$ by Theorem 4.13(a). Then β is *not* a limit point of K, i.e., there exists a $\epsilon > 0$ such that
$$(\beta - \epsilon, \beta + \epsilon) \cap K = \varnothing. \tag{4.29}$$
In other words, we have
$$(\beta - \epsilon, \beta + \epsilon) \subseteq \mathbb{R} \setminus K.$$
If $\beta - \frac{\epsilon}{2}$ is not an upper bound of K, then the definition of supremum shows that there is a $x \in K$ such that $\beta - \frac{\epsilon}{2} \leq x$ which implies that
$$x \in \left[\beta - \frac{\epsilon}{2}, \beta + \frac{\epsilon}{2}\right] \subset (\beta - \epsilon, \beta + \epsilon),$$
but this definitely contradicts the result (4.29). Then $\beta - \frac{\epsilon}{2}$ is an upper bound of K which is a contradiction. Hence we conclude that $\beta \in K$. The case for proving $\alpha \in K$ is very similar, so we omit the details here. Now we complete the proof of the problem. ∎

Problem 4.28

(⋆) Suppose that $E_0 = [0,1]$. Let E_1 be the set obtained from E_0 by removing the "middle third" $(\frac{1}{3}, \frac{2}{3})$. Next, let E_2 be the set obtained from E_1 by removing the "middle thirds" $(\frac{1}{9}, \frac{2}{9})$ and $(\frac{7}{9}, \frac{8}{9})$. In fact, E_n is given by
$$E_n = \bigcup_{k=0}^{2^{n-1}-1} \left(\left[\frac{3k+0}{3^n}, \frac{3k+1}{3^n}\right] \cup \left[\frac{3k+2}{3^n}, \frac{3k+3}{3^n}\right]\right), \tag{4.30}$$
where $n \geq 1$. The intersection
$$C = \bigcap_{n=1}^{\infty} E_n \tag{4.31}$$
is the well-known **Cantor set**. Prove that C is compact.

Proof. It is easy to see from the definition (4.30) that each E_n is a union of 2^n closed intervals in $\mathbb{R}$, so Theorem 4.7(b) implies that E_n is closed in $\mathbb{R}$. Next, we see from the definition (4.31) that C is the intersection of closed subsets in $\mathbb{R}$, so it is also closed in $\mathbb{R}$ by Theorem 4.7(a). It is clear that C is a bounded set. Hence we deduce from the Heine-Borel Theorem that C is compact, finishing the proof of the problem. ∎

4.5 Connected Sets

Problem 4.29

(⋆) Show that $\mathbb{Q}$ is not connected.

Proof. Let $a, b \in \mathbb{Q}$ with $a < b$. By Problem 2.9, we see that there is an irrational number θ such that $a < \theta < b$. Since $\theta \notin \mathbb{Q}$, it follows from Theorem 4.16 that $\mathbb{Q}$ is not connected. This completes the proof of the problem. ∎

> ### Problem 4.31
>
> (★)(★) Suppose that the pair U and V forms a separation of X and E is a connected subset of X. Prove that E is contained either in U or V.

Proof. Assume that $E \cap U \neq \varnothing$ and $E \cap V \neq \varnothing$. By Theorem 4.10, we deduce that both $E \cap U$ and $E \cap V$ are open in E. Since U and V are disjoint, it is easy to see that

$$(E \cap U) \cap (E \cap V) = E \cap (U \cap V) = E \cap \varnothing = \varnothing$$

so that $E \cap U$ and $E \cap V$ are also disjoint. Furthermore, we have

$$(E \cap U) \cup (E \cap V) = E.$$

By Definition 4.15, $E \cap U$ and $E \cap V$ form a separation of E which is a contradiction. Hence we have either

$$E \subseteq U \quad \text{or} \quad E \subseteq V.$$

We complete the proof of the problem. ∎

> ### Problem 4.31
>
> (★)(★) Suppose that $\{E_\alpha\}$ is a family of connected subsets of X and $\bigcap_\alpha E_\alpha \neq \varnothing$. Prove that
>
> $$E = \bigcup_\alpha E_\alpha$$
>
> is connected.

Proof. Let $p \in \bigcap_\alpha E_\alpha$. Assume that E had a separation, i.e., $E = U \cup V$, where U and V are disjoint nonempty open subsets of X. Now we have either $p \in U$ or $p \in V$. Let $p \in U$. Since each E_α is connected, it must be contained in either U or V by Problem 4.30. Since $p \in E_\alpha$ for every α and $p \in U$, we have $E_\alpha \subseteq U$ for every α which implies that $E \subseteq U$ or equivalently $V = \varnothing$. This is clearly a contradiction, so we have the desired result that E is connected. We complete the proof of the problem. ∎

> ### Problem 4.32
>
> (★)(★) Let U and V form a separation of E. Prove that
>
> $$\overline{U} \cap V = U \cap \overline{V} = \varnothing.$$

4.5. Connected Sets

Proof. By Definition 4.15, U and V are disjoint nonempty open sets in X and $X = U \cup V$. Since $U^c = X \setminus U = V$, U is also closed in X by Theorem 4.6. Then it follows from Theorem 4.9(b) that $\overline{U} = U$. Recall that $U \cap V = \varnothing$, so we have

$$\overline{U} \cap V = U \cap V = \varnothing.$$

The case for $U \cap \overline{V} = \varnothing$ can be proven similarly, so we omit the proof here. This completes the proof of the problem. ∎

Problem 4.33

(★)(★) Suppose that E is connected. Prove that $\overline{E}$ is connected.

Proof. Assume that the pair U and V was a separation of $\overline{E}$. By Problem 4.30, we have either $E \subseteq U$ or $E \subseteq V$. Without loss of generality, we may assume that $E \subseteq U$. By the proof of Problem 4.32, we see that U is closed in $\overline{E}$. Then we establish from Theorem 4.9(c) that

$$\overline{E} \subseteq U$$

and this means that $V = \varnothing$, a contradiction. Hence $\overline{E}$ is connected and this completes the proof of the problem. ∎

Problem 4.34

(★)(★) If $E \subseteq \mathbb{R}$ is connected, prove that E° is also connected.

Proof. Assume that E° was not connected. By Theorem 4.16, there exist $x, y \in E^\circ$ and a number z with $x < z < y$ but $z \notin E^\circ$. Since $E^\circ \subseteq E$, we always have $x, y \in E$. Since E is connected, Theorem 4.16 implies that $(x, y) \subseteq E$. In particular, we have

$$z \in (x, y) \subseteq E. \tag{4.32}$$

By Theorem 2.2 (Density of Rationals) and the set relation (4.32), there exists a $\epsilon > 0$ such that $N_\epsilon(z) \subseteq (x, y) \subseteq E$, therefore

$$z \in E^\circ$$

which contradicts to the assumption. Hence E° is connected and we end the proof of the problem. ∎

CHAPTER 5

Sequences in Metric Spaces

5.1 Fundamental Concepts

In fundamental calculus, we have learnt the concepts of sequences of real (or complex) numbers. In fact, those concepts can be generalized and explained without difficulties in a general setting of sequences in metric spaces. The main references for this part are [3, §4.1 - §4.4, pp. 70 - 74], [5, §3.1 - §3.6], [6, Chap. 2], [23, Chap. 3] and [27, Chap. 5 and 6].

5.1.1 Convergent Sequences in Metric Spaces

Let $\{p_n\}$ be a sequence of points in a metric space X with the metric d. We say that it **converges** if there is a point $p \in X$ such that *for every $\epsilon > 0$, there exists* a positive integer N such that

$$d(p_n, p) < \epsilon$$

for all $n \geq N$. In notation, we write $p_n \to p$ as $n \to \infty$[a] or

$$\lim_{n \to \infty} p_n = p. \tag{5.1}$$

The point p in the limit (5.1) is called the **limit** of $\{p_n\}$. If $\{p_n\}$ does not converge, then we say that it is **diverge**. A sequence $\{p_n\}$ is called **bounded** if there exists a positive number M and a (fixed) point $q \in X$ such that

$$d(p_n, q) \leq M$$

for all $n \in \mathbb{N}$.

In the following, $\{a_n\}, \{b_n\}, \{p_n\}, \{q_n\}, \{x_y\}, \{y_n\}, \ldots$ will always denote sequences in a metric space X. Some basic facts about convergent sequences are given as follows:

Theorem 5.1 (Uniqueness of Limits of Sequences). *A sequence $\{p_n\}$ can converge to at most one point in X.*

Theorem 5.2. *Let $\{p_n\}$ be a sequence in X and $p_n \to p$, where $p \in X$. Then $\{p_n\}$ is bounded and p is a limit point of the set $E = \{p_1, p_2, \ldots\}$.*

---
[a]Or simply $p_n \to p$.

Given a set $\{n_1, n_2, \ldots\}$ of positive integers such that $n_1 < n_2 < \cdots$. Then the sequence $\{p_{n_k}\}$ is a **subsequence** of the original sequence $\{p_n\}$. The following result characterizes the convergence of $\{p_n\}$ and convergence of its subsequences $\{p_{n_k}\}$:

Theorem 5.3. *In a metric space X, $\{p_n\}$ converges to p if and only if every subsequence $\{p_{n_k}\}$ converges to p.*

Next, the following result can be treated as a converse to the second assertion in Theorem 5.2:

Theorem 5.4. *Suppose that $E \subseteq X$. If p is a limit point of E, then there is a sequence $\{p_n\}$ in E such that the result (5.1) holds.*

5.1.2 Sequences in $\mathbb{R}^n$ and Some Well-known Sequences

If the metric space X is taken to be $\mathbb{R}^n$ for some $n \in \mathbb{N}$, then the convergence of sequences in $\mathbb{R}^n$ and the algebraic operations of points in $\mathbb{R}^n$ can be summarized as follows:

Theorem 5.5. *Suppose that $\mathbf{p}_k \in \mathbb{R}^n$, where $k \in \mathbb{N}$ and*

$$\mathbf{p}_k = (p_{1k}, p_{2k}, \ldots, p_{nk}).$$

Then $\mathbf{p}_k \to \mathbf{p} = (p_1, \ldots, p_n)$ as $k \to \infty$ if and only if

$$\lim_{k \to \infty} p_{ik} = p_i,$$

where $i = 1, 2, \ldots, n$. Furthermore, if $\mathbf{p}_k, \mathbf{q}_k \in \mathbb{R}^n$ and $c_k \in \mathbb{R}$ for $k \in \mathbb{N}$ and $\mathbf{p}_k \to \mathbf{p}$, $\mathbf{q}_k \to \mathbf{q}$, $c_k \to c$ as $k \to \infty$, then we have

$$\lim_{k \to \infty} (\mathbf{p}_k + \mathbf{q}_k) = \mathbf{p} + \mathbf{q}, \quad \lim_{k \to \infty} (\mathbf{p}_k \cdot \mathbf{q}_k) = \mathbf{p} \cdot \mathbf{q} \quad \text{and} \quad \lim_{k \to \infty} c_k \mathbf{p}_k = c\mathbf{p}.$$

We say that a sequence $\{p_n\}$ of real numbers is **monotonically increasing** (resp. **monotonically decreasing**) if $p_n \leq p_{n+1}$ (resp. $p_n \geq p_{n+1}$) for all $n \in \mathbb{N}$. A **monotonic sequence** is either a monotonically increasing or monotonically decreasing sequence. Now we have the so-called **Monotonic Convergence Theorem**:

The Monotonic Convergence Theorem. *Every monotonic and bounded sequence $\{p_n\}$ is convergent.*

Finally, the following two theorems state the **Squeeze Theorem for Convergent Sequences**[b] and some well-known limits of sequences in $\mathbb{R}$ and we are going to use them frequently in the problems and solutions.

Theorem 5.6 (Squeeze Theorem for Convergent Sequences). *Suppose that $\{a_n\}$, $\{b_n\}$ and $\{c_n\}$ are sequences of real numbers and $a_n \leq c_n \leq b_n$ for all but finitely many positive integers n. If $\lim_{n \to \infty} a_n = \lim_{n \to \infty} b_n = \ell$, then we have*

$$\lim_{n \to \infty} c_n = \ell.$$

---

[b]This is also called the **Sandwich Theorem**.

5.1. Fundamental Concepts

Theorem 5.7.

(a) If $k > 0$, then $\lim\limits_{n\to\infty} \dfrac{1}{n^k} = 0$ and $\lim\limits_{n\to\infty} \sqrt[n]{k} = 1$.

(b) $\lim\limits_{n\to\infty} \sqrt[n]{n} = 1$.

(c) If $|x| < 1$, then $\lim\limits_{n\to\infty} x^n = 0$.

5.1.3 Upper Limits and Lower Limits

The **extended real number system**, denoted by $[-\infty, +\infty]$, consists of $\mathbb{R}$ and the two symbols $+\infty$ and $-\infty$.[c] Then it is obvious that $+\infty$ is an upper bound of every subset of $[-\infty, +\infty]$ and every nonempty subset of it has the least upper bound.[d] This set preserves the original operations of $\mathbb{R}$ when we have $x, y \in \mathbb{R}$. When x is real and y is either $+\infty$ or $-\infty$, then we have the following conventions:

(a) $x + (+\infty) = +\infty$, $x + (-\infty) = -\infty$, $x - (+\infty) = -\infty$ and $x - (-\infty) = +\infty$, $\dfrac{x}{\pm\infty} = 0$.

(b) If $x > 0$, then $x \cdot (+\infty) = +\infty$ and $x \cdot (-\infty) = -\infty$.

(c) If $x < 0$, then $x \cdot (+\infty) = -\infty$ and $x \cdot (-\infty) = +\infty$.

(d) $(+\infty) + (+\infty) = (+\infty) \cdot (+\infty) = (-\infty) \cdot (-\infty) = +\infty$.

(e) $(-\infty) + (-\infty) = (+\infty)(-\infty) = -\infty$.

Definition 5.8. Suppose that $\{x_n\}$ is a sequence in $[-\infty, +\infty]$. Let E be the set of all **subsequential limits**, i.e., if $x \in E$, then $\{x_n\}$ has a subsequence $\{x_{n_k}\}$ such that

$$\lim_{k\to\infty} x_{n_k} = x.$$

It is allowed that E contains $+\infty$ and $-\infty$.

By the above discussion, we know that $\sup E$ and $\inf E$ exists in $[-\infty, +\infty]$. They are called the **upper limit** and **lower limit** of $\{x_n\}$ respectively. In notation, we have

$$\limsup_{n\to\infty} x_n = \sup E \quad \text{and} \quad \liminf_{n\to\infty} x_n = \inf E. \qquad (5.2)$$

The following theorem shows some important properties of $\sup E$. An analogous result is also valid for $\inf E$, so we won't repeat the statement here.

Theorem 5.9. *The upper limit $\sup E$ is the only number satisfying the following properties:*

(a) $\sup E \in E$.

---

[c] We remark that $[-\infty, +\infty]$ is *not* a field.

[d] The case for the definition of the greatest lower bound is exactly the same as the least upper bound, so we won't repeat here.

(b) If $x > \sup E$, then there exists a positive integer N such that

$$x_n < x$$

for all $n \geq N$.

Next, it is quite obvious that the following theorem holds:

Theorem 5.10. *Suppose that $\{x_n\}$ is a real sequence. Then $\lim\limits_{n\to\infty} x_n = x$ if and only if*

$$\limsup_{n\to\infty} x_n = \liminf_{n\to\infty} x_n = x.$$

Sometimes we need to compare upper limits or lower limits of two different sequences and the following result is very useful to deal with this question:

Theorem 5.11. *Suppose that N is a positive integer. If $a_n \leq b_n$ for all $n \geq N$, then we have*

$$\liminf_{n\to\infty} a_n \leq \liminf_{n\to\infty} b_n \quad \text{and} \quad \limsup_{n\to\infty} a_n \leq \limsup_{n\to\infty} b_n.$$

5.1.4 Cauchy Sequences and Complete Metric Spaces

Definition 5.12. *A sequence $\{p_n\}$ in a metric space X with metric d is called a **Cauchy sequence** if for every $\epsilon > 0$, there exists a positive integer N such that*

$$d(p_m, p_n) < \epsilon$$

whenever $m, n \geq N$.

The relationship between convergent sequences and Cauchy sequences in X can be characterized in the following theorem:

Theorem 5.13. *Let X be a metric space and $\{p_n\}$ be a sequence of X.*

(a) *If $\{p_n\}$ converges, then it is Cauchy.*

(b) *Suppose that X is compact or $\mathbb{R}^n$. If $\{p_n\}$ is Cauchy, then it is convergent.*

In other words, Theorem 5.13(a) says that the condition of a convergent sequence is stronger than that of a Cauchy sequence in *any* metric space. This kind of difference is due to the nature of the space the sequence lies and Theorem 5.13(b) indicates that they are equivalent if X is compact or the Euclidean space $\mathbb{R}^n$.

We remark that the converse of Theorem 5.13(a) is not true, see Problem 5.19 for an example. Thus, inspired by Theorem 5.13(b), it is natural to search more spaces where Cauchy sequences and convergent sequences are equivalent. In fact, such a metric space is called a **complete metric space**. For instances, every compact metric space and $\mathbb{R}^n$ is complete.

5.2 Convergence of Sequences

Problem 5.1

(★) Evaluate the limit
$$\lim_{n \to \infty} \Big(\frac{1}{2} + \frac{3}{2^2} + \cdots + \frac{2n-1}{2^n}\Big).$$

Proof. Suppose that
$$p_n = \frac{1}{2} + \frac{3}{2^2} + \cdots + \frac{2n-1}{2^n}. \tag{5.3}$$

Then we have
$$\frac{1}{2}p_n = \frac{1}{2^2} + \frac{3}{2^3} + \cdots + \frac{2n-1}{2^{n+1}}. \tag{5.4}$$

The subtraction of the expressions (5.3) and (5.4) gives
$$\frac{1}{2}p_n = \frac{1}{2} + \Big(\frac{1}{2} + \frac{1}{2^2} + \cdots + \frac{1}{2^{n-1}}\Big) - \frac{2n-1}{2^{n+1}} = \frac{1}{2} + 1 - \frac{1}{2^{n-1}} - \frac{2n-1}{2^{n+1}}$$

and then
$$p_n = 3 - \frac{1}{2^{n-2}} - \frac{2n-1}{2^n}$$

Hence, it follows from Theorem 5.5 and then Theorem 5.7(a) that
$$\lim_{n \to \infty} p_n = 3.$$

This completes the proof of the problem. ∎

Problem 5.2

(★) Evaluate the limit
$$\lim_{n \to \infty} \Big[\frac{1}{1 \cdot 2} + \frac{1}{2 \cdot 3} + \cdots + \frac{1}{n(n+1)}\Big].$$

Proof. Since
$$\frac{1}{1 \cdot 2} + \frac{1}{2 \cdot 3} + \cdots + \frac{1}{n(n+1)} = \Big(1 - \frac{1}{2}\Big) + \Big(\frac{1}{2} - \frac{1}{3}\Big) + \cdots + \Big(\frac{1}{n} - \frac{1}{n+1}\Big)$$
$$= 1 - \frac{1}{n+1},$$

we obtain from Theorem 5.5 and then Theorem 5.7(a) that
$$\lim_{n \to \infty} \Big[\frac{1}{1 \cdot 2} + \frac{1}{2 \cdot 3} + \cdots + \frac{1}{n(n+1)}\Big] = 1,$$

completing the proof of the problem. ∎

> **Problem 5.3**
>
> (★)(★) Given that $\theta > 1$ and $k \in \mathbb{N}$. Prove that
> $$\lim_{n \to \infty} \frac{n^k}{\theta^n} = 0.$$

Proof. Since $\theta > 1$, there exists a $\delta > 0$ such that $\theta = 1 + \delta$. For a *fixed* k, if $n > k$, then the Binomial Theorem implies that
$$\theta^n = (1+\delta)^n > \frac{n(n-1)\cdots(n-k)}{(k+1)!}\delta^{k+1}$$
so that
$$0 < \frac{n^k}{\theta^n} < \frac{n^k(k+1)!}{n(n-1)\cdots(n-k)\delta^{k+1}} = \frac{(k+1)!}{n(1-\frac{1}{n})(1-\frac{2}{n})\cdots(1-\frac{k}{n})\delta^{k+1}}. \tag{5.5}$$
By Theorem 5.7(a), we have $\lim_{n \to \infty}\left(1 - \frac{j}{n}\right) = 1$ for $1 \le j \le k$, we see that
$$\lim_{n \to \infty} \frac{(k+1)!}{n(1-\frac{1}{n})(1-\frac{2}{n})\cdots(1-\frac{k}{n})\delta^{k+1}} = 0.$$
Finally, we apply Theorem 5.6 (Squeeze Theorem for Convergent Sequences) to the inequalities (5.5) to get
$$\lim_{n \to \infty} \frac{n^k}{\theta^n} = 0$$
which is our desired result. This finishes the proof of the problem. ∎

> **Problem 5.4**
>
> (★) Given that $a_k \ge 0$ for $k = 1, 2, \ldots, m$. Evaluate the limit
> $$\lim_{n \to \infty} \sqrt[n]{a_1^n + \cdots + a_m^n}.$$

Proof. By rearrangement if necessary, we may assume that $a_1 \le a_2 \le \cdots \le a_m$. Thus we have
$$a_m^n \le a_1^n + \cdots + a_m^n \le m a_m^n$$
and then
$$a_m \le \sqrt[n]{a_1^n + \cdots + a_m^n} \le m^{\frac{1}{n}} a_m. \tag{5.6}$$
By Theorem 5.7(a), we have $\lim_{n \to \infty} m^{\frac{1}{n}} = 1$. Apply Theorem 5.6 (Squeeze Theorem for Convergent Sequences) to the inequalities (5.6), we establish that
$$\lim_{n \to \infty} \sqrt[n]{a_1^n + \cdots + a_m^n} = a_m,$$
completing the proof of the problem. ∎

5.2. Convergence of Sequences

> **Problem 5.5**
>
> (★) Given that $\ln(1+\frac{1}{n}) < \frac{1}{n}$ for every $n \in \mathbb{N}$. Suppose that
> $$p_n = \left(1+\frac{1}{3}\right)\left(1+\frac{1}{9}\right)\cdots\left(1+\frac{1}{3^n}\right), \qquad (5.7)$$
> where $n = 1, 2, \ldots$. Prove that $\{p_n\}$ converges.

Proof. Since each $1 + \frac{1}{3^n} > 0$ for all $n \in \mathbb{N}$, we see easily that $\{p_n\}$ is monotonic increasing. Furthermore, we can take the logarithm to both sides of (5.7) to get

$$\ln p_n = \ln\left(1+\frac{1}{3}\right) + \ln\left(1+\frac{1}{3^2}\right) + \cdots + \ln\left(1+\frac{1}{3^n}\right). \qquad (5.8)$$

By the given hint, we get the following inequality from the expression (5.8) that

$$\ln p_n < \frac{1}{3} + \frac{1}{3^2} + \cdots + \frac{1}{3^n} < 1 + \frac{1}{3} + \cdots = \frac{3}{2}$$

which means that $\{p_n\}$ is bounded above by $\frac{3}{2}$. By the Monotonic Convergence Theorem, we know that $\{p_n\}$ is convergent and this completes the proof of the problem. ∎

> **Problem 5.6**
>
> (★) Define $p_1 = \sqrt{3}$ and for $n = 2, 3, \ldots$, we have
> $$p_n = \underbrace{\sqrt{3 + \sqrt{3 + \cdots + \sqrt{3}}}}_{n \text{ square root symbols}}.$$
> Prove that $\{p_n\}$ converges.

Proof. It can be shown by induction that $\{p_n\}$ is monotonically increasing and bounded above by 3. By the Monotonic Convergence Theorem, we see that $\{p_n\}$ is convergent. We finish the proof of the problem. ∎

> **Problem 5.7**
>
> (★)(★) Suppose that $\lim\limits_{n\to\infty} \cos \delta_n = 1$ if $\lim\limits_{n\to\infty} \delta_n = 0$. For every positive integer n, we define
> $$p_n = \cos(2n!e\pi).$$
> Prove that
> $$\lim_{n\to\infty} p_n = 2\pi.$$

Proof. By definition, we have

$$e = \left(1 + \frac{1}{1!} + \cdots + \frac{1}{n!}\right) + \left[\frac{1}{(n+1)!} + \cdots\right].$$

so that

$$2n!e = 2n!\left(1 + \frac{1}{1!} + \cdots + \frac{1}{n!}\right) + 2\left[\frac{1}{n+1} + \frac{1}{(n+1)(n+2)} + \cdots\right] = 2N_n + \delta_n, \quad (5.9)$$

where $N_n = n!\left(1 + \frac{1}{1!} + \cdots + \frac{1}{n!}\right)$ and $\delta_n = 2\left[\frac{1}{n+1} + \frac{1}{(n+1)(n+2)} + \cdots\right]$. Since

$$\frac{2}{n+1} < \delta_n < 2\left[\frac{1}{n+1} + \frac{1}{(n+1)^2} + \cdots\right] = \frac{2}{n},$$

Theorem 5.6 (Squeeze Theorem for Convergent Sequences) shows that

$$\lim_{n \to \infty} \delta_n = 0.$$

Note that N_n is always a positive integer for every $n \in \mathbb{N}$, so we follow from the expression (5.9) and the fact $\cos(2n\pi + \theta) = \cos\theta$ that

$$\cos(2n!e\pi) = \cos(2N_n\pi + \delta_n\pi) = \cos\delta_n\pi.$$

Hence we conclude from the given hint that

$$\lim_{n \to \infty} \cos(2n!e\pi) = \lim_{n \to \infty} \cos\delta_n\pi = 1,$$

completing the proof of the problem. ■

Problem 5.8

(⋆) Suppose that $\{p_n\}$ is a sequence of nonnegative real numbers with $p_2 = 0$. If $p_{m+n} \le p_m + p_n$ for all $m, n \in \mathbb{N}$, prove that

$$\lim_{n \to \infty} \frac{p_n}{n} = 0 \quad (5.10)$$

Proof. Take $m = 2$. For any large positive integer n, we have $n = 2q + r$, where $q \in \mathbb{N}$ and $0 \le r \le 1$. By the hypothesis, we see that

$$0 \le p_n = p_{2q+r} \le p_{2+2+\cdots+2} + p_r \le qp_2 + p_r \le p_r$$

which implies that

$$0 \le \frac{p_n}{n} \le \frac{p_r}{n}. \quad (5.11)$$

Since $\lim_{n \to \infty} \frac{p_r}{n} = 0$, we apply Theorem 5.6 (Squeeze Theorem for Convergent Sequences) to the inequalities (5.11) to obtain the required result (5.10). This ends the proof of the problem. ■

5.2. Convergence of Sequences

Problem 5.9

(★) Given that $\{p_n\}$ is a sequence of real numbers. If $p_n \to p$ as $n \to \infty$, prove that $|p_n| \to |p|$ as $n \to \infty$.

Proof. Given $\epsilon > 0$. Since $p_n \to p$, there exists a positive integer N such that $|p_n - p| < \epsilon$ for all $n \geq N$. With this same N, it deduces from Problem 2.5 that

$$||p_n| - |p|| \leq |p_n - p| < \epsilon$$

for all $n \geq N$. Hence we have completed the proof of the problem. ∎

Problem 5.10

(★)(★) Suppose that $\{A_n\}$ and $\{B_n\}$ are sequences of real numbers converging to A and B respectively. Let

$$C_n = \frac{A_0 B_n + A_1 B_{n-1} + \cdots + A_n B_0}{n+1}.$$

Prove that $\lim_{n \to \infty} C_n = AB$.

Proof. By Theorem 5.2, $\{A_n\}$ and $\{B_n\}$ are bounded. Let M be a positive number such that

$$|A_n| \leq M \quad \text{and} \quad |B_n| \leq M \tag{5.12}$$

for all $n \in \mathbb{N}$. Since $A_n \to A$ and $B_n \to B$, for every $\epsilon > 0$, we know from the definition that there exists a positive integer N such that

$$|A_n - A| < \frac{\epsilon}{2(M + |A|)} \quad \text{and} \quad |B_n - B| < \frac{\epsilon}{2(M + |A|)} \tag{5.13}$$

for all $n \geq N$. Now we deduce from the inequalities (5.12) that

$$\begin{aligned}|A_k B_{n-k} - AB| &= |A_k B_{n-k} - AB_{n-k} + AB_{n-k} - AB| \\ &\leq |A_k - A||B_{n-k}| + |A||B_{n-k} - B| \\ &\leq M|A_k - A| + |A||B_{n-k} - B|. \end{aligned} \tag{5.14}$$

for all $k \in \mathbb{N}$.

Consider the positive integers n and k with $n \geq N^2$ and $[\sqrt{n}] \leq k \leq n - [\sqrt{n}]$.[e] By these, we know that

$$k \geq [\sqrt{n}] \geq N \quad \text{and} \quad n - k \geq \sqrt{n} \geq N.$$

Thus we follow from the inequalities (5.13) and (5.14) that

$$|A_k B_{n-k} - AB| < \frac{\epsilon}{2} \tag{5.15}$$

for $n \geq N^2$ and $[\sqrt{n}] \leq k \leq n - [\sqrt{n}]$.

---
[e] Recall that $[x]$ denotes the greatest integer less than or equal to x.

For $0 \leq k \leq \sqrt{n}$, we have

$$\sum_{k=0}^{[\sqrt{n}]} |A_k B_{n-k} - AB| \leq \sum_{k=0}^{[\sqrt{n}]} (M^2 + |AB|) \leq (M^2 + |AB|)([\sqrt{n}] + 1) \tag{5.16}$$

and similarly, for $n - [\sqrt{n}] \leq k \leq n$, we have

$$\sum_{k=n-[\sqrt{n}]}^{n} |A_k B_{n-k} - AB| \leq \sum_{k=n-[\sqrt{n}]}^{n} (M^2 + |AB|) \leq (M^2 + |AB|)([\sqrt{n}] + 1). \tag{5.17}$$

Finally, by combining the inequalities (5.15), (5.16) and (5.17), we derive that

$$\begin{aligned}
|C_n - AB| &\leq \Big| \frac{1}{n+1} \sum_{k=0}^{n} A_k B_{n-k} - AB \Big| \\
&\leq \frac{1}{n+1} \Big| \sum_{k=0}^{n} A_k B_{n-k} - (n+1)AB \Big| \\
&= \frac{1}{n+1} \Big| \sum_{k=0}^{n} (A_k B_{n-k} - AB) \Big| \\
&\leq \frac{1}{n+1} \sum_{k=0}^{[\sqrt{n}]} |A_k B_{n-k} - AB| + \frac{1}{n+1} \sum_{k=[\sqrt{n}]}^{n-[\sqrt{n}]} |A_k B_{n-k} - AB| \\
&\quad + \frac{1}{n+1} \sum_{k=n-[\sqrt{n}]}^{n} |A_k B_{n-k} - AB| \\
&\leq \frac{2(M^2 + |AB|)([\sqrt{n}] + 1)}{n+1} + \frac{\epsilon}{2}.
\end{aligned} \tag{5.18}$$

Hence, if we take n large enough[f], then we can make

$$\frac{2(M^2 + |AB|)([\sqrt{n}] + 1)}{n+1} < \frac{\epsilon}{2}.$$

By this and the inequality (5.18), we conclude that

$$|C_n - AB| < \epsilon$$

for large enough n and this proves the desired result. This completes the proof of the problem. ∎

Problem 5.11

(★)(★) Suppose that $\{p_n\}$ is a sequence of real numbers such that $p_n - p_{n-2} \to 0$. Prove that

$$\lim_{n \to \infty} \frac{p_n - p_{n-1}}{n} = 0. \tag{5.19}$$

---

[f]Recall that once we choose ϵ, the positive integer N is fixed and then we can take any large integer n satisfying $n \geq N$.

5.3. Upper and Lower Limits

Proof. Given $\epsilon > 0$, since $p_n - p_{n-2} \to 0$, there exists a positive integer N such that

$$|p_n - p_{n-2}| < \frac{\epsilon}{2} \tag{5.20}$$

for $n \geq N$. We fix this N. It is clear that

$$\begin{aligned}
p_n - p_{n-1} &= (p_n - p_{n-2} + p_{n-2} - p_{n-1}) \\
&= (p_n - p_{n-2}) - (p_{n-1} - p_{n-3} + p_{n-3} - p_{n-2}) \\
&= (p_n - p_{n-2}) - (p_{n-1} - p_{n-3}) - (p_{n-3} - p_{n-4} + p_{n-4} - p_{n-2}) \\
&= (p_n - p_{n-2}) - (p_{n-1} - p_{n-3}) + (p_{n-2} - p_{n-4}) - (p_{n-3} - p_{n-4}) \\
&= (p_n - p_{n-2}) - (p_{n-1} - p_{n-3}) + (p_{n-2} - p_{n-4}) - \cdots \\
&\quad + (-1)^N (p_N - p_{N-2} + p_{N-2} - p_{N-1}).
\end{aligned} \tag{5.21}$$

Thus we deduce from the inequalities (5.20) and (5.21) that

$$\begin{aligned}
|p_n - p_{n-1}| &\leq |p_n - p_{n-2}| + |p_{n-1} - p_{n-3}| + |p_{n-2} - p_{n-4}| + \cdots \\
&\quad + |p_N - p_{N-2}| + |p_{N-2} - p_{N-1}| \\
&< (n - N + 1)\epsilon + |p_{N-2} - p_{N-1}|
\end{aligned}$$

for $n \geq N$ and thus

$$\left|\frac{p_n - p_{n-2}}{n}\right| < \frac{(n - N + 1)\epsilon}{2n} + \frac{|p_{N-2} - p_{N-1}|}{n} < \frac{\epsilon}{2} + \frac{\epsilon}{2} = \epsilon$$

for large enough n because N is *fixed*. By definition, we have the limit (5.19) which completes the proof of the problem. ∎

5.3 Upper and Lower Limits

Problem 5.12

(★) Suppose that $a_n = (-1)^n \left(2 + \frac{3}{n}\right)$ for $n = 1, 2, \ldots$. Find

$$\limsup_{n \to \infty} a_n \quad \text{and} \quad \liminf_{n \to \infty} a_n.$$

Proof. It is clear that

$$a_{2n} = 2 + \frac{3}{2n} \quad \text{and} \quad a_{2n-1} = -2 - \frac{3}{2n - 1}$$

for $n = 1, 2, \ldots$. Therefore, we have

$$\limsup_{n \to \infty} a_n = \lim_{n \to \infty} a_{2n} = 2 \quad \text{and} \quad \liminf_{n \to \infty} a_n = \lim_{n \to \infty} a_{2n-1} = -2.$$

Hence we complete the proof of the problem. ∎

Problem 5.13

(★)(★) If $\{a_n\}$ is a sequence of positive real numbers, prove that
$$\liminf_{n\to\infty} \frac{a_{n+1}}{a_n} \leq \liminf_{n\to\infty} \sqrt[n]{a_n} \leq \limsup_{n\to\infty} \sqrt[n]{a_n} \leq \limsup_{n\to\infty} \frac{a_{n+1}}{a_n}.$$

Proof. We only prove the case
$$\limsup_{n\to\infty} \sqrt[n]{a_n} \leq \limsup_{n\to\infty} \frac{a_{n+1}}{a_n}$$
because the case $\liminf_{n\to\infty} \frac{a_{n+1}}{a_n} \leq \liminf_{n\to\infty} \sqrt[n]{a_n}$ is very similar. Let $A = \limsup_{n\to\infty} \frac{a_{n+1}}{a_n}$. If $A = \infty$, then there is nothing to prove.

Suppose that $A < \infty$. Given $\epsilon > 0$. By Theorem 5.9(b), there exists a positive integer N such that
$$\frac{a_{n+1}}{a_n} < A + \epsilon \tag{5.22}$$
for all $n \geq N$. For $n > N$, we deduce from the inequality (5.22) that
$$0 < a_n = \frac{a_n}{a_{n-1}} \times \frac{a_{n-1}}{a_{n-2}} \times \cdots \times \frac{a_{N+1}}{a_N} \times a_N < (A+\epsilon)^{n-N} a_N = (A+\epsilon)^n \times \frac{a_N}{(A+\epsilon)^N}$$
so that
$$0 < \sqrt[n]{a_n} < (A+\epsilon) \times \sqrt[n]{\frac{a_N}{(A+\epsilon)^N}}. \tag{5.23}$$
Since $\frac{a_N}{(A+\epsilon)^N} > 0$, it follows from Theorem 5.7(a) that
$$\lim_{n\to\infty} \sqrt[n]{\frac{a_N}{(A+\epsilon)^N}} = 1$$
and we deduce from the inequalities (5.23) and Theorem 5.11 that
$$\limsup_{n\to\infty} \sqrt[n]{a_n} \leq (A+\epsilon) \limsup_{n\to\infty} \sqrt[n]{\frac{a_N}{(A+\epsilon)^N}} = (A+\epsilon) \lim_{n\to\infty} \sqrt[n]{\frac{a_N}{(A+\epsilon)^N}} = A+\epsilon.$$
Since ϵ is arbitrary, we must have
$$\limsup_{n\to\infty} \sqrt[n]{a_n} \leq A = \limsup_{n\to\infty} \frac{a_{n+1}}{a_n},$$
completing the proof of the problem. ∎

Problem 5.14

(★)(★) Prove that
$$\lim_{n\to\infty} \frac{\sqrt[n]{n!}}{n} = \frac{1}{e}.$$

5.3. Upper and Lower Limits

Proof. Put $a_n = \frac{n!}{n^n}$ into Problem 5.13, we have

$$\frac{a_{n+1}}{a_n} = \frac{(n+1)!}{(n+1)^{n+1}} \times \frac{n^n}{n!} = \left(\frac{n}{n+1}\right)^n \quad \text{and} \quad \sqrt[n]{a_n} = \frac{\sqrt[n]{n!}}{n}.$$

Recall the definition[g]

$$e^x = \lim_{n \to \infty} \left(1 + \frac{x}{n}\right)^n$$

for every $x \in \mathbb{R}$. By this, we have

$$\limsup_{n \to \infty} \frac{a_{n+1}}{a_n} = \liminf_{n \to \infty} \frac{a_{n+1}}{a_n} = \lim_{n \to \infty} \left(\frac{n}{n+1}\right)^n = \lim_{n \to \infty} \left(1 - \frac{1}{n+1}\right)^n = e^{-1}. \tag{5.24}$$

Thus it follows from the result (5.24) and Theorem 5.6 (Squeeze Theorem for Convergent Sequences) that

$$\limsup_{n \to \infty} \sqrt[n]{a_n} = \liminf_{n \to \infty} \sqrt[n]{a_n} = \frac{1}{e}.$$

Hence we obtain from Theorem 5.10 that

$$\lim_{n \to \infty} \frac{\sqrt[n]{n!}}{n} = \lim_{n \to \infty} \sqrt[n]{a_n} = \frac{1}{e},$$

completing the proof of the problem. ∎

Problem 5.15

(★)(★) *Suppose that $\{x_n\}$ is a convergent real sequence. Let*

$$a_n = \frac{x_1 + \cdots + x_n}{n}$$

for $n = 1, 2, \ldots$. Prove that

$$\lim_{n \to \infty} a_n = \lim_{n \to \infty} x_n. \tag{5.25}$$

Proof. Suppose that $\lim_{n \to \infty} x_n = x$. Let $x_n = x + \xi_n$, where $\xi_n \to 0$ as $n \to \infty$. It is easy to see that

$$a_n - x = \frac{x_1 + \cdots + x_n}{n} - x$$
$$= \frac{1}{n}[(x_1 - x) + (x_2 - x) + \cdots + (x_n - x)]$$
$$= \frac{1}{n}(\xi_1 + \xi_2 + \cdots + \xi_n). \tag{5.26}$$

Since $\xi_n \to 0$ as $n \to \infty$, given $\epsilon > 0$, there exists a positive integer N such that

$$-\epsilon < \xi_n < \epsilon$$

for all $n > N$. We *fix* this N, so we deduce from the inequality (5.26) that

$$\frac{1}{n}(\xi_1 + \cdots + \xi_N) - \epsilon < a_n - x < \frac{1}{n}(\xi_1 + \cdots + \xi_N) + \epsilon$$

---

[g]See, for example, [8, §215, Eqn. (1)].

for all $n > N$. By Theorem 5.11, we have

$$-\epsilon \leq \liminf_{n \to \infty}(a_n - x) \leq \limsup_{n \to \infty}(a_n - x) \leq \epsilon. \qquad (5.27)$$

Since ϵ is arbitrary, the inequality (5.27) means that

$$\liminf_{n \to \infty}(a_n - x) = \limsup_{n \to \infty}(a_n - x) = 0$$

which proves the desire inequality (5.25) by Theorem 5.10. This ends the proof of the problem. ∎

Problem 5.16

(★)(★) Let $\{a_n\}$ be a sequence of positive numbers. If

$$\limsup_{n \to \infty} a_n \times \limsup_{n \to \infty} \frac{1}{a_n} = 1, \qquad (5.28)$$

prove that $\{\frac{1}{a_n}\}$ is convergent.

Proof. We know that the left-hand side of the hypothesis (5.28) cannot be $0 \cdot \infty$, so the limits must be nonzero real numbers. Let $\alpha = \limsup_{n \to \infty} a_n$. By Theorem 5.9(a), *there exists* a subsequence $\{a_{n_k}\}$ such that

$$a_{n_k} \to \alpha \qquad (5.29)$$

as $k \to \infty$. Let

$$\liminf_{n \to \infty} \frac{1}{a_n} = \beta.$$

By the definition (5.2), we have

$$\beta \leq \lim_{j \to \infty} \frac{1}{a_{n_j}}$$

for all subsequences $\{a_{n_j}\}$. In particular, we know from the limit (5.29) that $\beta \leq \frac{1}{\alpha}$. Assume that $\beta < \frac{1}{\alpha}$. Then there exists a $\delta > 0$ such that

$$\beta < \beta + \delta < \frac{1}{\alpha}.$$

Since $\beta + \delta$ is *not* a lower bound of the set of all subsequential limits of $\{\frac{1}{a_n}\}$ anymore, we can find a subsequence $\{\frac{1}{a_{n_j}}\}$ such that

$$\lim_{j \to \infty} \frac{1}{a_{n_j}} < \beta + \delta < \frac{1}{\alpha}$$

which implies that

$$\alpha < \lim_{j \to \infty} a_{n_j}$$

but this contradicts the definition that α is the least upper bound of the set of all subsequential limits of $\{a_n\}$. Therefore, we must have $\beta = \frac{1}{\alpha}$, i.e.,

$$\limsup_{n \to \infty} a_n \times \liminf_{n \to \infty} \frac{1}{a_n} = 1. \qquad (5.30)$$

5.3. Upper and Lower Limits

By comparing the results (5.28) and (5.30), we get

$$\limsup_{n\to\infty} \frac{1}{a_n} = \liminf_{n\to\infty} \frac{1}{a_n} = \frac{1}{\alpha}.$$

Hence, by Theorem 5.10, we have proven that $\{\frac{1}{a_n}\}$ converges and this completes the proof of the problem. ∎

Problem 5.17

(⋆)(⋆)(⋆) Suppose that $a_n \geq 0$ and $b_n \geq 0$ for all $n = 1, 2, \ldots$. Show that

$$\liminf_{n\to\infty} a_n \times \liminf_{n\to\infty} b_n \leq \liminf_{n\to\infty}(a_n \times b_n) \leq \liminf_{n\to\infty} a_n \times \limsup_{n\to\infty} b_n.$$

Proof. Let $A = \liminf\limits_{n\to\infty} a_n$. We prove the inequality

$$\liminf_{n\to\infty}(a_n \times b_n) \leq \liminf_{n\to\infty} a_n \times \limsup_{n\to\infty} b_n \tag{5.31}$$

first. By the definition (5.2), there exists a subsequence $\{a_{n_k}\}$ such that

$$\lim_{k\to\infty} a_{n_k} = A. \tag{5.32}$$

For the corresponding subsequence $\{b_{n_k}\}$, let $B = \limsup\limits_{k\to\infty} b_{n_k}$. Then there exists a subsequence $\{b_{n_{k_j}}\}$ (of $\{b_{n_k}\}$) such that

$$\lim_{j\to\infty} b_{n_{k_j}} = B. \tag{5.33}$$

It is clear that

$$B = \limsup_{k\to\infty} b_{n_k} \leq \limsup_{n\to\infty} b_n. \tag{5.34}$$

Apply Theorem 5.10 to the limit (5.32), we know that

$$\lim_{j\to\infty} a_{n_{k_j}} = A, \tag{5.35}$$

where $\{a_{n_{k_j}}\}$ is *any* subsequence of $\{a_{n_k}\}$. Combining the limits (5.33) and (5.35), we achieve

$$\lim_{j\to\infty}(a_{n_{k_j}} \times b_{n_{k_j}}) = AB.$$

By the definition (5.2) again, we have

$$\liminf_{n\to\infty}(a_n \times b_n) \leq AB. \tag{5.36}$$

Since $a_n \geq 0$ and $b_n \geq 0$ for all $n = 1, 2, \ldots$, A and B must be nonnegative. Substitute the inequality (5.34) into the inequality (5.36), we are able to obtain the desired inequality (5.31).

Next, we prove the inequality

$$\liminf_{n\to\infty} a_n \times \liminf_{n\to\infty} b_n \leq \liminf_{n\to\infty}(a_n \times b_n). \tag{5.37}$$

Let $C = \liminf\limits_{n\to\infty}(a_n \times b_n)$. If $A = 0$, then there is nothing to prove. Thus we may assume that $A > 0$. By the analogous result of Theorem 5.9, there exists a positive integer N such that

$$a_n > 0 \tag{5.38}$$

for all $n \geq N$. Again the analogous result of Theorem 5.9 implies that a subsequence $\{a_{n_k} b_{n_k}\}$ of $\{a_n b_n\}$ exists such that

$$\lim_{k\to\infty} (a_{n_k} b_{n_k}) = C. \tag{5.39}$$

For the sequence $\{a_{n_k}\}$, we can find a subsequence $\{a_{n_{k_j}}\}$ such that

$$\lim_{j\to\infty} a_{n_{k_j}} = A' = \liminf_{k\to\infty} a_{n_k}. \tag{5.40}$$

By the definition of A, A' and the lower limit, we have

$$0 < A = \liminf_{n\to\infty} a_n \leq \liminf_{k\to\infty} a_{n_k} = A'. \tag{5.41}$$

Then it follows from the inequality (5.38), the limits (5.39) and (5.40) with the application of Theorem 5.10 to the limit (5.39) that

$$\lim_{j\to\infty} b_{n_{k_j}} = \lim_{j\to\infty} \left[(a_{n_{k_j}} b_{n_{k_j}}) \times \frac{1}{a_{n_{k_j}}}\right] = \frac{C}{A'}.$$

By the definition (5.2), this implies that

$$\liminf_{n\to\infty} b_n \leq \frac{C}{A'}. \tag{5.42}$$

Hence, after substituting the definition of A' and C back into the inequality (5.42) and then using the inequality (5.41), we obtain the expected inequality (5.37), completing the proof of the problem. ∎

> **Remark 5.1**
>
> Similar to Problem 5.17, if $a_n \geq 0$ and $b_n \geq 0$ for all $n = 1, 2, \ldots$, then we can prove that
>
> $$\liminf_{n\to\infty} a_n \times \limsup_{n\to\infty} b_n \leq \limsup_{n\to\infty}(a_n \times b_n) \leq \limsup_{n\to\infty} a_n \times \limsup_{n\to\infty} b_n.$$

> **Problem 5.18**
>
> (★)(★) *Reprove Problem 5.16 by using Problem 5.17 and Remark 5.1.*

Proof. Since $a_n > 0$ for all $n = 1, 2, \ldots$, we have $\frac{1}{a_n} > 0$ for all $n = 1, 2, \ldots$. In other words, the sequences $\{a_n\}$ and $\{\frac{1}{a_n}\}$ satisfy the hypotheses of Problem 5.17 and Remark 5.1. Put $b_n = \frac{1}{a_n}$ into Problem 5.17 and Remark 5.1, we have

$$\liminf_{n\to\infty} a_n \times \liminf_{n\to\infty} \frac{1}{a_n} \leq 1 \leq \liminf_{n\to\infty} a_n \times \limsup_{n\to\infty} \frac{1}{a_n} \tag{5.43}$$

and
$$\liminf_{n\to\infty} a_n \times \limsup_{n\to\infty} \frac{1}{a_n} \leq 1 \leq \limsup_{n\to\infty} a_n \times \limsup_{n\to\infty} \frac{1}{a_n} \quad (5.44)$$
respectively. Combining the inequalities (5.43) and (5.44), it is easy to see that
$$1 \leq \liminf_{n\to\infty} a_n \times \limsup_{n\to\infty} \frac{1}{a_n} \leq 1$$
which means that
$$\liminf_{n\to\infty} a_n \times \limsup_{n\to\infty} \frac{1}{a_n} = 1.$$
By this and the hypothesis (5.28), we have $\limsup_{n\to\infty} \frac{1}{a_n} \neq 0$ and then we derive that
$$\liminf_{n\to\infty} a_n = \limsup_{n\to\infty} a_n.$$
Hence the sequence $\{a_n\}$ and then the sequence $\{\frac{1}{a_n}\}$ are convergent and we finish the proof of the problem. ∎

5.4 Cauchy Sequences and Complete Metric Spaces

Problem 5.19

(⋆) Suppose that $X = \mathbb{R}^+$, the set of all positive real numbers, and the metric d of X is given by
$$d(x,y) = |x - y| \quad (5.45)$$
for all $x, y \in X$. Prove that if $p_n = \frac{1}{n}$ for all $n = 1, 2, \ldots$, then $\{p_n\}$ is Cauchy but not convergent in X.

Proof. Given that $\epsilon > 0$. By Theorem 2.1 (The Archimedean Property), there exists a positive integer N such that
$$\frac{1}{N} < \frac{\epsilon}{2}. \quad (5.46)$$
It is clear that the inequality (5.46) holds for every $n \geq N$. Therefore, if $m, n \geq N$, then we have
$$d(p_n, p_m) = \left|\frac{1}{n} - \frac{1}{m}\right| \leq \frac{1}{n} + \frac{1}{m} < \frac{\epsilon}{2} + \frac{\epsilon}{2} = \epsilon.$$
By Definition 5.12, the sequence $\{p_n\}$ is Cauchy. However, since its limit point is $0 \notin X$, it does not satisfy the definition of a convergent sequence (see §5.1.1). Hence we have completed the proof of the problem. ∎

Remark 5.2

In other words, Problem 5.19 shows that the space $\mathbb{R}^+$ with the metric d given by (5.45) is *not* complete.

Problem 5.20

(⋆) *Suppose that $\{a_n\}$ is a bounded real sequence and $|x| < 1$. We define $\{p_n\}$ by*

$$p_n = a_n x^n + \cdots + a_1 x + a_0$$

for $n = 1, 2, \ldots$. Prove that $\{p_n\}$ is Cauchy.

Proof. Since $\{a_n\}$ is bounded, there exists a positive constant M such that

$$|a_n| \leq M$$

for all $n = 1, 2, \ldots$. Given $\epsilon > 0$. Let N be a positive integer such that

$$N > \frac{1}{\log |x|} \times \log \frac{\epsilon(1 - |x|)}{M} - 1. \tag{5.47}$$

Then for all $m > n \geq N$, we follow from the inequality (5.47) that

$$\begin{aligned}
|p_m - p_n| &= |a_m x^m + \cdots + a_{n+1} x^{n+1}| \\
&\leq M(|x|^m + |x|^{m-1} + \cdots + |x|^{n+1}) \\
&\leq M|x|^{n+1}(1 + |x| + \cdots + |x|^{m-n-1}) \\
&< M|x|^{n+1}(1 + |x| + \cdots) \\
&= \frac{M|x|^{n+1}}{1 - |x|} \\
&\leq \frac{M|x|^{N+1}}{1 - |x|} \\
&< \epsilon.
\end{aligned}$$

Hence $\{p_n\}$ is Cauchy by Definition 5.12 and we finish the proof of the problem. ∎

Problem 5.21

(⋆) *Suppose that $\{p_n\}$ is a real sequence defined by*

$$p_n = \frac{n+1}{n-2}$$

for $n = 3, 4, \ldots$. Prove that $\{p_n\}$ is Cauchy.

Proof. We note that

$$\begin{aligned}
|p_m - p_n| &= \left| \frac{m+1}{m-2} - \frac{n+1}{n-2} \right| \\
&= \left| \frac{3(n-m)}{(m-2)(n-2)} \right| \\
&\leq \left| \frac{3n}{(m-2)(n-2)} \right| + \left| \frac{3m}{(m-2)(n-2)} \right|. \tag{5.48}
\end{aligned}$$

5.4. Cauchy Sequences and Complete Metric Spaces

Since $\frac{3n}{n-2}$ and $\frac{3m}{m-2}$ are obviously bounded, we can find a positive constant M such that

$$\left|\frac{3n}{n-2}\right| \leq M \quad \text{and} \quad \left|\frac{3m}{m-2}\right| \leq M.$$

Put these into the inequality (5.48), we achieve

$$|p_m - p_n| \leq M\left(\frac{1}{m-2} + \frac{1}{n-2}\right). \tag{5.49}$$

Therefore, if we let N to be a positive integer such that $N > \frac{2M}{\epsilon} + 2$, then for $m, n \geq N$, we derive from the inequality (5.49) that

$$|p_m - p_n| \leq M\left(\frac{1}{m-2} + \frac{1}{n-2}\right) < \frac{\epsilon}{2} + \frac{\epsilon}{2} = \epsilon.$$

Hence the sequence $\{p_n\}$ is Cauchy and we finish the proof of the problem. ∎

Problem 5.22

(⋆) Suppose that $\{p_n\}$ is a real sequence defined by

$$p_n = \frac{\sin(1!)}{1 \times 2} + \frac{\sin(2!)}{2 \times 3} + \cdots + \frac{\sin(n!)}{n \times (n+1)}$$

for $n = 1, 2, \ldots$. Determine the convergence of $\{p_n\}$.

Proof. Without the knowledge of the limit (if it exists) of $\{p_n\}$, we prove that it is Cauchy in $\mathbb{R}$ so that it converges in $\mathbb{R}$ by Theorem 5.13(b). To this end, given $\epsilon > 0$. Let N be a positive integer such that

$$N > \frac{1}{\epsilon} - 1. \tag{5.50}$$

Then for $m > n \geq N$, we obtain from the inequality (5.50) and the fact $|\sin x| \leq 1$ that

$$\begin{aligned}
|p_m - p_n| &= \left|\frac{\sin(m!)}{m(m+1)} + \frac{\sin[(m-1)!]}{(m-1)m} + \cdots + \frac{\sin[(n+1)!]}{(n+1)(n+2)}\right| \\
&\leq \frac{1}{m(m+1)} + \frac{1}{(m-1)m} + \cdots + \frac{1}{(n+1)(n+2)} \\
&= \left(\frac{1}{m} - \frac{1}{m+1}\right) + \left(\frac{1}{m-1} - \frac{1}{m}\right) + \cdots + \left(\frac{1}{n+1} - \frac{1}{n+2}\right) \\
&= \frac{1}{n+1} - \frac{1}{m+1} \\
&< \frac{1}{n+1} \\
&< \epsilon
\end{aligned}$$

By Definition 5.12, $\{p_n\}$ is Cauchy and by Theorem 5.13(b), it is convergent in $\mathbb{R}$. This completes the proof of the problem. ∎

Problem 5.23

(★) Suppose that $\{p_n\}$ is a real sequence defined by
$$p_n = 1 + \frac{1}{2} + \cdots + \frac{1}{n}$$
for $n = 1, 2, \ldots$. Determine the convergence of $\{p_n\}$.

Proof. Take $\epsilon = \frac{1}{2}$. Consider
$$\begin{aligned} |p_{2n} - p_n| &= \frac{1}{n+1} + \frac{1}{n+2} + \cdots + \frac{1}{2n} \\ &> \underbrace{\frac{1}{2n} + \frac{1}{2n} + \cdots + \frac{1}{2n}}_{n \text{ terms}} \\ &= \frac{1}{2}. \end{aligned}$$

Thus there is *no* integer N such that $|p_m - p_n| < \frac{1}{2}$ for all $m, n \geq N$. By Definition 5.12, $\{p_n\}$ is not Cauchy and by Theorem 5.13, it is not convergent in $\mathbb{R}$. This ends the proof of the problem. ∎

Problem 5.24

(★)(★) Suppose that $\{p_n\}$ is Cauchy in a metric space X with metric d. If a subsequence $\{p_{n_k}\}$ converges, prove that $\{p_n\}$ is convergent in X.

Proof. Suppose that $\lim\limits_{k \to \infty} p_{n_k} = p$. Given $\epsilon > 0$. Since $p_{n_k} \to p$, there exists a positive integer N_1 such that $k \geq N_1$ implies
$$d(p_{n_k}, p) < \frac{\epsilon}{2}. \tag{5.51}$$
Since $\{p_n\}$ is Cauchy, there is a positive integer N_2 such that $m, n \geq N_2$ implies that
$$d(p_m, p_n) < \frac{\epsilon}{2}. \tag{5.52}$$
Note that if $k \to \infty$, then $n_k \to \infty$. Therefore, k can be chosen large enough so that $n_k \geq N_2$. Then for $n \geq N_2$, we deduce from the inequalities (5.51) and (5.52) that
$$d(p_n, p) \leq d(p_n, p_{n_k}) + d(p_{n_k}, p) < \frac{\epsilon}{2} + \frac{\epsilon}{2} = \epsilon.$$
By definition, $\{p_n\}$ also converges. This ends the proof of the problem. ∎

Problem 5.25 (The Bolzano-Weierstrass Theorem)

(★)(★) Suppose that $\{x_n\}$ is a bounded infinite sequence of real numbers. Prove that there is a convergent subsequence $\{x_{n_k}\}$.

5.4. Cauchy Sequences and Complete Metric Spaces

Proof. Without loss of generality, we may assume that $\{x_n\} \subseteq [0,1]$. Since $[0,1] = [0,\frac{1}{2}] \cup [\frac{1}{2},1]$ and $\{x_n\}$ is infinite, at least one of $[0,\frac{1}{2}]$ and $[\frac{1}{2},1]$ contains infinitely many elements of $\{x_n\}$. Let this interval be I_1 and $\{x_{n^{(1)}}\}$ be the subsequence of $\{x_n\}$ such that

$$\{x_{n^{(1)}}\} \subseteq I_1$$

and the length of I_1 is $\frac{1}{2}$. Next, since

$$I_1 = \left[0, \frac{1}{4}\right] \cup \left[\frac{1}{4}, \frac{1}{2}\right] \quad \left(\text{or } I_1 = \left[\frac{1}{2}, \frac{3}{4}\right] \cup \left[\frac{3}{4}, 1\right]\right)$$

and $\{x_{n^{(1)}}\}$ is infinite, at least one of $[0, \frac{1}{4}]$ and $[\frac{1}{4}, \frac{1}{2}]$ (or one of $[\frac{1}{2}, \frac{3}{4}]$ and $[\frac{3}{4}, 1]$) contains infinitely many elements of $\{x_{n^{(1)}}\}$. Let this interval be I_2 and $\{x_{n^{(2)}}\}$ be the subsequence of $\{x_{n^{(1)}}\}$ such that

$$\{x_{n^{(2)}}\} \subseteq I_2$$

and the length of I_2 is $\frac{1}{4}$. This process can be continued inductively, we are able to construct a sequence of intervals

$$I_1 \supseteq I_2 \supseteq \cdots$$

and a sequence of infinite subsequences of the original sequence $\{x_n\}$

$$\{x_{n^{(1)}}\} \supseteq \{x_{n^{(2)}}\} \supseteq \cdots,$$

where the length of each I_k is $\frac{1}{2^k}$ and I_k contains $\{x_{n^{(k)}}\}$ for $k = 1, 2, \ldots$. Now we may pick one term y_k from each infinite subset $\{x_{n^{(k)}}\}$ and consider the subsequence $\{y_k\}$ of $\{x_n\}$.

We claim that $\{y_k\}$ is Cauchy. To this end, given $\epsilon > 0$, Theorem 2.1 (The Archimedean Property) implies that there exists a positive integer N such that $\frac{1}{2^N} < \epsilon$. Thus for this N, if $k, j > N$, then we have

$$|y_k - y_j| \leq \max\left(\frac{1}{2^k}, \frac{1}{2^j}\right) < \frac{1}{2^N} < \epsilon.$$

This proves the claim. By Theorem 5.13(b), the subsequence $\{y_k\}$ is convergent and this completes the proof of the problem. ∎

Problem 5.26

(★)(★) *If the real sequence $\{x_n\}$ is unbounded, prove that there is an unbounded subsequence $\{x_{n_k}\}$.*

Proof. Since $\{x_n\}$ is unbounded, there exists a term x_{n_1} such that

$$|x_{n_1}| > 1.$$

Split the sequence $\{x_n\}$ into two subsequences $\{x_1, \ldots, x_{n_1-1}\}$ and $\{x_{n_1}, x_{n_1+1}, \ldots\}$. The first subsequence is obviously bounded, so the second subsequence must be unbounded. Thus there exists a term x_{n_2} in $\{x_{n_1}, x_{n_1+1}, \ldots\}$ such that

$$|x_{n_2}| > 2.$$

We may continue this way to obtain a subsequence $\{x_{n_k}\}$ satisfying the condition

$$|x_{n_k}| > k$$

for each positive integer k. By definition, the subsequence $\{x_{n_k}\}$ is unbounded. This finishes the proof of the problem. ∎

Problem 5.27

(★) Let X be a complete metric space. Prove that if a subset E of X is closed in X, then E is also a complete metric space with the same metric as X.

Proof. Suppose that E is closed in X. Let $\{p_n\}$ be a Cauchy sequence of E. Since X is complete, there is a $p \in X$ such that $p_n \to p$. In other words, p is a limit point of $\{p_n\}$ by Theorem 5.2. Since E is closed in X, E must contain p. Since E is also a metric space (see §4.1.4), it follows from the definition in §5.1.4 that E is a complete metric space. Hence we have finished the proof of the problem. ∎

Problem 5.28

(★)(★) Suppose that $\{p_n\}$ and $\{q_n\}$ are sequences in a metric space X with metric d. Let $\{p_n\}$ be Cauchy and $d(p_n, q_n) \to 0$ as $n \to \infty$. Prove that $\{q_n\}$ is Cauchy.

Proof. Given that $\epsilon > 0$. Then there is a positive integer N_1 such that $m, n \geq N_1$ implies

$$d(p_m, p_n) < \frac{\epsilon}{3}. \tag{5.53}$$

Furthermore, since $d(p_n, q_n) \to 0$ as $n \to \infty$, there exists a positive integer N_2 such that

$$d(p_n, q_n) < \frac{\epsilon}{3} \tag{5.54}$$

for all $n \geq N_2$. Let $N = \max(N_1, N_2)$. Thus for $m, n \geq N$, we follow from the inequalities (5.53) and (5.54) that

$$d(q_m, q_n) \leq d(q_m, p_m) + d(p_m, p_n) + d(p_n, q_n) < \frac{\epsilon}{3} + \frac{\epsilon}{3} + \frac{\epsilon}{3} = \epsilon.$$

Hence $\{q_n\}$ is Cauchy and we complete the proof of the problem. ∎

5.5 Recurrence Relations

Problem 5.29

(★) Let $0 < A < 2$. Suppose that $\{p_n\}$ is a real sequence such that

$$p_{n+1} = A p_n + (1 - A) p_{n-1}, \tag{5.55}$$

where $n = 1, 2, \ldots$. Prove that $\{p_n\}$ converges.

Proof. Rewrite the relation (5.55) as

$$p_n - p_{n-1} = (A - 1)(p_{n-1} - p_{n-2})$$

5.5. Recurrence Relations

for $n = 2, 3, \ldots$. Then it becomes

$$p_n - p_{n-1} = (A-1)^{n-1}(p_1 - p_0) \tag{5.56}$$

for $n = 1, 2, \ldots$. By the formula (5.56), we get

$$\begin{aligned} p_n - p_0 &= (p_n - p_{n-1}) + (p_{n-1} - p_{n-2}) + \cdots + (p_1 - p_0) \\ &= [(A-1)^{n-1} + (A-1)^{n-2} + \cdots + (A-1) + 1](p_1 - p_0) \\ &= \frac{1 - (A-1)^n}{1 - (A-1)}(p_1 - p_0) \\ &= \frac{1 - (A-1)^n}{2 - A}(p_1 - p_0). \end{aligned} \tag{5.57}$$

Since $0 < A < 2$, $-1 < A - 1 < 1$ and then Theorem 5.7(c) gives $\lim_{n \to \infty}(1-A)^n = 0$. By this, we obtain from the equation (5.57) that

$$\lim_{n \to \infty} p_n = p_0 + \lim_{n \to \infty} \frac{1 - (A-1)^n}{2 - A}(p_1 - p_0) = p_0 + \frac{1}{2 - A}(p_1 - p_0).$$

Hence $\{p_n\}$ converges and it finishes the proof of the problem. ∎

Problem 5.30

(★) Suppose that $\{p_n\}$ is a real sequence defined by

$$p_1 = 1 \quad \text{and} \quad p_{n+1} = \frac{1}{2}\left(p_n + \frac{1}{p_n}\right). \tag{5.58}$$

Find $\lim_{n \to \infty} p_n$ if it exists.

Proof. It is clear that $p_n \geq 0$ for all $n = 1, 2, \ldots$. By the A.M. $\geq$ G.M., we can show further that

$$p_n \geq 1 \tag{5.59}$$

for all $n = 1, 2, \ldots$. Thus the sequence $\{p_n\}$ is bounded below by 1. By the inequality (5.59), we see that

$$p_{n+1} - p_n = \frac{1}{2}\left(p_n + \frac{1}{p_n}\right) - p_n = \frac{1}{2}\left(\frac{1 - p_n^2}{p_n}\right) \leq 0$$

for every $n = 1, 2, \ldots$. Therefore, the sequence $\{p_n\}$ is monotonically decreasing. By the Monotonic Convergence Theorem, we know that $\{p_n\}$ is convergent. Let

$$\lim_{n \to \infty} p_n = p$$

for some real p. By the recurrence relation (5.58), we gain that

$$p = \frac{1}{2}\left(p + \frac{1}{p}\right)$$
$$p^2 = 1$$
$$p = \pm 1.$$

By the inequality (5.59), we must have $p \geq 1$, so $p = 1$. This ends the proof of the problem. ∎

> **Problem 5.31**
>
> (★)(★) Define the real sequence $\{p_n\}$ by $p_1 = p_2 = 1$ and
>
> $$p_n = \frac{p_{n-1}^2 + 2}{p_{n-2}} \qquad (5.60)$$
>
> for every positive integer $n \geq 3$. Suppose that $p_n = Ax^n + By^n$ for some constants A, B, x and y. Prove that $p_n \in \mathbb{N}$.

Proof. By the hypothesis $p_1 = p_2 = 1$, we have

$$Ax + By = 1 \quad \text{and} \quad Ax^2 + By^2 = 1. \qquad (5.61)$$

Solving the equations (5.61), we have

$$Ax + \left(\frac{1 - Ax^2}{y^2}\right)y = 1$$

which implies that

$$A = \frac{y-1}{x(y-x)}. \qquad (5.62)$$

By substituting the constant (5.62) back to one of the equations (5.61), we gain

$$B = \frac{1-x}{y(y-x)}. \qquad (5.63)$$

Next, we put $p_n = Ax^n + By^n$ into the recurrence relation (5.60) to get

$$(Ax^n + By^n)(Ax^{n-2} + By^{n-2}) = (Ax^{n-1} + By^{n-1})^2 + 2$$
$$A^2 x^{2n-2} + ABx^n y^{n-2} + ABx^{n-2} y^n + B^2 y^{2n-2} = A^2 x^{2n-2} + 2ABx^{n-1} y^{n-1} + B^2 y^{2n-2} + 2.$$

After simplification, we have

$$ABx^n y^{n-2} + ABx^{n-2} y^n = 2ABx^{n-1} y^{n-1} + 2$$
$$ABx^n y^{n-2} - 2ABx^{n-1} y^{n-1} + ABx^{n-2} y^n = 2$$
$$ABx^{n-2} y^{n-2}(x^2 - 2xy + y^2) = 2$$
$$ABx^{n-2} y^{n-2}(x - y)^2 = 2. \qquad (5.64)$$

By substituting the constants (5.62) and (5.63) into (5.64), we achieve that

$$(y-1)(1-x)(xy)^{n-3} = 2. \qquad (5.65)$$

Since the equation (5.65) is valid for all $n = 3, 4, 5, \ldots$, we must have $xy = 1$ and then we have $(y-1)(1-x) = 2$ or equivalently

$$x + y = 4.$$

Now the conditions $xy = 1$ and $x + y = 4$ show that x and y are solutions of the equation

$$z^2 - 4z + 1 = 0,$$

5.5. Recurrence Relations

so $x = 2 + \sqrt{3}$ and $y = 2 - \sqrt{3}$. Therefore, it follows from the constants (5.62) and (5.63) that

$$A = \frac{3 - \sqrt{3}}{6(2 + \sqrt{3})} \quad \text{and} \quad B = \frac{3 + \sqrt{3}}{6(2 - \sqrt{3})}.$$

Thus we conclude that

$$p_n = Ax^n + By^n = \frac{3 - \sqrt{3}}{6} \times (2 + \sqrt{3})^{n-1} + \frac{3 + \sqrt{3}}{6} \times (2 - \sqrt{3})^{n-1}$$

for every $n = 3, 4, \ldots$. Now it can be shown by induction that p_n is an integer for $n = 3, 4, \ldots$. Since $p_1 = p_2 = 1$, we establish that $p_n \in \mathbb{N}$ for all $n = 1, 2, \ldots$. This completes the proof of the problem. ∎

CHAPTER 6

Series of Numbers

6.1 Fundamental Concepts

In this section, we summarize some basic results about series of real or complex numbers. The main references for the current topic are [3, Chap. 8], [5, §3.7, pp. 94 - 101], [23, §3.21 - §3.55, pp. 58 - 78] and [27, Chap. 7].

6.1.1 Definitions and Addition of Series

Suppose that $\{a_n\}$ is a sequence of real or complex numbers. We form a new sequence $\{s_n\}$, where

$$s_n = a_1 + a_2 + \cdots + a_n = \sum_{k=1}^{n} a_k \tag{6.1}$$

for each $n = 1, 2, \ldots$. Here we use the symbol

$$a_1 + a_2 + \cdots$$

to mean the **infinite series**

$$\sum_{n=1}^{\infty} a_n. \tag{6.2}$$

The numbers s_n are called the n-**th partial sums** of the series (6.2).

Definition 6.1. *The series (6.2) is said to* **converge** *or* **diverge** *according to the sequence* $\{s_n\}$ *is convergent or divergent. In the case of convergence, we write*

$$\sum_{n=1}^{\infty} a_n = s = \lim_{n \to \infty} s_n.$$

Here we call s the **sum** of the series (6.2).

The following result describes addition of convergent series.

Theorem 6.2. Suppose that $s = \sum_{n=1}^{\infty} a_n$ and $t = \sum_{n=1}^{\infty} b_n$ are convergent series. Suppose, further, that $\alpha, \beta \in \mathbb{C}$. Then we obtain

$$\sum_{n=1}^{\infty} (\alpha a_n \pm \beta b_n) = \alpha s \pm \beta t.$$

6.1.2 Tests for Convergence of Series

Many tests have been developed for determining the convergence of a given series. Some of them are given as follows:[a]

Theorem 6.3. Suppose that $a_n \geq 0$ for $n = 1, 2, \ldots$. Then the series $\sum a_n$ converges if and only if the sequence $\{s_n\}$ is bounded, where s_n is the n-th partial sum (6.1).

Theorem 6.4 (Cauchy Criterion). The series $\sum a_n$ converges if and only if for every $\epsilon > 0$, there exists a positive integer N such that $m \geq n \geq N$ implies

$$\left| \sum_{k=n}^{m} a_k \right| < \epsilon.$$

Theorem 6.5. Suppose that $\{a_n\}$ is monotonically decreasing and bounded below by 0. Then the series $\sum a_n$ converges if and only if the series

$$\sum 2^n a_{2^n}$$

converges.

The following three theorems are the famous **comparison test**, the **root test** and the **ratio test**.

Theorem 6.6 (Comparison Test). Let N be a fixed positive integer.

(a) Suppose that $|a_n| \leq b_n$ for all $n \geq N$. If $\sum b_n$ converges, then $\sum a_n$ converges.

(b) Suppose that $a_n \geq b_n \geq 0$ for all $n \geq N$. If $\sum b_n$ is divergent, then $\sum a_n$ is divergent.

Theorem 6.7 (Root Test). Let $\sum a_n$ be a series of complex numbers and let

$$\alpha = \limsup_{n \to \infty} \sqrt[n]{|a_n|}.$$

(a) If $\alpha < 1$, then $\sum a_n$ converges.

(b) If $\alpha > 1$, then $\sum a_n$ diverges.

(c) If $\alpha = 1$, then no conclusion can be drawn.

---
[a] Other useful tests can be found in [3, Theorems 8.10, 8.21 & 8.23, pp. 186, 190 & 191].

6.1. Fundamental Concepts

Theorem 6.8 (Ratio Test). Let $\sum a_n$ be a series of complex numbers and let
$$R = \limsup_{n \to \infty} \left| \frac{a_{n+1}}{a_n} \right|.$$

(a) If $R < 1$, then $\sum a_n$ converges.

(b) If $\left| \frac{a_{n+1}}{a_n} \right| \geq 1$ for all $n \geq N$, where N is a fixed positive integer, then $\sum a_n$ diverges.

> **Remark 6.1**
>
> We notice that the conclusions of the Root Test and the Ratio Test can be stronger. Firstly, we introduce the concept of **absolute convergence**: A series $\sum a_n$ is called **converges absolutely** if the series $\sum |a_n|$ converges. Then the absolute convergence of $\sum a_n$ implies the convergence of $\sum a_n$ and the conclusions of the Root Test and the Ratio Test can be changed from convergence to absolute convergence. Another importance property of absolute convergence will be given in §6.1.6.

6.1.3 Three Special Series

The following two results are very useful and important in many real problems. In fact, Theorem 6.9 (Geometric Series) can tell us the exact value of a certain type of series, called **geometric series**. Theorem 6.10 (p-series) is commonly applied with Theorem 6.6 (Comparison Test) to determine the convergence of other series.

Theorem 6.9 (Geometric Series). If $-1 < x < 1$, then we have
$$\sum_{n=0}^{\infty} x^n = \frac{1}{1-x}.$$

Theorem 6.10 (p-series). The series
$$\sum_{n=1}^{\infty} \frac{1}{n^p}$$
converges if $p > 1$ and diverges if $p \leq 1$.

The next result discusses about the convergence of the so-called **alternating series**. We have the precise definition first.

Definition 6.11. If $a_n > 0$ for each $n = 1, 2, \ldots$, we call the series
$$\sum_{n=1}^{\infty} (-1)^n a_n \tag{6.3}$$
an **alternating series**.

Theorem 6.12 (Alternating Series Test). Suppose that $\{a_n\}$ is a monotonically decreasing sequence and $\lim_{n \to \infty} a_n = 0$. Then the series (6.3) converges.

6.1.4 Series in the form $\sum_{n=1}^{\infty} a_n b_n$

Sometimes we have to cope with series in the form

$$\sum_{n=1}^{\infty} a_n b_n. \tag{6.4}$$

The main tool for studying the convergence of the series (6.4) is the so-called **partial summation formula**.

Theorem 6.13. *Suppose that $\{a_n\}$ and $\{b_n\}$ are two sequences of complex numbers. Denote*

$$A_n = a_1 + \cdots + a_n.$$

Then we have the following identity

$$\sum_{k=1}^{n} a_k b_k = A_n b_{n+1} - \sum_{k=1}^{n} A_k (b_{k+1} - b_k).$$

By this tool, we can prove two famous results: **Dirichlet's Test** and **Abel's Test**.

Theorem 6.14 (Dirichlet's Test). *Suppose that $\sum a_n$ is a series of complex numbers whose partial sums $\{A_n\}$ is bounded. If $\{b_n\}$ is a monotonically decreasing sequence and $\lim_{n \to \infty} b_n = 0$, then the series (6.4) converges.*

Theorem 6.15 (Abel's Test). *Suppose that $\sum a_n$ is convergent and $\{b_n\}$ is a monotonic bounded sequence. Then the series (6.4) converges.*

6.1.5 Multiplication of Series

By Theorem 6.2, we see easily that two convergent series may be added term by term and the resulting series will converge to the sum of the two series. However, the situation is a little bit complicated when multiplication of series is considered. Let's define a product of two series first. It is called the **Cauchy product**.[b]

Definition 6.16. *Suppose that we have two series $\sum_{n=0}^{\infty} a_n$ and $\sum_{n=0}^{\infty} b_n$. Define*

$$c_n = \sum_{k=0}^{n} a_k b_{n-k},$$

*where $n = 0, 1, 2, \ldots$. Then the series $\sum_{n=0}^{\infty} c_n$ is the **Cauchy product** of the two series.*

Theorem 6.17 (Mertens' Theorem). *Suppose that $\sum_{n=0}^{\infty} a_n$ converges absolutely, $\sum_{n=0}^{\infty} a_n = A$ and $\sum_{n=0}^{\infty} b_n$ converges with sum B. Then the Cauchy product of the two series converges to AB, i.e.,*

$$\sum_{n=0}^{\infty} c_n = AB.$$

---

[b]There is another way to define multiplication of series which is known to be **Dirichlet product**.

6.1.6 Rearrangement of Series

It may happen that different rearrangements of a convergent series give different numbers. One example can be found in [23, Example 3.53, p. 76]. To make sure different rearrangements give the same sum, the series must converge absolutely:

Theorem 6.18. *If $\sum a_n$ is a series of complex numbers and it converges absolutely, then every rearrangement of $\sum a_n$ converges to the same sum.*

6.1.7 Power Series

Definition 6.19. *An infinite series of the form*

$$\sum_{n=0}^{\infty} c_n z^n \tag{6.5}$$

*is called a **power series**. Here all c_n and z are complex numbers. The numbers c_n are called the **coefficients** of the series (6.5).*

It is well-known that with every power series there is associated a circle, namely the **circle of convergence**, such that the series (6.5) converges for every z in the interior of the circle and diverges if z is outside the circle. The following theorem provides us an easy method to establish the circle of convergence.

Theorem 6.20. *Given a power series in the form (6.5). Take*

$$\alpha = \limsup_{n \to \infty} \sqrt[n]{|c_n|} \quad \text{and} \quad R = \frac{1}{\alpha}.$$

*Then the series (6.5) converges if $|z| < R$ and diverges if $|z| > R$. The number R is called the **radius of convergence** of the power series (6.5).*

6.2 Convergence of Series of Nonnegative Terms

Unless otherwise specified the terms in the series are assumed to be nonnegative.

> **Problem 6.1**
>
> (★) Prove that $\displaystyle\sum_{n=1}^{\infty} \frac{n}{n^4 - n^2 + 1}$ converges.

Proof. By simple algebra, we can show that

$$0 < \frac{n}{n^4 - n^2 + 1} < \frac{2}{n^3}$$

because $(n^2-1)^2 + 1 > 0$ for every positive integer n. By Theorem 6.10 (p-series),

$$\sum_{n=1}^{\infty} \frac{1}{n^3}$$

converges. Then Theorem 6.6 (Comparison Test) implies that

$$\sum_{n=1}^{\infty} \frac{n}{n^4 - n^2 + 1}$$

converges which completes the proof of the problem. ∎

Problem 6.2

(⋆)(⋆) *Let $a_n \geq 0$ for $n = 1, 2, \ldots$. Suppose that $\sum_{n=1}^{\infty} n a_n$ converges. Prove that $\sum_{n=1}^{\infty} a_n$ converges.*

Proof. Let $s_n = a_1 + 2a_2 + \cdots + na_n$ for $n = 1, 2, \ldots$. Since $\sum_{n=1}^{\infty} n a_n$ converges, Theorem 6.3 implies that $\{s_n\}$ is bounded, i.e., there exists a positive constant M such that

$$|s_n| \leq M$$

for all positive integers n. For $k = 1, 2, \ldots$, we note that

$$a_k = \frac{s_k - s_{k-1}}{k}. \tag{6.6}$$

Given that $\epsilon > 0$, for positive integers m, n with $m > n > \frac{2M}{\epsilon}$, we deduce from the equation (6.6) that

$$\left| \sum_{k=n}^{m} a_k \right| = \left| \frac{s_n - s_{n-1}}{n} + \frac{s_{n+1} - s_n}{n+1} + \cdots + \frac{s_m - s_{m-1}}{m} \right|$$

$$= \left| -\frac{s_{n-1}}{n} + \left(\frac{1}{n} - \frac{1}{n+1} \right) s_n + \cdots + \left(\frac{1}{m-1} - \frac{1}{m} \right) s_{m-1} + \frac{s_m}{m} \right|$$

$$\leq M \left[\frac{1}{n} + \left(\frac{1}{n} - \frac{1}{n+1} \right) + \left(\frac{1}{n+1} - \frac{1}{n+2} \right) + \cdots + \left(\frac{1}{m-1} - \frac{1}{m} \right) + \frac{1}{m} \right]$$

$$= \frac{2M}{n}$$

$$< \epsilon.$$

By Theorem 6.4 (Cauchy Criterion), the series $\sum_{n=1}^{\infty} a_n$ converges. This completes the proof of the problem. ∎

6.2. Convergence of Series of Nonnegative Terms

Problem 6.3

($\star$) Suppose that $a_1, a_2, \ldots$ are complex numbers and $\sum_{n=1}^{\infty} a_n$ converges. Prove that

$$\lim_{n \to \infty} a_n = 0. \tag{6.7}$$

Proof. If we take $m = n$ in Theorem 6.4 (Cauchy Criterion), then we have

$$|a_n| < \epsilon$$

for all $n \geq N$. Thus it follows from the definition that the limit (6.7) holds. We complete the proof of the problem. ∎

Problem 6.4

($\star$)($\star$) Prove that $\sum_{n=1}^{\infty} \dfrac{1}{n}$ is divergent without using Theorem 6.10 (*p*-series).

Proof. Let $s_n = 1 + \frac{1}{2} + \cdots + \frac{1}{n}$ for each positive integer n. By the fact $e^x > 1 + x > 0$ for $x > 0$, we have

$$\begin{aligned}
e^{s_n} &= \exp\left(1 + \frac{1}{2} + \cdots + \frac{1}{n}\right) \\
&= e^1 \times e^{\frac{1}{2}} \times \cdots \times e^{\frac{1}{n}} \\
&> (1+1)\left(1 + \frac{1}{2}\right) \times \cdots \times \left(1 + \frac{1}{n}\right) \\
&= 2 \times \frac{3}{2} \times \cdots \times \frac{n}{n-1} \times \frac{n+1}{n} \\
&= n+1.
\end{aligned}$$

Thus $\{e^{s_n}\}$ is unbounded and hence $\{s_n\}$ is also unbounded. By Theorem 6.3, the series

$$\sum_{n=1}^{\infty} \frac{1}{n}$$

is divergent. This completes the proof of the problem. ∎

Remark 6.2

The series in Problem 6.4 is called the **harmonic series**. There are many proofs of Problem 6.4 and Theorem 6.10 (*p*-series) is clearly one of them.

> **Problem 6.5**
>
> (★) Suppose that $a_n \geq 0$ for $n = 1, 2, \ldots$ and $\sum_{n=1}^{\infty} a_n$ converges. Prove that the series
> $$\sum_{n=1}^{\infty} a_n^2$$
> converges. Is the converse true?

Proof. Since $\sum_{n=1}^{\infty} a_n$ converges, it follows from Problem 6.3 that $a_n \to 0$ as $n \to \infty$. Therefore, there exists a positive constant M such that
$$0 \leq a_n < M$$
for $n = 1, 2, \ldots$. Then we have
$$0 \leq a_n^2 < M a_n$$
for $n = 1, 2, \ldots$. By Theorem 6.2, the series $\sum_{n=1}^{\infty} M a_n$ converges. By Theorem 6.6 (Comparison Test), the series
$$\sum_{n=1}^{\infty} a_n^2$$
converges.

The converse is false. For examples, the series $\sum_{n=1}^{\infty} \frac{1}{n^2}$ converges by Theorem 6.10 (p-series). However, Theorem 6.10 (p-series) or Problem 6.4 shows that the series $\sum_{n=1}^{\infty} \frac{1}{n}$ is divergent. Thus we have completed the proof of the problem. ∎

> **Problem 6.6**
>
> (★) Prove that the series $\sum_{n=2}^{\infty} \frac{1}{n(\log n)}$ diverges.

Proof. Since the sequence $\{\log n\}$ is increasing, the sequence $\{\frac{1}{n(\log n)}\}$ is decreasing and it is clear that $\{\frac{1}{n(\log n)}\}$ is bounded below by 0 for $n = 2, 3, \ldots$. Consider
$$\sum_{k=2}^{n} 2^k \times \frac{1}{2^k (\log 2^k)} = \frac{1}{\log 2} \sum_{k=2}^{n} \frac{1}{k}.$$
By Theorem 6.10 (p-series),
$$\sum_{n=2}^{\infty} \frac{1}{n}$$

6.2. Convergence of Series of Nonnegative Terms

diverges. Thus it follows from Theorem 6.6 (Comparison Test) that

$$\sum_{n=2}^{\infty} \frac{1}{n(\log n)}$$

is divergent. Hence we finish the proof of the problem. ∎

Problem 6.7

(★) Prove that the series $\sum_{n=1}^{\infty} \frac{n^n}{7^{n^2}}$ converges.

Proof. Since

$$\limsup_{n \to \infty} \sqrt[n]{\frac{n^n}{7^{n^2}}} = \lim_{n \to \infty} \frac{n}{7^n} = 0,$$

we know from Theorem 6.7 (Root Test) that the series converges. We have completed the proof of the problem. ∎

Problem 6.8

(★) Prove that both series

$$\sum_{n=1}^{\infty} \frac{1}{n!} \quad \text{and} \quad \sum_{n=1}^{\infty} \frac{1}{n^n} \qquad (6.8)$$

converge.

Proof. Put $a_n = \frac{1}{n!}$ so that

$$\frac{a_{n+1}}{a_n} = \frac{1}{n+1}$$

which gives

$$R = \limsup_{n \to \infty} \left| \frac{a_{n+1}}{a_n} \right| = \lim_{n \to \infty} \frac{1}{n+1} = 0.$$

By Theorem 6.8 (Ratio Test), we see that the first series in (6.8) converges.

Since $\log k \leq \log n$ for all $k = 1, 2, \ldots, n$, we must have

$$0 \leq \log(n!) = \sum_{k=1}^{n} \log k \leq \sum_{k=1}^{n} \log n = \log(n^n) \qquad (6.9)$$

for every positive integer n. By the inequalities (6.9), we deduce that $n! \leq n^n$ for each $n = 1, 2, \ldots$ and then

$$0 < \frac{1}{n^n} \leq \frac{1}{n!}.$$

By our first assertion and Theorem 6.6 (Comparison Test), we conclude that

$$\sum_{n=1}^{\infty} \frac{1}{n^n}$$

also converges, completing the proof of the problem. ∎

> **Problem 6.9**
>
> (⋆) (⋆) *Discuss the convergence of the series*
>
> $$\sum_{n=2}^{\infty} \frac{1}{(\log n)^{\log n}} \quad \text{and} \quad \sum_{n=3}^{\infty} \frac{1}{(\log \log n)^{\log \log n}}. \tag{6.10}$$

Proof. Since $a^{\log b} = b^{\log a}$ holds for $a > 0$ and $b > 0$, we have

$$(\log n)^{\log n} = n^{\log \log n} \tag{6.11}$$

for every positive integer $n \geq 2$. Furthermore, when $n \geq e^{e^2}$, we have $\log \log n \geq 2$ so it follows from the expression (6.11) that

$$(\log n)^{\log n} = n^{\log \log n} \geq n^2 > 0$$

or equivalently,

$$0 < \frac{1}{(\log n)^{\log n}} \leq \frac{1}{n^2}$$

for $n \geq e^{e^2}$. By Theorem 6.10 (*p*-series), we know that $\displaystyle\sum_{n \geq e^{e^2}}^{\infty} \frac{1}{n^2}$ converges. Thus Theorem 6.6 (Comparison Test) shows that

$$\sum_{n \geq e^{e^2}}^{\infty} \frac{1}{(\log n)^{\log n}}$$

is convergent and the convergence of the first series in (6.10) follows from this fact immediately.

We claim that the second series in (6.10) is divergent. To this end, we first note that for $n \geq 3$, we always have $\log n > \log \log n > 0$ so that

$$\frac{1}{(\log \log n)^{\log \log n}} > \frac{1}{(\log n)^{\log \log n}} \tag{6.12}$$

for $n \geq 3$. Furthermore, we follow from the fact $a^b = e^{b \log a}$ for $a > 0$ and $b \in \mathbb{R}$ that

$$(\log n)^{\log \log n} = e^{(\log \log n) \times (\log \log n)}. \tag{6.13}$$

By the fact[c] that $\frac{x}{e^x} \to 0$ as $x \to +\infty$, we obtain

$$\frac{x}{e^{\frac{x}{2}}} \to 0$$

as $x \to +\infty$. In other words, for large x, we must have

$$0 < x < e^{\frac{x}{2}}$$

or equivalently

$$0 < 2 \log x < x. \tag{6.14}$$

---

[c] See [23, Theorem 8.6(f), p. 180].

6.2. Convergence of Series of Nonnegative Terms

By substituting $x = \log\log n$ for large enough n into the inequalities (6.14), we see that
$$0 < 2\log\log\log n < \log\log n$$
and they imply that
$$0 < (\log\log n)^2 < \log n. \tag{6.15}$$
Combining the inequalities (6.15) and the equation (6.13), we can show that
$$\frac{1}{(\log n)^{\log\log n}} = \frac{1}{e^{(\log\log n)^2}} > \frac{1}{e^{\log n}} = \frac{1}{n}. \tag{6.16}$$
Next, we achieve from the inequalities (6.12) and (6.16) that
$$\frac{1}{(\log\log n)^{\log\log n}} > \frac{1}{n}.$$
Finally, by using Theorem 6.10 (*p*-series) and then Theorem 6.6 (Comparison Test), we conclude that
$$\sum_{n=3}^{\infty} \frac{1}{(\log\log n)^{\log\log n}}$$
is divergent. This completes the proof of the problem. ∎

Problem 6.10

(⋆)(⋆)(⋆) Suppose that, for $n = 1, 2, \ldots$, we have
$$0 < a_n \leq \sum_{k=n+1}^{\infty} a_k \quad \text{and} \quad \sum_{n=1}^{\infty} a_n = A \tag{6.17}$$
for some constant A. Prove that there is either a finite subsequence $\{n_1, n_2, \ldots, n_j\}$ such that $a_{n_1} + \cdots + a_{n_j} = p$ or an infinite subsequence $\{n_j\}$ such that $\sum_{j=1}^{\infty} a_{n_j} = p$, where $0 < p < 1$.

Proof. Fix $p \in (0,1)$. Let n_1 be the *least* positive integer such that
$$a_{n_1} \leq p. \tag{6.18}$$
Such an n_1 must exist, otherwise,
$$\sum_{n=1}^{m} a_n > \sum_{n=1}^{m} p = mp > A$$
for some positive integer m, a contradiction. Furthermore, the inequality (6.18) implies that
$$\sum_{k=n_1}^{\infty} a_k \geq p.$$

Otherwise, the first hypothesis in (6.17) shows that

$$a_{n_1-1} \leq \sum_{k=n_1}^{\infty} a_k < p$$

which obviously contradicts the definition of n_1 in (6.18).

If $\sum_{k=n_1}^{\infty} a_k = p$, then an infinite subsequence $\{n_1, n_1+1, n_1+2, \ldots\}$ exists. Otherwise, we have $\sum_{k=n_1}^{\infty} a_k > p$ and let n_2 be the *greatest* integer such that

$$\sum_{k=n_1}^{n_2} a_k \leq p. \tag{6.19}$$

The existence of such an n_2 is guaranteed by the condition (6.18). If the equality holds, then a finite subsequence $\{n_1, n_1+1, \ldots, n_2\}$ exists. Otherwise, we have

$$\sum_{k=n_1}^{n_2} a_k < p$$

and then there exists an integer n_3 which is the *least* integer greater than n_2 such that

$$\sum_{k=n_1}^{n_2} a_k + a_{n_3} \leq p. \tag{6.20}$$

We remark that $n_3 > n_2 + 1$, otherwise, $n_3 = n_2 + 1$ which contradicts the definition of n_2 given by the inequality (6.19). In addition, we must have

$$\sum_{k=n_1}^{n_2} a_k + \sum_{k=n_3}^{\infty} a_k \geq p.$$

Otherwise, the first hypothesis in (6.17) implies that

$$p > \sum_{k=n_1}^{n_2} a_k + \sum_{k=n_3}^{\infty} a_k \geq \sum_{k=n_1}^{n_2} a_k + a_{n_3-1}$$

which obviously contradicts the definition of n_3 given by the inequality (6.20). If the equality holds, then an infinite subsequence $\{n_1, \ldots, n_2, n_3, n_3+1, \ldots\}$ exists. Otherwise, we have

$$\sum_{k=n_1}^{n_2} a_k + \sum_{k=n_3}^{\infty} a_k > p$$

and then we let n_4 be the *greatest* integer such that

$$\sum_{k=n_1}^{n_2} a_k + \sum_{k=n_3}^{n_4} a_k \leq p.$$

6.3. Alternating Series and Absolute Convergence

Similar to n_2, the existence of n_4 is guaranteed by the condition (6.20). If the equality holds, then a finite subsequence $\{n_1, \ldots, n_2, n_3, \ldots, n_4\}$ exists.

Now we may see that the above process can be continued. If the process terminates after finitely many steps, then we have either

$$a_{n_1} + \cdots + a_{n_j} = p \quad \text{or} \quad \sum_{j=1}^{\infty} a_{n_j} = p.$$

If the process does not terminate, then we must have

$$\sum_{k=n_1}^{n_2} a_k + \cdots + \sum_{k=n_{2j-1}}^{n_{2j}} a_k + \sum_{k=n_{2j}+1}^{\infty} a_k > p \tag{6.21}$$

and

$$\sum_{k=n_1}^{n_2} a_k + \cdots + \sum_{k=n_{2j-1}}^{n_{2j}} a_k + \sum_{k=n_{2j}+1}^{n_{2j+2}} a_k < p \tag{6.22}$$

where $n_{2j+1} > n_{2j} + 1$ for every positive integer j. Combining the inequalities (6.21) and (6.22) to get

$$0 < p - \left[\sum_{k=n_1}^{n_2} a_k + \cdots + \sum_{k=n_{2j-1}}^{n_{2j}} a_k\right] < \sum_{k=n_{2j}+1}^{\infty} a_k.$$

Since $n_{2j+1} \to +\infty$ as $j \to +\infty$, the second hypothesis in (6.17) and Problem 6.3 ensure that

$$\lim_{j \to \infty} \sum_{k=n_{2j}+1}^{\infty} a_k = 0.$$

By this and Theorem 5.6 (Squeeze Theorem for Convergent Sequences), an infinite subsequence $\{n_1, n_2, n_3, n_4, \ldots\}$ (after renaming) exists such that

$$\sum_{j=1}^{\infty} a_{n_j} = p.$$

This completes the proof of the problem. ∎

6.3 Alternating Series and Absolute Convergence

Problem 6.11

(★) *Determine the convergence of the series*

$$\sum_{n=1}^{\infty} \frac{\cos n\pi}{n^{\frac{4}{5}}}. \tag{6.23}$$

Proof. Recall the fact that
$$\cos n\pi = (-1)^n$$
for $n = 1, 2, \ldots$. Then we have
$$\sum_{n=1}^{\infty} \frac{\cos n\pi}{n^{\frac{4}{5}}} = \sum_{n=1}^{\infty} (-1)^n \frac{1}{n^{\frac{4}{5}}}$$
which is an alternating series. It is clear that the sequence $\{n^{-\frac{4}{5}}\}$ is monotonically decreasing and
$$\lim_{n \to \infty} \frac{1}{n^{\frac{4}{5}}} = 0.$$
By Theorem 6.12 (Alternating Series Test), the series (6.23) is convergent and this completes the proof of the problem. ∎

Problem 6.12

(⋆) *Discuss the convergence of the series*
$$\sum_{n=1}^{\infty} \frac{(-1)^{n-1}}{n^p}. \tag{6.24}$$

Proof. If $p < 0$, then $n^{-p} \to +\infty$ as $n \to +\infty$. By Problem 6.3, the series (6.24) is divergent in this case. Similarly, the series (6.24) is also divergent if $p = 0$. When $p > 0$, we know that
$$\frac{1}{n^p} > \frac{1}{(n+1)^p}$$
for $n = 1, 2, \ldots$ so that $\{n^{-p}\}$ is monotonically decreasing. Besides, we have $\frac{1}{n^p} \to 0$ as $n \to \infty$. Thus Theorem 6.12 (Alternating Series Test) implies that the series (6.24) converges in this case. This ends the proof of the problem. ∎

Problem 6.13

(⋆) *In Remark 6.1, we know that absolute convergence implies convergence. Is the converse true?*

Proof. The converse is false. For example, the series
$$\sum_{n=1}^{\infty} (-1)^n \frac{1}{n}$$
converges by Theorem 6.12 (Alternating Series Test), but $\sum_{n=1}^{\infty} \frac{1}{n}$ is divergent by Problem 6.4. This finishes the proof of the problem. ∎

6.3. Alternating Series and Absolute Convergence

> **Problem 6.14**
>
> (★) Let a_n be real numbers for $n = 1, 2, \ldots$. Suppose that $\sum_{n=1}^{\infty} a_n$ converges absolutely. Prove that $\sum_{n=1}^{\infty} a_n^2$ converges absolutely.

Proof. Since $\sum_{n=1}^{\infty} a_n$ converges, Problem 6.3 implies that $a_n \to 0$ as $n \to \infty$. Thus there exists a positive integer N such that $n \geq N$ implies

$$|a_n| < 1.$$

Then we must have

$$0 \leq a_n^2 \leq |a_n|$$

for $n \geq N$. Since $\sum_{n=1}^{\infty} |a_n|$ converges, Theorem 6.6 (Comparison Test) implies that

$$\sum_{n=N}^{\infty} a_n^2 \qquad (6.25)$$

converges. Since $\sum_{n=1}^{N-1} a_n^2$ is finite, we follow from this and the convergence of (6.25) that

$$\sum_{n=1}^{\infty} a_n^2$$

converges. Since $|a_n^2| = a_n^2$ for each $n = 1, 2, \ldots$, the series actually converges absolutely. Hence this completes the proof of the problem. ■

> **Problem 6.15**
>
> (★)(★) Let
>
> $$\sum_{n=3}^{\infty} \frac{a^n}{n^b (\log n)^c}, \qquad (6.26)$$
>
> where $a, b, c \in \mathbb{R}$. Prove that the series (6.26) converges absolutely if $|a| < 1$.

Proof. If $a = 0$, then each term in the series (6.26) is zero, so it converges absolutely. Suppose that $a \neq 0$. Now we have

$$\limsup_{n \to \infty} \left| \frac{a^{n+1}}{(n+1)^b [\log(n+1)]^c} \right| \times \left| \frac{n^b (\log n)^c}{a^n} \right| = |a| \lim_{n \to \infty} \left\{ \left(\frac{n}{n+1} \right)^b \times \left[\frac{\log n}{\log(n+1)} \right]^c \right\}. \qquad (6.27)$$

Since
$$\lim_{n\to\infty} \frac{n}{n+1} = 1 \quad \text{and} \quad \lim_{n\to\infty} \frac{\log n}{\log(n+1)} = 1,$$
we deduce from the expression (6.27) that
$$\limsup_{n\to\infty} \left\| \frac{a^{n+1}}{(n+1)^b [\log(n+1)]^c} \right| \times \left| \frac{n^b (\log n)^c}{a^n} \right\| = |a|.$$

Therefore, we follow from Theorem 6.8 (Ratio Test) that the series (6.26) converges absolutely when $|a| < 1$ and $a = 0$. As we have shown in the previously that it converges absolutely when $a = 0$, we gain our desired result that it converges absolutely if $|a| < 1$. We have finished the proof of the problem. ∎

6.4 The Series $\sum_{n=1}^{\infty} a_n b_n$ and Multiplication of Series

Problem 6.16

(⋆)(⋆) Suppose that $\sum_{n=1}^{\infty} a_n^2$ and $\sum_{n=1}^{\infty} b_n^2$ converge, $a_n \geq 0$ and $b_n \geq 0$ for every $n = 1, 2, \ldots$. Prove that
$$\sum_{n=1}^{\infty} a_n b_n$$
converges absolutely.

Proof. Let
$$\sum_{n=1}^{\infty} a_n^2 = A \quad \text{and} \quad \sum_{n=1}^{\infty} b_n^2 = B$$
respectively. Recall the **Schwarz inequality** that if $a_1, a_2, \ldots, a_n$ and $b_1, b_2, \ldots, b_n$ are complex numbers, then we have
$$\Big| \sum_{k=1}^{n} a_k \overline{b}_k \Big|^2 \leq \Big(\sum_{k=1}^{n} |a_k|^2 \Big) \times \Big(\sum_{k=1}^{n} |b_k|^2 \Big). \tag{6.28}$$
Since a_n and b_n are nonnegative for every $n \in \mathbb{N}$, we have
$$\Big| \sum_{k=1}^{n} a_k b_k \Big| = \sum_{k=1}^{n} a_k b_k = \sum_{k=1}^{n} |a_k b_k| \tag{6.29}$$
for all $n \in \mathbb{N}$. Now we substitute the expression (6.29) into the inequality (6.28) to obtain
$$\sum_{k=1}^{n} |a_k b_k| \leq \Big(\sum_{k=1}^{n} a_k^2 \Big)^{\frac{1}{2}} \times \Big(\sum_{k=1}^{n} b_k^2 \Big)^{\frac{1}{2}} < \sqrt{AB}$$

6.4. The Series $\Sigma_{n=1}^{\infty} a_n b_n$ and Multiplication of Series

for every $n \in \mathbb{N}$. Thus the partial sums of the series $\displaystyle\sum_{n=1}^{\infty} |a_n b_n|$ is bounded. Hence it follows from Theorem 6.3 that $\displaystyle\sum_{n=1}^{\infty} |a_n b_n|$ converges, i.e., $\displaystyle\sum_{n=1}^{\infty} a_n b_n$ converges absolutely by definition. This completes the proof of the problem. ∎

Problem 6.17

(★) Suppose that $0 < \theta < 1$. Prove that the series $\displaystyle\sum_{n=1}^{\infty} \frac{e^{2\pi i n \theta}}{n}$ converges.

Proof. Let $a_n = e^{2\pi i n \theta}$ and $b_n = \frac{1}{n}$ for each $n = 1, 2, \ldots$. It is clear that $\{b_n\}$ is monotonically decreasing and
$$\lim_{n \to \infty} b_n = \lim_{n \to \infty} \frac{1}{n} = 0.$$
Since $0 < \theta < 1$, we have $e^{2\pi i \theta} \neq 1$. Thus we have
$$A_n = \sum_{k=1}^{n} a_k = \sum_{k=1}^{n} e^{2\pi i k \theta} = \frac{e^{2\pi i \theta} - e^{2\pi i (n+1) \theta}}{1 - e^{2\pi i \theta}}. \tag{6.30}$$
Fix the θ. Since $|e^{2\pi i \theta}| = 1$, we deduce from the expression (6.30) that
$$|A_n| \leq \frac{2}{|1 - e^{2\pi i \theta}|}$$
which means that $\{A_n\}$ is bounded. Hence we apply Theorem 6.14 (Dirichlet's Test) to conclude that the series is convergent and we have finished the proof of the problem. ∎

Problem 6.18

(★) Prove that the series
$$\sum_{n=1}^{\infty} \frac{n^4 \sin(\frac{1}{n^3})}{e^n (n+1)} \tag{6.31}$$
converges.

Proof. For each $n = 1, 2, \ldots$, let
$$a_n = \frac{n^4}{e^n (n+1)} \quad \text{and} \quad b_n = \sin\left(\frac{1}{n^3}\right).$$
Consider
$$\begin{aligned}\limsup_{n \to \infty} \left|\frac{a_{n+1}}{a_n}\right| &= \lim_{n \to \infty} \frac{(n+1)^4}{e^{n+1}(n+2)} \times \frac{e^n(n+1)}{n^4} \\ &= \lim_{n \to \infty} \frac{1}{e} \times \left(\frac{n+1}{n}\right)^4 \times \left(\frac{n+1}{n+2}\right)\end{aligned}$$

$$= \frac{1}{e}.$$

By Theorem 6.8, we see that the series $\sum_{n=1}^{\infty} a_n$ converges. It is clear that $\{b_n\}$ is bounded by 1. Furthermore, since $\frac{1}{n^3} \to 0$ as $n \to \infty$ and $\sin x$ is decreasing on the interval $[0, \frac{\pi}{2}]$, $\{b_n\}$ is monotonically decreasing. Hence we follow from Theorem 6.15 (Abel's Test) that the series (6.31) is convergent. This completes the proof of the problem. ∎

Problem 6.19

(★)(★) *Suppose that $\sum_{n=0}^{\infty} a_n$ and $\sum_{n=0}^{\infty} b_n$ converge absolutely. Prove that their Cauchy product also converges absolutely.*

Proof. We have to prove that

$$C_n = \sum_{n=0}^{\infty} |c_n|$$

converges. Since $\sum_{n=0}^{\infty} a_n$ and $\sum_{n=0}^{\infty} b_n$ converge absolutely, we let

$$A = \sum_{n=0}^{\infty} |a_n|, \quad A_n = \sum_{k=0}^{n} |a_k|, \quad B = \sum_{n=0}^{\infty} |b_n| \quad \text{and} \quad B_n = \sum_{k=0}^{n} |b_k|.$$

Now for all $n \geq 0$, $|a_n|$ and $|b_n|$ are nonnegative terms of A_n and B_n respectively. This implies that

$$0 \leq A_n \leq A \quad \text{and} \quad 0 \leq B_n \leq B$$

for all $n \geq 0$. Therefore, we achieve that

$$|C_n| = \sum_{n=0}^{\infty} |c_n|$$
$$\leq |a_0||b_0| + (|a_0||b_1| + |a_1||b_0|) + \cdots + (|a_0||b_n| + |a_1||b_{n-1}| + \cdots + |a_n||b_0|)$$
$$= |a_0|B_n + |a_1|B_{n-1} + \cdots + |a_n|B_0$$
$$\leq (|a_0| + |a_1| + \cdots + |a_n|)B$$
$$= A_n B$$
$$\leq AB.$$

Hence $\{C_n\}$ is bounded and it follows from Theorem 6.3 that $\sum_{n=1}^{\infty} |c_n|$ converges, i.e., the series

$$\sum_{n=0}^{\infty} c_n$$

converges absolutely. ∎

6.4. The Series $\Sigma_{n=1}^{\infty} a_n b_n$ and Multiplication of Series

Problem 6.20

(⋆)(⋆) Does the Cauchy product of the series $\displaystyle\sum_{n=0}^{\infty} \frac{(-1)^{n+1}}{\sqrt{n+1}}$ with itself converge?

Proof. Let $a_n = \frac{(-1)^{n+1}}{\sqrt{n+1}}$. By Definition 6.16, we have

$$\begin{aligned} c_n &= \sum_{k=0}^{n} a_k a_{n-k} \\ &= \sum_{k=0}^{n} \frac{(-1)^{k+1}}{\sqrt{k+1}} \times \frac{(-1)^{n-k+1}}{\sqrt{n-k+1}} \\ &= (-1)^n \sum_{k=0}^{n} \frac{1}{\sqrt{k+1}\sqrt{n-k+1}}. \end{aligned} \tag{6.32}$$

Since $k+1$ and $n-k+1$ are nonnegative for all $n, k = 0, 1, \ldots$ with $0 \leq k \leq n$, the A.M. $\geq$ G.M. implies that

$$\sqrt{(k+1)(n-k+1)} \leq \frac{k+1+n-k+1}{2} = \frac{n+2}{2}. \tag{6.33}$$

Now, by combining the expression (6.32) and the inequality (6.33), we derive that

$$|c_n| = \sum_{k=0}^{n} \frac{1}{\sqrt{k+1}\sqrt{n-k+1}} \geq \sum_{k=0}^{n} \frac{2}{n+2} = \frac{2n+2}{n+2}.$$

Since $\frac{2n+2}{n+2} \to 2$ as $n \to \infty$, it means that there exists a positive integer N such that

$$|c_n| > 1$$

for all $n \geq N$. Thus it follows from Problem 6.3 that the series

$$\sum_{n=0}^{\infty} c_n$$

is divergent. Hence we complete the proof of the problem. ∎

Remark 6.3

It is clear that

$$\sum_{n=0}^{\infty} \frac{(-1)^{n+1}}{\sqrt{n+1}}$$

converges by Theorem 6.12 (Alternating Series Test), Problem 6.20 tells us that absolute convergence of one of the two series in Theorem 6.17 (Mertens' Theorem) cannot be dropped.

6.5 Power Series

Problem 6.21

(★) Determine the radius of convergence of each of the following series:

(a) $\sum_{n=1}^{\infty} (\log 2n) z^n$.

(b) $\sum_{n=1}^{\infty} \dfrac{z^n}{(2n)!}$.

Proof.

(a) Since $1 \leq \sqrt[n]{\log 2n} \leq \sqrt[n]{n}$ for $n \geq 2$, we deduce from Theorems 5.7(b) and 6.6 (Comparison Test) that
$$\lim_{n \to \infty} \sqrt[n]{\log 2n} = 1.$$
By Theorem 6.20, the radius of convergence of the power series is 1.

(b) For $n = 1, 2, \ldots$, we have
$$0 < \frac{1}{(2n)!} \leq \frac{1}{n!}. \tag{6.34}$$
By Problem 5.14, we must have
$$\lim_{n \to \infty} \sqrt[n]{n!} = +\infty. \tag{6.35}$$
Now we apply Theorem 6.6 (Comparison Test) and the limit (6.35) to the inequalities (6.34) to conclude that
$$\lim_{n \to \infty} \frac{1}{\sqrt[n]{(2n)!}} = 0.$$
By Theorem 6.20, the radius of convergence of the power series is $+\infty$. ∎

Problem 6.22

(★)(★) Suppose that the radius of convergence of the power series $\sum_{n=0}^{\infty} c_n z^n$ is 3. Prove that 1 is the radius of convergence of the power series
$$\sum_{n=0}^{\infty} c_n z^{n^3}. \tag{6.36}$$

Proof. By the hypothesis, we know that
$$\limsup_{n \to \infty} \sqrt[n]{|c_n|} = \frac{1}{3}.$$

6.5. Power Series

We consider
$$\limsup_{n\to\infty} \sqrt[n]{|c_n z^{n^3}|} = \limsup_{n\to\infty}(\sqrt[n]{|c_n|} \times |z|^{n^2}).$$

By Remark 5.1, we obtain
$$\liminf_{n\to\infty} |z|^{n^2} \times \limsup_{n\to\infty} \sqrt[n]{|c_n|} \leq \limsup_{n\to\infty}(\sqrt[n]{|c_n|} \times |z|^{n^2}) \leq \limsup_{n\to\infty} \sqrt[n]{|c_n|} \times \limsup_{n\to\infty} |z|^{n^2}. \quad (6.37)$$

By Theorem 5.10, we have
$$\liminf_{n\to\infty} |z|^{n^2} = \limsup_{n\to\infty} |z|^{n^2} = \lim_{n\to\infty} |z|^{n^2}. \quad (6.38)$$

Thus we follow from the inequalities (6.37) and (6.38) that
$$\limsup_{n\to\infty} \sqrt[n]{|c_n z^{n^3}|} = \lim_{n\to\infty} |z|^{n^2} \times \limsup_{n\to\infty} \sqrt[n]{|c_n|} = \frac{1}{3} \lim_{n\to\infty} |z|^{n^2}.$$

By Theorem 6.7 (Root Test), the power series (6.36) converges if
$$\frac{1}{3} \lim_{n\to\infty} |z|^{n^2} < 1. \quad (6.39)$$

It is obvious that the inequality (6.39) holds if $|z| < 1$. However, if $|z| > 1$, then we must have
$$\lim_{n\to\infty} |z|^{n^2} = +\infty$$

so that the inequality (6.39) does not hold. Hence the radius of convergence is 1 and we end the proof of the problem. ∎

Problem 6.23

$(\star)(\star)$ Let R be the radius of convergence of the power series $\sum_{n=0}^{\infty} c_n z^n$, where $c_n \neq 0$ for all $n = 1, 2, \ldots$. Prove that

$$\liminf_{n\to\infty} \left|\frac{c_n}{c_{n+1}}\right| \leq R \leq \limsup_{n\to\infty} \left|\frac{c_n}{c_{n+1}}\right|. \quad (6.40)$$

Proof. If $0 < R < +\infty$, then we know from Theorem 6.20 that
$$\frac{1}{R} = \limsup_{n\to\infty} \sqrt[n]{|c_n|}. \quad (6.41)$$

Since $|c_n| > 0$ for all $n = 1, 2, \ldots$, it follows from Problem 5.13 that
$$\liminf_{n\to\infty} \left|\frac{c_{n+1}}{c_n}\right| \leq \limsup_{n\to\infty} \sqrt[n]{|c_n|} \leq \limsup_{n\to\infty} \left|\frac{c_{n+1}}{c_n}\right|. \quad (6.42)$$

It is obvious that
$$\frac{1}{\liminf_{n\to\infty} \left|\frac{c_{n+1}}{c_n}\right|} = \limsup_{n\to\infty} \left|\frac{c_n}{c_{n+1}}\right| \quad \text{and} \quad \frac{1}{\limsup_{n\to\infty} \left|\frac{c_{n+1}}{c_n}\right|} = \liminf_{n\to\infty} \left|\frac{c_n}{c_{n+1}}\right|, \quad (6.43)$$

Hence we obtain the desired result (6.40) by combining the inequality (6.42) and the expressions (6.41) and (6.43). This completes the proof of the problem. ∎

CHAPTER 7

Limits and Continuity of Functions

7.1 Fundamental Concepts

A real or complex function f is a function whose values $f(x)$ are real or complex respectively. The main references for this chapter are [3, §4.5 - §4.23, pp. 74 - 95], [5, Chap. 4 & 5, pp. 102 - 159], [6, Chap. 3], [23, Chap. 4, pp. 83 - 98] and [27, Chap. 9, pp. 211 - 250].

7.1.1 Limits of Functions

Suppose that X and Y are metric spaces with metrics d_X and d_Y respectively. Let $E \subseteq X$, $f : E \to Y$ be a function, p be a limit point of E and $q \in Y$. The notation[a]

$$\lim_{x \to p} f(x) = q \qquad (7.1)$$

means that *for every* $\epsilon > 0$, *there exists* a $\delta > 0$ such that

$$d_Y(f(x), q) < \epsilon$$

for all points $x \in E$ with

$$0 < d_X(x, p) < \delta. \qquad (7.2)$$

By the inequality (7.2), it is clear that $x \neq p$. Furthermore, we note that the number δ depends on ϵ. The notation (7.1) is read "the **limit of** $f(x)$ as x tends to p."

The following theorem relates limits of functions and limits of convergent sequences:

Theorem 7.1. *The limit (7.1) holds if and only if*

$$\lim_{n \to \infty} f(p_n) = q$$

for every sequence $\{p_n\} \subseteq E \setminus \{p\}$ *such that* $p_n \to p$.

---
[a] Sometimes we write $f(x) \to q$ as $x \to p$.

By Theorems 5.1 (Uniqueness of Limits of Sequences) and 7.1, we know that if f has a limit at the point p, then this limit must be **unique**.

Suppose that $f, g : X \to Y$ are two functions from X to Y. As usual, the **sum** $f + g$ is defined to be the function whose value at $x \in X$ is $f(x) + g(x)$. The **difference** $f - g$, the **product** fg and the **quotient** $\frac{f}{g}$ can be defined similarly. Similar to Theorem 5.5, we have the following algebra of limits of functions.

Theorem 7.2. *Suppose that X and Y are metric spaces, $E \subseteq X$ and p is a limit point of E. Let $f, g : E \to Y$,*
$$\lim_{x \to p} f(x) = A \quad \text{and} \quad \lim_{x \to p} g(x) = B.$$

Then we have

(a) $\lim_{x \to p} (f \pm g)(x) = A \pm B$.

(b) $\lim_{x \to p} (fg)(x) = AB$.

(c) $\lim_{x \to p} \left(\frac{f}{g}\right)(x) = \frac{A}{B}$ *provided that* $B \neq 0$.

In Chapter 5, we have the **Squeeze Theorem for Convergent Sequences** (Theorem 5.6). Similarly, we have a corresponding result for limits of functions (see [2, Theorem 3.3, p. 133]):

Theorem 7.3 (Squeeze Theorem for Limits of Functions). *Let $f, g, h : (a, b) \to \mathbb{R}$ and $p \in (a, b)$. Suppose that $g(x) \leq f(x) \leq h(x)$ for all $x \in (a, b) \setminus \{p\}$ and*
$$\lim_{x \to p} g(x) = \lim_{x \to p} h(x).$$

Then we have
$$\lim_{x \to p} f(x).$$

7.1.2 Continuity and Uniform Continuity of Functions

Definition 7.4 (Continuity). Suppose that X and Y are metric spaces with metrics d_X and d_Y respectively. Let $p \in E \subseteq X$ and $f : E \to Y$. The function f is said to be **continuous at** p if for every $\epsilon > 0$, there exists a $\delta > 0$ such that
$$d_Y(f(x), f(p)) < \epsilon$$

for all $x \in E$ with $d_X(x, p) < \delta$. If f is continuous at every point of E, then f is said to be **continuous on** E.

If p is a **limit point** of E, then Definition 7.4 (Continuity) implies that
$$\lim_{x \to p} f(x) = f(p).$$

In view of Theorem 7.1, the following result, which can be treated as an equivalent way of formulating the definition of continuity, is obvious:

7.1. Fundamental Concepts

Theorem 7.5. *Suppose that X and Y are metric spaces. Let $p \in E \subseteq X$ and $f : E \to Y$. The function f is **continuous at** p if and only if, for every sequence $\{x_n\} \subseteq E \setminus \{p\}$ converging to p, the corresponding sequence $\{f(x_n)\} \subseteq Y$ converging to $f(p)$, i.e.,*

$$\lim_{n \to \infty} f(x_n) = f\left(\lim_{n \to \infty} x_n\right).$$

> **Remark 7.1**
>
> It is well-known that every polynomial is continuous on its domain and a rational function is continuous at points wherever the denominator is nonzero. Furthermore, it follows from the triangle inequality that the function $f : E \subseteq \mathbb{R}^n \to \mathbb{R}$ defined by $f(\mathbf{x}) = |\mathbf{x}|$ is also continuous on E.

For compositions of functions, we see that a continuous function of a continuous function is always continuous. The detailed description is given in the following result:

Theorem 7.6. *Given that X, Y and Z are metric spaces. Let $E \subseteq X$. Suppose that $f : E \to Y$, $g : f(E) \to Z$ and $h = g \circ f : X \to Z$, i.e.,*

$$h(x) = g(f(x))$$

for every $x \in E$. If f is continuous at $p \in E$ and g is continuous at $f(p)$, then h is continuous at p.

Another useful characterization of the continuity of a function depends on the "topology" (open sets and closed sets) of the metric spaces X and Y. In fact, we have

Theorem 7.7. *Let $f : X \to Y$ be a function. Then f is continuous on X if and only if $f^{-1}(V)$ is open in X for every open set V in Y.*

> **Remark 7.2**
>
> The property "open set" in Theorem 7.7 can be replaced by the property "closed set".

Besides the continuity of functions, **uniform continuity of functions** is another important concept. The definition is as follows:

Definition 7.8 (Uniform Continuity). *Suppose that X and Y are metric spaces with metrics d_X and d_Y respectively. Then the function $f : X \to Y$ is said to be **uniformly continuous** on X if for every $\epsilon > 0$, there exists a $\delta > 0$ such that*

$$d_Y(f(x), f(y)) < \epsilon$$

for all $x, y \in X$ with $d_X(x, y) < \delta$.

If we compare Definitions 7.4 (Continuity) and 7.8 (Uniform Continuity), then we can see their differences easily. In fact, continuity of a function f is defined at a point $x \in X$, but uniform continuity of f is defined on the whole set X. In addition, the chosen δ in Definitions 7.4 (Continuity) depends on two factors: the point p and the initial value ϵ, while that in Definition 7.8 (Uniform Continuity) relies solely on the initial value ϵ and different points won't give different δ. **Obviously, a uniform continuous function f is continuous.**

7.1.3 The Extreme Value Theorem

There are many nice properties when the domain of a continuous function is a compact set. In fact, the next theorem shows that the image of a compact set under a continuous function f is always compact.

Theorem 7.9 (Continuity and Compactness). *Let $f : X \to Y$ be a continuous function from the metric space X to the metric space Y. If $E \subseteq X$ is compact, then $f(E) \subseteq Y$ is compact.*

We say that a function $\mathbf{f} : E \subseteq X \to \mathbb{R}^k$ is **bounded** on E if there exists a positive constant M such that
$$|\mathbf{f}(x)| \leq M$$
for all $x \in E$. Thus, Theorem 7.9 (Continuity and Compactness) and the Heine-Borel Theorem imply that $\mathbf{f}(E)$ is bounded on E. In particular, when $k = 1$ (i.e., f is a real function), the Completeness Axiom implies that both
$$\sup f(E) \quad \text{and} \quad \inf f(E)$$
exist in $\mathbb{R}$.

The following famous result, namely the **Extreme Value Theorem**, shows that a continuous real function $f : E \to \mathbb{R}$ takes on the values $\sup f(E)$ and $\inf f(E)$.

The Extreme Value Theorem. *Suppose that $f : E \to \mathbb{R}$ is continuous, where E is compact. Let*
$$M = \sup_{x \in E} f(x) \quad \text{and} \quad m = \inf_{x \in E} f(x).$$
Then there exists $p, q \in E$ such that
$$f(p) = M \quad \text{and} \quad f(q) = m.$$

In §7.1.2, we know that a uniform continuous function f is continuous, but the converse is not always true. Interestingly, this is true when the domain is a compact set.

Theorem 7.10. *Suppose that X and Y are metric spaces and X is compact. If $f : X \to Y$ is continuous, then f is uniformly continuous on X.*

If f is bijective, then we can say more:

Theorem 7.11. *Suppose that X and Y are metric spaces and X is compact. If $f : X \to Y$ is bijective and continuous, then the **inverse function** $f^{-1} : Y \to X$ is continuous.*

7.1.4 The Intermediate Value Theorem

Similar to functions with compact domains, there are also many nice properties when the domain of the function f is connected. Actually, the following result says that continuity "preserves" connectedness.

Theorem 7.12 (Continuity and Connectedness). *Let $f : X \to Y$ be a continuous function from the metric space X to the metric space Y. If $E \subseteq X$ is connected, then $f(E) \subseteq Y$ is connected.*

7.1. Fundamental Concepts

Besides the **Extreme Value Theorem**, we have another important theorem for continuous functions. It is called the **Intermediate Value Theorem**.

The Intermediate Value Theorem. *Suppose that $f : [a,b] \to \mathbb{R}$ is continuous. If we have $f(a) < c < f(b)$ or $f(b) < c < f(a)$, then there exists a point $x_0 \in (a,b)$ at which $f(x_0) = c$.*

7.1.5 Discontinuity of Functions

Suppose that $f : X \to Y$ is a function. If f is *not* continuous at $p \in X$, then we say that f is **discontinuous** at p. In this case, classifications of all the discontinuous points of a function should be considered.

Definition 7.13 (Left-hand Limits and Right-hand Limits). *Suppose that $f : (a,b) \to Y$ and $p \in [a,b)$, where Y is a metric space. We define $f(p+)$ the **right-hand limit** of f at p by*

$$f(p+) = q$$

*if $f(x_n) \to q$ as $n \to \infty$ for all sequences $\{x_n\} \subseteq (p,b)$ converging to p. Similarly, if $p \in (a,b]$, we define $f(p-)$ the **left-hand limit** of f at p by*

$$f(p-) = q$$

if $f(x_n) \to q$ as $n \to \infty$ for all sequences $\{x_n\} \subseteq (a,p)$ converging to p.

Theorem 7.14. *Suppose that $f : (a,b) \to Y$, where Y is a metric space. Then $\lim_{x \to p} f(x)$ exists if and only if*

$$f(p+) = f(p-) = \lim_{x \to p} f(x).$$

In particular, f is continuous at $p \in (a,b)$ if and only if

$$f(p+) = f(p-) = f(p).$$

By this result, we can determine the type of the discontinuity (if it exists) of a mapping f as follows:

Definition 7.15 (Types of Discontinuity). *Suppose that $f : (a,b) \to Y$, where Y is a metric space. Suppose that f is discontinuous at $p \in (a,b)$. Then one of the following conditions is satisfied:*

(a) *Both $f(p+)$ and $f(p-)$ exist, but $f(p+) = f(p-) \neq f(p)$.*

(b) *Both $f(p+)$ and $f(p-)$ exist, but $f(p+) \neq f(p-)$.*

(c) *Either $f(p+)$ and $f(p-)$ does not exist.*

The function f is said to have a **discontinuity of the first kind** at p (or a **simple discontinuity**) in Case (a) or Case (b). For Case (c), f is said to have a **discontinuity of the second kind** at p.

Theorem 7.16 (Countability of Simple Discontinuities). *The set of all simple discontinuities of the function $f : (a,b) \to \mathbb{R}$ is **at most countable**.*

7.1.6 Monotonic Functions

There is a close connection between real monotonic functions $f : (a, b) \to \mathbb{R}$ and types of its discontinuities in the sense of Definition 7.15 (Types of Discontinuity).

Definition 7.17 (Monotonic Functions)**.** *Suppose that $f : (a, b) \to \mathbb{R}$ is a function. We say f is **monotonically increasing** on (a, b) if*

$$f(x) \leq f(y) \tag{7.3}$$

*for every $x < y$ with $x, y \in (a, b)$. Similarly, f is **monotonically decreasing** on (a, b) if*

$$f(x) \geq f(y) \tag{7.4}$$

for every $x < y$ with $x, y \in (a, b)$.[b]

Theorem 7.18. *Suppose that $f : (a, b) \to \mathbb{R}$ is monotonically increasing. Then both $f(p+)$ and $f(p-)$ exist for every $p \in (a, b)$ and we have*

$$f(p-) \leq f(p) \leq f(p+).$$

In other words, monotonic functions does not have discontinuities of the second kind.

> **Remark 7.3**
>
> Notice that analogous results of Theorem 7.18 hold for monotonically decreasing functions.

Theorem 7.19 (Froda's Theorem)**.** *Suppose that the function $f : (a, b) \to \mathbb{R}$ is monotonic. Then the set of discontinuities is **at most countable**.*

7.1.7 Limits at Infinity

Sometimes we need to deal with limits at infinity, so we need the following definition:

Definition 7.20 (Limits at Infinity)**.** *Suppose that $f : E \subseteq \mathbb{R} \to \mathbb{R}$. We say that $f(x)$ converges to L as $x \to +\infty$ in E if for every $\epsilon > 0$, there exists a constant M such that*

$$|f(x) - L| < \epsilon$$

for all $x \in E$ such that $x > M$. Analogous definition evidently holds when $x \to -\infty$.

By this definition, we can do the same things with limits at infinity as what we have done with limits at a point p. In short, it means that the analogue of Theorem 7.2 also holds if the operations of the numbers there are well-defined.

---

[b]If the inequality sign in (7.3) or (7.4) is strict, then we say f is **strictly increasing** or **strictly decreasing** on (a, b) respectively.

7.2 Limits of Functions

> **Problem 7.1**
>
> (★) Suppose that $f : (a, b) \to \mathbb{R}$ and $x \in \mathbb{R}$. If $\lim_{h \to 0} |f(x+h) - f(x)| = 0$, prove that
>
> $$\lim_{h \to 0} |f(x+h) - f(x-h)| = 0.$$
>
> Is the converse true?

Proof. By Theorem 7.1, we have

$$\lim_{n \to \infty} |f(x + h_n) - f(x)| = 0 \tag{7.5}$$

for every sequence $\{h_n\}$ such that $h_n \neq 0$ and $h_n \to 0$. By the triangle inequality, we have

$$|f(x+h_n) - f(x-h_n)| = |f(x+h_n) - f(x) + f(x) - f(x-h_n)|$$
$$\leq |f(x+h_n) - f(x)| + |f(x) - f(x-h_n)| \tag{7.6}$$

Apply Theorem 5.6 (Squeeze Theorem for Convergent Sequences) to the inequality (7.6) and then using the limit (7.5), we deduce

$$\lim_{n \to \infty} |f(x + h_n) - f(x - h_n)| = 0$$

for every sequence $\{h_n\}$ such that $h_n \neq 0$ and $h_n \to 0$. By Theorem 7.1, it means that

$$\lim_{h \to 0} |f(x+h) - f(x-h)| = 0$$

holds.

The converse is false. For example, consider $f : \mathbb{R} \to \mathbb{R}$ to be defined by

$$f(x) = \begin{cases} x^2, & \text{if } x \neq 0; \\ 1, & \text{if } x = 0. \end{cases}$$

Then we have

$$\lim_{h \to 0} |f(0+h) - f(0-h)| = \lim_{h \to 0} |f(h) - f(-h)| = 0$$

but

$$\lim_{h \to 0} |f(0+h) - f(0)| = \lim_{h \to 0} |f(h) - 1| = 1.$$

This completes the proof of the problem. ∎

> **Problem 7.2**
>
> (★) Let $x \in \mathbb{R}$. Prove that
>
> $$\lim_{n \to \infty} \left[\lim_{k \to \infty} \cos^{2k}(n! \pi x) \right] = \begin{cases} 1, & \text{if } x \in \mathbb{Q}; \\ 0, & \text{if } x \in \mathbb{R} \setminus \mathbb{Q}. \end{cases}$$

Proof. Suppose that $x \in \mathbb{R} \setminus \mathbb{Q}$. Then $n!x \notin \mathbb{Z}$ for any $n \in \mathbb{N}$ and thus $n!\pi x$ is not a multiple of π for every $n \in \mathbb{N}$. Therefore, we have
$$0 \le |\cos(n!\pi x)| < 1$$
which implies that
$$0 \le |\cos(n!\pi x)| < \delta \tag{7.7}$$
for some $\delta < 1$. Now we apply Theorems 5.6 (Squeeze Theorem for Convergent Sequences) and 5.7(c) to (7.7) to obtain
$$\lim_{k \to \infty} \cos^{2k}(n!\pi x) = 0$$
and hence
$$\lim_{n \to \infty} \left[\lim_{k \to \infty} \cos^{2k}(n!\pi x) \right] = 0.$$

Next, we suppose that $x \in \mathbb{Q}$. Then $x = \frac{p}{q}$ for some $p, q \in \mathbb{Z}$ and $q \ne 0$. Take $n > q$ so that $n!x \in \mathbb{N}$ implying that
$$|\cos(n!\pi x)| = 1. \tag{7.8}$$
Now the expression (7.8) shows easily that
$$\lim_{k \to \infty} \cos^{2k}(n!\pi x) = 1$$
and hence
$$\lim_{n \to \infty} \left[\lim_{k \to \infty} \cos^{2k}(n!\pi x) \right] = 1.$$
This ends the proof of the problem. ∎

> **Remark 7.4**
>
> The limit in Problem 7.2 is the analytical form of the so-called the **Dirichlet function**.

> **Problem 7.3**
>
> (★) Suppose that $a_0, a_1, \ldots, a_n, b_0, b_1, \ldots, b_m \in \mathbb{R}$ and $a_0 b_0 \ne 0$. Let
> $$R(x) = \frac{a_0 x^n + a_1 x^{n-1} + \cdots + a_n}{b_0 x^m + b_1 x^{m-1} + \cdots + b_m}.$$
> **Prove that**
> $$\lim_{x \to \infty} R(x) = \begin{cases} +\infty, & \text{if } n > m; \\ \dfrac{a_0}{b_0}, & \text{if } n = m; \\ 0, & \text{if } n < m. \end{cases}$$

Proof. Express $R(x)$ in the following form
$$R(x) = \frac{a_0 x^{n-m} + a_1 x^{n-m-1} + \cdots + a_n x^{-m}}{b_0 + b_1 x^{-1} + \cdots + b_m x^{-m}}. \tag{7.9}$$

7.2. Limits of Functions

When $n > m$, the denominator and the numerator of (7.9) tend to b_0 and ∞ respectively as $x \to \infty$. Therefore, we have
$$\lim_{x \to \infty} R(x) = \infty.$$

When $n = m$, the term $a_0 x^{n-m}$ becomes a_0 and the other terms in the numerator still have the factor x^{-1}. Thus the denominator and the numerator of (7.9) tend to b_0 and a_0 respectively as $x \to \infty$. Hence we have
$$\lim_{x \to \infty} R(x) = \frac{a_0}{b_0}$$
in this case.

When $n < m$, since each term in the numerator involves the factor x^{-1}, each term tends to 0 as $x \to \infty$. In other words, we have
$$\lim_{x \to \infty} R(x) = 0$$
in this case, completing the proof of the problem. ∎

Problem 7.4

(★) Let $n \in \mathbb{N}$ and $a \in \mathbb{R}$, evaluate the limit
$$\lim_{x \to a} \frac{(x^n - a^n) - na^{n-1}(x - a)}{(x - a)^2}. \tag{7.10}$$

Proof. Put $x = a + y$. Now $x \to a$ if and only if $y \to 0$. By the Binomial Theorem, the limit (7.10) becomes
$$\lim_{y \to 0} \frac{(a+y)^n - a^n - na^{n-1}y}{y^2} = \lim_{y \to 0} \left(\frac{a^n + C_1^m a^{n-1}y + C_2^m a^{n-2}y^2 + \cdots + y^n - a^n - na^{n-1}y}{y^2} \right)$$
$$= \lim_{y \to 0} (C_2^m a^{n-2} + C_3^m a^{n-3} y + \cdots + y^{n-2})$$
$$= C_2^m a^{n-2}.$$

This finishes the proof of the problem. ∎

Problem 7.5

(★)(★) Suppose that $\lim_{x \to 0} f(x) = A$ and $\lim_{x \to A} g(x) = B$. Prove or disprove
$$\lim_{x \to 0} g(f(x)) = B.$$

Proof. Consider the following functions[c]
$$f(x) = \begin{cases} \frac{1}{q}, & \text{if } x = \frac{p}{q}, p \in \mathbb{Z}, q \in \mathbb{N} \text{ and } p, q \text{ are coprime}; \\ 0, & x \in \mathbb{R} \setminus \mathbb{Q} \end{cases}$$

---
[c] The function is called the **Riemann function** and some properties of this function will be given in Problem 9.8.

and
$$g(x) = \begin{cases} 1, & \text{if } x \neq 0; \\ -1, & \text{if } x = 0. \end{cases}$$

Now it is easy to check that
$$\lim_{x \to 0} f(x) = 0 \quad \text{and} \quad \lim_{x \to 0} g(x) = 1.$$

However, we claim that the limit
$$\lim_{x \to 0} g(f(x)) \tag{7.11}$$
does not exist. To see this, if $\{x_n\} \subseteq \mathbb{Q} \setminus \{0\}$ and $x_n \to 0$, then we have $f(x_n) \neq 0$ for every $n \in \mathbb{N}$ and this implies that
$$g(f(x_n)) = 1 \tag{7.12}$$
for every $n \in \mathbb{N}$. Next, if we choose $\{y_n\} \subseteq \mathbb{R} \setminus (\mathbb{Q} \cup \{0\})$ and $y_n \to 0$, then we have $f(y_n) = 0$ for every $n \in \mathbb{N}$. In this case, we have
$$g(f(x_n)) = -1 \tag{7.13}$$
for every $n \in \mathbb{N}$. Since the two values (7.12) and (7.13) are not equal, it follows from Theorem 7.1 that the limit (7.11) does not exist. This completes the proof of the problem. ∎

Problem 7.6

(⋆)(⋆) Prove that
$$\lim_{x \to \infty} \left(1 + \frac{1}{x}\right)^x = e.$$

Proof. Fix $x \in [1, \infty)$. Let n_x be the *unique* integer such that
$$n_x \leq x < n_x + 1.$$

Now we have
$$\left(1 + \frac{1}{x}\right)^x \leq \left(1 + \frac{1}{n_x}\right)^{n_x+1} = \left(1 + \frac{1}{n_x}\right)^{n_x} \left(1 + \frac{1}{n_x}\right) \tag{7.14}$$
and
$$\left(1 + \frac{1}{x}\right)^x \geq \left(1 + \frac{1}{n_x+1}\right)^{n_x} = \left(1 + \frac{1}{n_x+1}\right)^{n_x+1} \left(1 + \frac{1}{n_x+1}\right)^{-1}. \tag{7.15}$$

Combining the inequalities (7.14) and (7.15), we get
$$\left(1 + \frac{1}{n_x+1}\right)^{n_x+1} \left(1 + \frac{1}{n_x+1}\right)^{-1} \leq \left(1 + \frac{1}{x}\right)^x \leq \left(1 + \frac{1}{n_x}\right)^{n_x} \left(1 + \frac{1}{n_x}\right). \tag{7.16}$$

It is clear that $n_x \to \infty$ if and only if $x \to \infty$ so that
$$\lim_{n_x \to \infty} \left(1 + \frac{1}{x}\right)^x = \lim_{x \to \infty} \left(1 + \frac{1}{x}\right)^x.$$

Furthermore, since
$$\lim_{n_x \to \infty} \left(1 + \frac{1}{n_x}\right) = \lim_{n_x \to \infty} \left(1 + \frac{1}{n_x+1}\right)^{-1} = 1 \tag{7.17}$$

7.2. Limits of Functions

and[d]
$$\lim_{n_x \to \infty} \left(1 + \frac{1}{n_x}\right)^{n_x} = \lim_{n_x \to \infty} \left(1 + \frac{1}{n_x + 1}\right)^{n_x+1} = e, \quad (7.18)$$

we apply the results (7.17), (7.18) and Theorem 5.6 (Squeeze Theorem for Convergent Sequences) to the inequalities (7.16) to obtain the desired result

$$\lim_{x \to \infty} \left(1 + \frac{1}{x}\right)^x = \lim_{n_x \to \infty} \left(1 + \frac{1}{n_x}\right)^{n_x} = e.$$

This finishes the proof of the problem. ∎

Problem 7.7

(★)(★) Suppose that $f : (0,1) \to \mathbb{R}$ satisfies the conditions

$$\lim_{x \to 0} f(x) = 0 \quad \text{and} \quad \lim_{x \to 0} \frac{f(2x) - f(x)}{x} = 0.$$

Prove that

$$\lim_{x \to 0} \frac{f(x)}{x} = 0.$$

Proof. Given that $\epsilon > 0$. By the second condition, there exists a $\delta > 0$ such that

$$\left|\frac{f(2x) - f(x)}{x}\right| < \epsilon \quad (7.19)$$

for all $x \in (0, \delta)$. It is easy to see that $\frac{x}{2^n} \in (0, \delta)$ for every positive integer n. Therefore, it follows from the inequality (7.19) that

$$\left|f\left(\frac{x}{2^{n-1}}\right) - f\left(\frac{x}{2^n}\right)\right| < \epsilon \times \frac{x}{2^n} \quad (7.20)$$

for every positive integer n. By the triangle inequality and the inequality (7.20), we follow that

$$\left|f(x) - f\left(\frac{x}{2^n}\right)\right| \leq \underbrace{\left|f(x) - f\left(\frac{x}{2}\right)\right| + \left|f\left(\frac{x}{2}\right) - f\left(\frac{x}{2^2}\right)\right| + \cdots + \left|f\left(\frac{x}{2^{n-1}}\right) - f\left(\frac{x}{2^n}\right)\right|}_{n \text{ terms}}$$

$$< \epsilon x \left(\frac{1}{2} + \frac{1}{2^2} + \cdots + \frac{1}{2^n}\right)$$

$$< \frac{\epsilon x}{2}\left(1 + \frac{1}{2} + \frac{1}{2^2} + \cdots\right)$$

$$= \epsilon x \quad (7.21)$$

for all $x \in (0, \delta)$. By the first condition and Theorem 7.1, we have

$$\lim_{n \to \infty} \left|f\left(\frac{x}{2^n}\right)\right| = \lim_{x \to 0} |f(x)| = 0. \quad (7.22)$$

By the triangle inequality and the inequality (7.21), we know that

$$|f(x)| \leq \left|f(x) - f\left(\frac{x}{2^n}\right)\right| + \left|f\left(\frac{x}{2^n}\right)\right| < \epsilon x + \left|f\left(\frac{x}{2^n}\right)\right| \quad (7.23)$$

---
[d]See the proof of Problem 5.14.

for all $x \in (0, \delta)$. By taking $n \to \infty$ to both sides of the inequality (7.23) and then applying the result (7.22), we gain
$$|f(x)| \le \epsilon x$$
or equivalently,
$$\left|\frac{f(x)}{x}\right| \le \epsilon$$
for all $x \in (0, \delta)$. By definition, it means that
$$\lim_{x \to 0} \frac{f(x)}{x} = 0,$$
completing the proof of the problem. ∎

7.3 Continuity and Uniform Continuity of Functions

Problem 7.8

(★)(★) Suppose that $f : [0,1] \to \mathbb{R}$ is continuous and $f(x) = 0$ for every irrational x in $[0,1]$. Prove that $f \equiv 0$ on $[0,1]$.

Proof. By Problem 2.9 and Remark 2.1, we know that given any $x \in [0,1]$, there exists a sequence $\{\theta_n\} \subseteq [0,1]$ such that
$$\lim_{n \to \infty} \theta_n = x.$$
Since f is continuous on $[0,1]$, it follows from Theorem 7.5 that
$$f(x) = f\left(\lim_{n \to \infty} \theta_n\right) = \lim_{n \to \infty} f(\theta_n) = 0.$$
Hence we have $f \equiv 0$ on $[0,1]$. This completes the proof of the problem. ∎

Problem 7.9

(★)(★) Suppose that $f : \mathbb{R} \to \mathbb{R}$ and f is continuous at 0. If $f(x+y) = f(x) + f(y)$ for every $x, y \in \mathbb{R}$, prove that there exists a constant a such that
$$f(x) = ax$$
for all $x \in \mathbb{R}$.

Proof. Since f is continuous at 0, for every $\epsilon > 0$, there exists a $\delta > 0$ such that
$$|f(h) - f(0)| < \epsilon \tag{7.24}$$

7.3. Continuity and Uniform Continuity of Functions

for all $|h| < \delta$. By the hypothesis and the inequality (7.24), we have

$$|f(h)| = |[f(h) + f(0)] - f(0)| = |f(h+0) - f(0)| < \epsilon \qquad (7.25)$$

for all $|h| < \delta$. Let $p \in \mathbb{R}$. Then it follows from the hypothesis and the inequality (7.25) that

$$|f(p+h) - f(p)| = |f(p) + f(h) - f(p)| = |f(h)| < \epsilon$$

for all $|h| < \delta$. By Definition 7.4 (Continuity), the function f is continuous at every point $p \in \mathbb{R}$.

Let $a = f(1)$. Then it can be proven by induction that

$$f(n) = na \qquad (7.26)$$

for every positive integer n. Since $f(0) = f(0+0) = f(0) + f(0)$, we have

$$f(0) = 0. \qquad (7.27)$$

Furthermore, the hypothesis and the fact (7.27) imply that

$$f(-x) + f(x) = f(-x+x) = f(0) = 0$$

so that

$$f(-x) = -f(x) \qquad (7.28)$$

for every $x \in \mathbb{R}$.

Now, by combining the facts (7.26), (7.27) and (7.28), we deduce that the formula (7.26) holds for every integer n. If $r = \frac{p}{q}$, where $p \in \mathbb{Z}$ and $q \in \mathbb{N}$, then we see from the hypothesis and the expression (7.26) that

$$\underbrace{f\left(\frac{p}{q}\right) + \cdots + f\left(\frac{p}{q}\right)}_{q \text{ terms}} = f\left(q \cdot \frac{p}{q}\right) = f(p) = ap$$

which implies that

$$f\left(\frac{p}{q}\right) = a \times \frac{p}{q},$$

i.e., $f(r) = ar$ for every rational r. Finally, by an argument similar to the proof of Problem 7.8, we can show that

$$f(x) = ax$$

for all $x \in \mathbb{R}$. Hence we have finished the proof of the problem. ■

Problem 7.10

(★)(★) *Find all continuous functions $f : \mathbb{R} \to \mathbb{R}$ such that*

$$f(x) + f(3x) = 0 \qquad (7.29)$$

holds for all $x \in \mathbb{R}$.

Proof. The hypothesis implies that $f(0) = 0$. Furthermore, it is clear that $f \equiv 0$ satisfies the hypothesis (7.29). We claim that $f \equiv 0$ is the *only* function satisfying the requirement (7.29).

To this end, we prove that
$$f(x) = (-1)^n f\left(\frac{x}{3^n}\right) \qquad (7.30)$$
for every positive integer n and $x \in \mathbb{R}$. We use induction. The case $n = 1$ follows from the hypothesis directly. Assume that
$$f(x) = (-1)^k f\left(\frac{x}{3^k}\right) \qquad (7.31)$$
for some positive integer k and all $x \in \mathbb{R}$. If $n = k+1$, then we replace x by $\frac{x}{3^{k+1}}$ in the hypothesis (7.29) and we obtain
$$f\left(\frac{x}{3^{k+1}}\right) + f\left(\frac{x}{3^k}\right) = 0. \qquad (7.32)$$
By putting the assumption (7.31) into the expression (7.32), we see that for all $x \in \mathbb{R}$,
$$f\left(\frac{x}{3^{k+1}}\right) + (-1)^k f(x) = 0$$
$$f(x) = (-1)^{k+1} f\left(\frac{x}{3^{k+1}}\right).$$
Thus the statement (7.30) is true for $n = k+1$ if it is true for $n = k$. By induction, we are able to prove that the formula (7.30) is true for all $n \in \mathbb{N}$ and $x \in \mathbb{R}$.

Since f is continuous at 0, we deduce from Theorem 7.5 that
$$\lim_{n \to \infty} f\left(\frac{x}{3^n}\right) = f(0) = 0 \qquad (7.33)$$
for every $x \in \mathbb{R}$. Combining the formula (7.30) and the result (7.33), we derive that
$$f(x) = 0$$
for all $x \in \mathbb{R}$, i.e., $f \equiv 0$ which is the *only* function satisfying the hypothesis (7.29). We finish the proof of the problem. ∎

Problem 7.11

(⋆) Prove that the Dirichlet function $D(x)$ in Problem 7.2 is discontinuous on $\mathbb{R}$.

Proof. By the definition in Problem 7.2, we see that
$$D(x) = \begin{cases} 1, & \text{if } x \in \mathbb{Q}; \\ 0, & \text{if } x \in \mathbb{R} \setminus \mathbb{Q}. \end{cases}$$

Let $x_0 \in \mathbb{R}$. Then *for every* $\delta > 0$, the interval $(x_0 - \delta, x_0 + \delta)$ contains a rational number q and an irrational number θ.[e] Thus if $x_0 \in \mathbb{Q}$, then $D(x_0) = 1$ and we may take an irrational $\theta \in (x_0 - \delta, x_0 + \delta)$ so that
$$|D(x_0) - D(\theta)| = |1 - 0| = 1.$$

---

[e]See Theorem 2.2 (Density of Rationals) and Problem 2.9.

7.3. Continuity and Uniform Continuity of Functions

Otherwise, if $x_0 \in \mathbb{R} \setminus \mathbb{Q}$, then $D(x_0) = 0$ and we may take a rational $q \in (x_0 - \delta, x_0 + \delta)$ so that

$$|D(x_0) - D(q)| = |0 - 1| = 1.$$

In other words, the function $D(x)$ does not satisfy Definition 7.4 (Continuity). Therefore, $D(x)$ is discontinuous at x_0 and hence it is discontinuous on $\mathbb{R}$. We complete the proof of the problem. ∎

Problem 7.12

(★) Suppose that X and Y are metric spaces. Let $f : X \to Y$ and $g : X \to Y$ be continuous on X. Define $\varphi(x) : X \to Y$ and $\psi(x) : X \to Y$ by

$$\varphi(x) = \max(f(x), g(x)) \quad \text{and} \quad \psi(x) = \min(f(x), g(x))$$

for every $x \in X$ respectively. Prove that φ and ψ are continuous on X.

Proof. By the following identities

$$\varphi(x) = \max(f(x), g(x)) = \frac{1}{2}[f(x) + g(x) + |f(x) - g(x)|]$$

and

$$\psi(x) = \min(f(x), g(x)) = \frac{1}{2}[f(x) + g(x) - |f(x) - g(x)|],$$

we see from Theorem 7.2 and Remark 7.1 that φ and ψ are continuous at every point $x \in X$. Hence we have completed the proof of the problem. ∎

Problem 7.13

(★) Suppose that $f : X \to \mathbb{R}$ is continuous and $a \in \mathbb{R}$, where X is a metric space. Define $f_a : X \to \mathbb{R}$ by

$$f_a(x) = \begin{cases} a, & \text{if } f(x) > a; \\ f(x), & \text{if } |f(x)| \le a; \\ -a, & \text{if } f(x) < -a, \end{cases}$$

where $x \in X$. Prove that f_a is continuous on X.

Proof. It is clear from the definition that

$$f_a(x) = \max(\min(a, f(x)), -a).$$

Hence we deduce from Problem 7.12 that f_a is continuous on X. ∎

Problem 7.14

(⋆) Let X be a metric space with metric d_X and $f : X \to X$ be continuous. Suppose that
$$d_X(f(x), f(y)) \geq d_X(x, y) \tag{7.34}$$
for all $x, y \in X$. Prove that f is one-to-one and $f^{-1} : f(X) \to X$ is continuous.

Proof. If $x_1 \neq x_2$, then the inequality (7.34) implies that
$$d_X(f(x_1), f(x_2)) \geq d_X(x_1, x_2) > 0.$$
By definition (§1.1.2), we see that f is one-to-one. Recall that $f^{-1}(f(x)) = x$ for every $x \in X$. It follows from this and the inequality (7.34) that
$$d_X(f^{-1}(p), f^{-1}(q)) \leq d_X(f(f^{-1}(p)), f(f^{-1}(q))) = d_X(p, q)$$
for every $p, q \in f(X)$. Thus we conclude from Definition 7.4 (Continuity) that f^{-1} is continuous on $f(X)$, completing the proof of the problem. ∎

Problem 7.15 (The Sign-preserving Property)

(⋆)(⋆) Suppose that $f : E \subseteq \mathbb{R} \to \mathbb{R}$ is continuous at p and $f(p) \neq 0$. Then there exists a $\delta > 0$ such that f has the same sign as $f(p)$ in $(p - \delta, p + \delta)$.

Proof. By the continuity of f at p, for every $\epsilon > 0$, there exists a $\delta > 0$ such that
$$f(p) - \epsilon < f(x) < f(p) + \epsilon \tag{7.35}$$
whenever $x \in (p - \delta, p + \delta)$. Put $\epsilon = \frac{f(p)}{2}$ into the inequalities (7.35), we see that
$$\frac{1}{2}f(p) < f(x) < \frac{3}{2}f(p)$$
on $(p - \delta, p + \delta)$. Hence this shows that f has the same sign as $f(p)$ in $(p - \delta, p + \delta)$, completing the proof of the problem. ∎

Problem 7.16

(⋆) If $f(x) = x^4$ on $\mathbb{R}$, prove that f is not uniformly continuous on $\mathbb{R}$.

Proof. Fix $\epsilon = 1$. For every $\delta > 0$, Theorem 2.1 (The Archimedean Property) implies that there exists a positive integer N such that
$$N\delta^4 > 1. \tag{7.36}$$
Take $x = N\delta$ and $(N + \frac{1}{2})\delta$. Then it is easy to see that
$$|x - y| = \left|N\delta - \left(N + \frac{1}{2}\right)\delta\right| < \delta.$$

7.3. Continuity and Uniform Continuity of Functions

Now we have from the inequality (7.36) that

$$\begin{aligned}
|f(x) - f(y)| &= |x^4 - y^4| \\
&= |x-y||x+y||x^2 + y^2| \\
&= \frac{\delta}{2} \times \Big(2N + \frac{1}{2}\Big)\delta \times \Big[N^2 + \Big(N + \frac{1}{2}\Big)^2\Big]\delta^2 \\
&> \frac{1}{4}(4N+1)\delta^4 \\
&> N\delta^4 \\
&> 1.
\end{aligned}$$

Hence we follow from Definition 7.8 (Uniform Continuity) that f is not uniformly continuous. This ends the proof of the problem. ∎

Problem 7.17

(⋆) Construct a continuous real function f defined on $(0,1)$ but not uniformly continuous.

Proof. The function $f : (0,1) \to \mathbb{R}$ defined by

$$f(x) = \frac{1}{x}$$

is obviously continuous on $(0,1)$ because it is the quotient of two polynomials and the denominator does not vanish in $(0,1)$. See Remark 7.1.

Assume that f was uniformly continuous on $(0,1)$. We pick $\epsilon = 1$. *For every $\delta > 0$, if we take $x = \min(\delta, 1)$ and $y = \frac{x}{2}$,* then we always have $x, y \in (0,1)$ and

$$|x - y| = \frac{x}{2} < \delta,$$

but

$$\Big|\frac{1}{x} - \frac{1}{y}\Big| = \Big|\frac{1}{x} - \frac{2}{x}\Big| = \frac{1}{x} \geq 1 = \epsilon.$$

Hence it contradicts Definition 7.8 (Uniform Continuity) and so f is not uniformly continuous on $(0,1)$, finishing the proof of the problem. ∎

Remark 7.5

Problems 7.16 and 7.17 tell us that the condition "compact" in Theorem 7.10 cannot be dropped.

Problem 7.18

(⋆)(⋆) Let $a > 0$. If $f : [a, +\infty) \to \mathbb{R}$ is continuous and

$$\lim_{x \to +\infty} f(x)$$

exists, prove that f is uniformly continuous.

Proof. Given that $\epsilon > 0$. Since $\lim_{x \to +\infty} f(x)$ exists, there exists a $M > a$ such that $x, y > M$ implies[f]

$$|f(x) - f(y)| < \epsilon.$$

Since f is continuous on $[a, M+1]$, it is uniformly continuous there. Thus there exists a $\delta > 0$ such that $x, y \in [a, M+1]$ with $|x - y| < \delta$ implies

$$|f(x) - f(y)| < \epsilon.$$

Let $\delta' = \min(\delta, 1)$ and $x, y \in [a, +\infty)$ with $|x - y| < \delta' \leq \delta$. If $x \in [a, M]$ and $y \in (M+1, +\infty)$, then it is evident that

$$|x - y| \geq |y| - |x| > M + 1 - M = 1,$$

a contradiction. Thus if $x \in [a, M]$, then y is forced to lie in $[a, M+1]$. In this case, we have

$$x, y \in [a, M+1]. \tag{7.37}$$

Similarly, if $y \in (M+1, \infty)$, then x is forced to lie in $(M+1, +\infty)$ too. Therefore, we have the case

$$x, y \in (M+1, +\infty) \subset (M, +\infty). \tag{7.38}$$

Hence we deduce from the set relations (7.37) and (7.38) that

$$|f(x) - f(y)| < \epsilon.$$

By Definition 7.8 (Uniform Continuity), f is uniformly continuous on $[a, +\infty)$, completing the proof of the problem. ∎

Problem 7.19

(★)(★) Let $f : (a, b) \to \mathbb{R}$ be uniformly continuous. Prove that f is bounded on (a, b).

Proof. Take $\epsilon = 1$. Since f is uniformly continuous on (a, b), there is a $\delta > 0$ such that

$$|f(x) - f(y)| < 1$$

for all $x, y \in (a, b)$ with $|x - y| < \delta$. Fix this δ.

We note that *for every* $x \in (a, b)$, we have

$$\left| x - \frac{1}{2}(a+b) \right| < \frac{1}{2}(b-a) = \left(\frac{b-a}{2\delta} \right) \times \delta. \tag{7.39}$$

By Theorem 2.1 (The Archimedean Property), we can find a positive integer N such that

$$N > \frac{b-a}{2\delta}. \tag{7.40}$$

We *fix* this N. Combining the inequalities (7.39) and (7.40), we know that

$$\left| x - \frac{1}{2}(a+b) \right| < N\delta \tag{7.41}$$

---

[f]See Definition 7.20.

7.3. Continuity and Uniform Continuity of Functions

for every $x \in (a,b)$. Geometrically, the inequality (7.41) means that the distance between the point x and the mid-point of the interval (a,b) is always less than N times of the *fixed* length δ.

By the inequality (7.41), we can divide the interval with endpoints x and $\frac{1}{2}(a+b)$ into n subintervals each of length less than δ. It is clear that $n \le N$. More precisely, for every $x \in (a,b)$, we can construct a sequence $\{x_1, x_2, \ldots, x_n\} \subseteq (a,b)$ of $n(\le N)$ points such that

$$x_1 = \frac{1}{2}(a+b), \ldots, x_n = x \quad \text{or} \quad x_1 = x, \ldots, x_n = \frac{1}{2}(a+b),$$

where $|x_k - x_{k+1}| < \delta$ for $k = 1, 2, \ldots, n-1$. This construction and the triangle inequality give

$$\begin{aligned}
\left| f(x) - f\left(\frac{1}{2}(a+b)\right) \right| &= |f(x_n) - f(x_1)| \\
&\le |f(x_n) - f(x_{n-1})| + |f(x_{n-1}) - f(x_{n-2})| + \cdots + |f(x_2) - f(x_1)| \\
&< \underbrace{1 + 1 + \cdots + 1}_{n \text{ terms}} \\
&= n
\end{aligned} \qquad (7.42)$$

for every $x \in (a,b)$. Apply the triangle inequality again, it follows from the inequality (7.42) that

$$|f(x)| \le \left| f(x) - f\left(\frac{1}{2}(a+b)\right) \right| + \left| f\left(\frac{1}{2}(a+b)\right) \right| < n + f\left(\frac{1}{2}(a+b)\right)$$

holds for every $x \in (a,b)$. This completes the proof of the problem. ∎

Problem 7.20

(★) (★) Suppose that $f : E \subseteq \mathbb{R} \to \mathbb{R}$ is uniformly continuous on E. If $\{p_n\}$ is a Cauchy sequence in E, show that $\{f(p_n)\}$ is a Cauchy sequence in $\mathbb{R}$. Can the condition "uniformly continuous" be replaced by "continuous"?

Proof. By Definition 7.8 (Uniform Continuity), given $\epsilon > 0$, there exists a $\delta > 0$ such that

$$|f(x) - f(y)| < \epsilon \qquad (7.43)$$

for all $x, y \in E$ with $|x - y| < \delta$. Since $\{p_n\}$ is Cauchy, there exists a positive integer N such that $m, n \ge N$ implies that

$$|p_m - p_n| < \delta. \qquad (7.44)$$

Therefore, for $m, n \ge N$, we establish from the inequalities (7.43) and (7.44) that

$$|f(p_m) - f(p_n)| < \epsilon.$$

By Definition 5.12, $\{f(p_n)\}$ is also Cauchy.

We note that the conclusion does not hold anymore if the condition "uniformly continuous" is replaced by "continuous". For example, let $f : (0,1) \to \mathbb{R}$ be given by

$$f(x) = \frac{1}{x}$$

for $x \in (0,1)$. Then f is clearly continuous on $(0,1)$ and the sequence $\{\frac{1}{n}\} \subseteq (0,1)$ is Cauchy by Problem 5.19. However, $\{f(\frac{1}{n})\} = \{n\}$ which is a divergent sequence. Hence we have finished the proof of the problem. ∎

> **Problem 7.21**
>
> (★)(★) Suppose that X and Y are metric spaces and $f : X \to Y$. Prove that f is continuous on X if and only if
> $$f^{-1}(V^\circ) \subseteq [f^{-1}(V)]^\circ$$
> for every $V \subseteq Y$.

Proof. Suppose that f is continuous on X. Let $V \subseteq Y$. Since V° is open in Y, Theorem 7.7 tells us that $f^{-1}(V^\circ)$ is open in X. Since $f^{-1}(V^\circ) \subseteq f^{-1}(V)$ and we recall from Problem 4.11(c) $[f^{-1}(V)]^\circ$ is the largest open subset of $f^{-1}(V)$, we must have

$$f^{-1}(V^\circ) \subseteq [f^{-1}(V)]^\circ. \tag{7.45}$$

Conversely, we suppose that the set relation (7.45) holds for every $V \subseteq Y$. Let U be an open set in Y. Then Problem 4.11(b) implies that $U^\circ = U$ and so we follow from the set relation (7.45) that

$$f^{-1}(U) = f^{-1}(U^\circ) \subseteq [f^{-1}(U)]^\circ.$$

Since $[f^{-1}(U)]^\circ \subseteq f^{-1}(U)$, we see that

$$f^{-1}(U) = [f^{-1}(U)]^\circ$$

and it deduces from Problem 4.11(b) that $f^{-1}(U)$ is open in X. By Theorem 7.7, f is continuous on X, finishing the proof of the problem. ∎

7.4 The Extreme Value Theorem and the Intermediate Value Theorem

> **Problem 7.22**
>
> (★)(★) A function f is called **periodic** if, for some nonzero constant T, we have
> $$f(x+T) = f(x)$$
> for all values of x in the domain of f. Suppose that $f : \mathbb{R} \to \mathbb{R}$ is a nonempty, continuous and periodic function. Prove that f attains its supremum and infimum.

Proof. Let $T > 0$ be a period of f. Since f is continuous on $\mathbb{R}$, it is also continuous on $I = [0, T]$. By the Extreme Value Theorem, there exists $p, q \in I$ such that

$$f(p) = \sup_{x \in I} f(x) \quad \text{and} \quad f(q) = \inf_{x \in I} f(x). \tag{7.46}$$

For any $y \in \mathbb{R}$, we know that $y = x + nT$ for some integer n and $x \in I$, so the periodicity of f implies that

$$f(y) = f(x + nT) = f(x), \tag{7.47}$$

7.4. The Extreme Value Theorem and the Intermediate Value Theorem

where $y \in \mathbb{R}$ and $x \in I$. Thus we follow from the results (7.46) and (7.47) that

$$\sup_{y \in \mathbb{R}} f(y) = \sup_{x \in I} f(x) = f(p) \quad \text{and} \quad \inf_{y \in \mathbb{R}} f(y) = \inf_{x \in I} f(x) = f(q).$$

This ends the proof of the problem. ∎

Problem 7.23

(★) (★) A function $f : [a, b] \to \mathbb{R}$ is said to have a **local maximum** at p if there exists a neighborhood $N_\delta(p)$ of p such that $f(x) \leq f(p)$ for all $x \in N_\delta(p) \cap [a, b]$. The concept of **local minimum** at p can be defined similarly.
Suppose that $f : [a, b] \to \mathbb{R}$ is continuous and has a local minimum at p and at q, where $p < q$. Prove that f has a local maximum at r, where $r \in (p, q)$.

Proof. Let $E = [p, q] \subseteq [a, b]$. Since f is continuous on $[a, b]$, it is also continuous on E. By the Extreme Value Theorem, there exists a $r \in E$ such that

$$f(r) = \sup_{x \in E} f(x). \tag{7.48}$$

If $p < r < q$, then since the supremum (7.48) indicates that $f(r)$ is the maximum value of f on E, it is definitely a local maximum of f. If $r = p$, then since $f(p)$ is a local minimum, there exists a $\delta > 0$ such that $p < x < p + \delta < q$ implies

$$f(p) \leq f(x) \leq f(r),$$

i.e., $f(x) = f(p) = f(r)$ for all $x \in (p, p+\delta) \subseteq (p, q)$. Now we may pick any $r' \in (p, p+\delta) \subseteq (p, q)$ so that

$$f(r') = f(r) = \sup_{x \in E} f(x),$$

i.e., f has a local maximum at $r' \in (p, q)$. The case for $r = q$ is similar, so we omit the details here. Thus we have completed the proof of the problem. ∎

Problem 7.24

(★) (★) (★) Prove, without using Theorem 7.10, if $f : [a, b] \to \mathbb{R}$ is continuous, then f is uniformly continuous on $[a, b]$.

Proof. The following argument is due to Lüroth [14]. Given that $\epsilon > 0$. For each $p \in [a, b]$, define

$$E_p = \{\delta > 0 \mid |f(x) - f(y)| < \epsilon \text{ for all } x, y \in [p - \tfrac{\delta}{2}, p + \tfrac{\delta}{2}] \subseteq [a, b]\}$$

and

$$\delta(p) = \sup E_p.$$

In other words, $\delta(p)$ is the length of the *largest* interval $I(p)$ centred at p and

$$|f(x) - f(y)| < \epsilon$$

for all $x, y \in I(p)$. Since f is continuous at p, Definition 7.4 (Continuity) shows that $E_p \neq \emptyset$ for every $p \in [a, b]$. There are two cases for consideration:

- **Case (1):** $\delta(p) = \infty$ **for some** $p \in [a,b]$. Then any $\delta \in E_p$ satisfies Definition 7.8 (Uniform Continuity). We are done in this case.

- **Case (2):** $\delta(p) < \infty$ **for all** $p \in [a,b]$. We claim that $\delta(p)$ is continuous. To this end, consider the intervals $I(p)$ and $I(p+\omega)$ centred at points p and $p+\omega$, where $\omega > 0$. If $p + \omega - \frac{\delta(p+\omega)}{2} \leq p - \frac{\delta(p)}{2}$, then we have

$$p + \frac{\delta(p)}{2} \leq p + \omega + \frac{\delta(p)}{2} \leq p + \frac{\delta(p+\omega)}{2} \leq p + \omega + \frac{\delta(p+\omega)}{2}.$$

In other words, these imply that

$$I(p) \subseteq I(p+\omega).$$

In this case, $I(p)$ is *not* the largest closed interval anymore, a contradiction. Thus we have

$$p + \omega - \frac{\delta(p+\omega)}{2} > p - \frac{\delta(p)}{2}$$

which is equivalent to

$$\delta(p+\omega) - \delta(p) < 2\omega. \tag{7.49}$$

Similarly, if $p + \frac{\delta(p)}{2} \geq p + \omega + \frac{\delta(p+\omega)}{2}$, then we have

$$p + \omega - \frac{\delta(p+\omega)}{2} \geq p - \frac{\delta(p+\omega)}{2} \geq p + \omega - \frac{\delta(p)}{2} \geq p - \frac{\delta(p)}{2}.$$

In other words, these mean that

$$I(p+\omega) \subseteq I(p)$$

which is a contradiction again. Thus we have

$$p + \frac{\delta(p)}{2} < p + \omega + \frac{\delta(p+\omega)}{2}$$

which is equivalent to

$$\delta(p+\omega) - \delta(p) > -2\omega. \tag{7.50}$$

Combining the inequalities (7.49) and (7.50), we obtain

$$|\delta(p+\omega) - \delta(p)| < 2\omega$$

for every $\omega > 0$. By Definition 7.4 (Continuity), $\delta(p)$ is continuous. This proves our claim.

Since $\delta(p)$ is continuous on the compact set $[a,b]$, the Extreme Value Theorem ensures that there exists a $c \in [a,b]$ such that

$$0 < \delta(c) \leq \delta(p)$$

for every $p \in [a,b]$. Now this $\delta(c)$ evidently satisfies Definition 7.8 (Uniform Continuity).

We have completed the proof of the problem. ∎

Problem 7.25

(★)(★) *Suppose that $I = [0,1]$ and $f : I \to I$ is a continuous function such that $f \circ f = f$. Let $E = \{x \in I \mid f(x) = x\}$. Prove that E is connected.*

7.4. The Extreme Value Theorem and the Intermediate Value Theorem

Proof. Let $y = f(x) \in f(I) \subseteq I$. Then the hypothesis shows that
$$f(f(x)) = f(x),$$
i.e., $f(y) = y$ or $y \in E$. In other words, we have $f(I) \subseteq E$. On the other hand, if $x \in E$, then we have $f(x) = x$ so that $x \in f(I)$. In conclusion, we have
$$E = f(I). \tag{7.51}$$

Since I is connected and f is continuous on I, Theorem 7.12 (Continuity and Connectedness) implies that $f(I)$ is also connected. Hence our desired result follows from the expression (7.51) immediately. This ends the proof of the problem. ∎

Problem 7.26

(⋆)(⋆) *Suppose that $f, g : [a, b] \to \mathbb{R}$ are continuous functions satisfying*
$$g(x) < f(x) \tag{7.52}$$
for all $x \in [a, b]$. Prove that there exists a $\delta > 0$ such that
$$g(x) + \delta < f(x)$$
for all $x \in [a, b]$.

Proof. Define the function $h : [a, b] \to \mathbb{R}$ by
$$h(x) = f(x) - g(x)$$
for all $x \in [a, b]$. Since f and g are continuous on $[a, b]$, h is also continuous on $[a, b]$. Thus it deduces from the Extreme Value Theorem that h attains a minimum value $h(q)$ *for some* $q \in [a, b]$, i.e., $h(x) \geq h(q)$ for all $[a, b]$. By the hypothesis (7.52), we know that $h(x) > 0$ for all $x \in [a, b]$ so that $h(q) > 0$. Let $\delta = \frac{h(q)}{2}$. Then we have
$$h(x) \geq h(q) > \delta$$
for all $x \in [a, b]$. This finished the proof of the problem. ∎

Problem 7.27

(⋆) *Let $f : X \to Y$ be a function from a metric space X to another metric space Y. If $f(p) = p$ for some $p \in X$, then we call p a **fixed point** of f.*
Suppose that $f : [a, b] \to \mathbb{R}$ is continuous, $f(a) \leq a$ and $f(b) \geq b$. Prove that f has a fixed point in $[a, b]$.

Proof. If $f(a) = a$ or $f(b) = b$, then f has a fixed point a or b and we are done. Thus without loss of generality, we may assume that $f(a) < a$ and $f(b) > b$. Define $g : [a, b] \to \mathbb{R}$ by
$$g(x) = f(x) - x$$

for $x \in [a,b]$. Since f is continuous on $[a,b]$, g is also continuous on $[a,b]$. Since we have $g(a) = f(a) - a < 0$ and $g(b) = f(b) - b > 0$, it follows from the Intermediate Value Theorem that there exists a $p \in (a,b)$ such that $g(p) = 0$, i.e.,

$$f(p) = p.$$

We finish the proof of the problem. ∎

Problem 7.28

(⋆) For each $n = 1, 2, \ldots$, suppose that $f_n : \mathbb{R} \to \mathbb{R}$ is defined by

$$f_n(x) = Ax^n + x^{n-1} + \cdots + x - 1,$$

where $A > 1$. Prove that $f_n(x)$ has a *unique* positive root α_n for each $n = 1, 2, \ldots$.

Proof. If $n = 1$, then $f_1(x) = Ax - 1$ which has a unique positive root $\frac{1}{A}$ obviously. Without loss of generality, we may assume that $n \geq 2$. By Remark 7.1, each $f_n(x)$ is continuous on $\mathbb{R}$. Furthermore, we have

$$f_n(0) = -1 < 0 \quad \text{and} \quad f_n(1) = A + (n-1) - 1 > n - 1 > 0.$$

By the Intermediate Value Theorem, we know that there exists a $\alpha_n \in (0,1)$ such that

$$f_n(\alpha_n) = 0.$$

If $x, y \in (0, +\infty)$ with $x > y$, then we have

$$\begin{aligned}
f_n(x) - f_n(y) &= Ax^n + x^{n-1} + \cdots + x - 1 - (Ay^n + y^{n-1} + \cdots + y - 1) \\
&= A(x^n - y^n) + (x^{n-1} - y^{n-1}) + \cdots + (x - y) \\
&= (x - y)\Big[A(x^{n-1} + x^{n-2}y + \cdots + xy^{n-2} + y^{n-1}) \\
&\quad + (x^{n-2} + x^{n-3}y + \cdots + xy^{n-3} + y^{n-2}) + \cdots + (x + y) + 1\Big]
\end{aligned} \quad (7.53)$$

Since $x > y > 0$, we have $x - y > 0$ and every term in the expression (7.53) is positive. Thus $f_n(x) > f_n(y)$ and this means that f_n is strictly increasing on $(0, +\infty)$. Hence the positive root α_n of $f_n(x)$ is unique. This ends the proof of the problem. ∎

7.5 Discontinuity of Functions

Problem 7.29

(⋆) We say that a function f is **right continuous** at p if the **right-hand limit** of f at p exists and equals $f(p)$. (**Left continuous** at p can be defined similarly.) Construct a function which is *only* right continuous in its domain but unbounded.

7.5. Discontinuity of Functions

Proof. Define $f : (-1, 0] \to \mathbb{R}$ by

$$f(x) = \begin{cases} \dfrac{1}{x}, & \text{if } x \in (-1, 0); \\ 0, & \text{if } x = 0. \end{cases}$$

We check Definition 7.13 (Left-hand Limits and Right-hand Limits). Let $p \in (-1, 0)$ and $\{x_n\} \subseteq (p, 0)$ be such that $x_n \to p$ as $n \to \infty$. Then we have

$$f(p) = \frac{1}{p} \quad \text{and} \quad f(x_n) = \frac{1}{x_n}$$

for each $n = 1, 2, \ldots$. It is clear that $f(x_n) \to f(p)$, so f is right continuous at p.[g] Besides, f is not left continuous at 0 because the left-hand limit of f at 0 does not exit. In fact, f is unbounded on $(-1, 0]$ because

$$f(x_n) \to -\infty$$

for all sequences $\{x_n\} \subseteq (-1, 0)$ such that $x_n \to 0$ as $n \to \infty$, completing the proof of the problem. ∎

Problem 7.30

(★) Let $f : \mathbb{R} \to \mathbb{R}$ be the **greatest integer function**, i.e., $f(x) = [x]$. What are the type(s) of the discontinuities of f?

Proof. For each $n \in \mathbb{Z}$, if $n \leq x < n + 1$, then we have

$$f(x) = [x] = n.$$

This implies that f is continuous on $(n, n+1)$ for every $n \in \mathbb{Z}$. However, we have

$$f(n+) = \lim_{\substack{x \to n \\ x > n}} f(x) = n = f(n) \quad \text{and} \quad f(n-) = \lim_{\substack{x \to n \\ x < n}} f(x) = n - 1 \neq f(n).$$

By Definition 7.15 (Types of Discontinuity), f has simple discontinuities at every integer. This completes the proof of the problem. ∎

Problem 7.31

(★) Check the types of the discontinuities of the function $f : \mathbb{R} \to \mathbb{R}$ defined by

$$f(x) = \begin{cases} \sin \dfrac{1}{x}, & \text{if } x \neq 0; \\ 1, & \text{if } x = 0. \end{cases}$$

---

[g] In fact, f is continuous at every $p \in (-1, 0)$.

Proof. For $x \neq 0$, it is clear that f is continuous at x, so the only possible discontinuity of f is the origin. Let $x_n = \frac{1}{2n\pi}$ and $y_n = \frac{1}{(2n\pi + \frac{\pi}{2})}$. Then it is clear that $\{x_n\}, \{y_n\} \subseteq (0, +\infty)$, $x_n \to 0$ and $y_n \to 0$ as $n \to \infty$. Besides, we have

$$\lim_{n \to \infty} f(x_n) = \lim_{n \to \infty} \sin \frac{1}{x_n} = \lim_{n \to \infty} \sin 2n\pi = 0$$

and

$$\lim_{n \to \infty} f(y_n) = \lim_{n \to \infty} \sin \frac{1}{y_n} = \lim_{n \to \infty} \sin \left(2n\pi + \frac{\pi}{2}\right) = 1.$$

By Definition 7.13 (Left-hand Limits and Right-hand Limits), $f(0+)$ does not exist. By Definition 7.15 (Types of Discontinuity), we see that f has a discontinuity of the second kind at 0, completing the proof of the problem. ∎

7.6 Monotonic Functions

Problem 7.32

(★)(★) Suppose that $f : \mathbb{R} \to \mathbb{R}$ is a function. Prove that if f is strictly monotonic in $\mathbb{R}$, then f is one-to-one in $\mathbb{R}$.

Proof. We just prove the case when the function f is strictly increasing. If $p, q \in \mathbb{R}$ and $p \neq q$, then either $p < q$ or $p > q$. If $p < q$, then since f is strictly increasing in $\mathbb{R}$, we have $f(p) < f(q)$. Similarly, if $p > q$, then since f is strictly increasing in $\mathbb{R}$, we have $f(p) > f(q)$. In both cases, we have

$$f(p) \neq f(q).$$

By definition (see §1.1.2), we conclude that f is one-to-one in $\mathbb{R}$, finishing the proof of the problem. ∎

Problem 7.33

(★)(★) Prove that if $f : \mathbb{R} \to \mathbb{R}$ is a continuous and one-to-one function, then it is strictly monotonic.

Proof. Assume that f was not strictly monotonic. Then one can find $a < b < c$ such that either $f(a) < f(b) > f(c)$ or $f(a) > f(b) < f(c)$. Suppose that $f(a) < f(b) > f(c)$.[h] Then it implies that

$$f(b) > \max(f(a), f(c)).$$

Since f is continuous on $\mathbb{R}$, it is also continuous on $[a, b]$ and $[b, c]$. By the Intermediate Value Theorem, there exists $p \in (a, b)$ and $q \in (b, c)$ such that

$$f(p) = f(q).$$

However, $p < b < q$ which contradicts the hypothesis that f is one-to-one. Hence we have completed the proof of the problem. ∎

---

[h]The case for $f(a) > f(b) < f(c)$ is similar, so we omit the details here.

7.6. Monotonic Functions

Remark 7.6

Combining Problems 7.32 and 7.33, for a continuous function $f : \mathbb{R} \to \mathbb{R}$, we establish the fact that f is one-to-one if and only if f is strictly monotonic.

Problem 7.34

(⋆)(⋆)(⋆) Let $f : (a,b) \to \mathbb{R}$. Suppose that for each $p \in (a,b)$, there exists a neighborhood $(p-\delta, p+\delta)$ of p in (a,b) such that f is decreasing in $(p-\delta, p+\delta)$. Show that f is decreasing in (a,b).

Proof. Assume that f was not decreasing in (a,b). Thus there exist $p,q \in (a,b)$ with $p < q$ such that $f(p) < f(q)$. By the hypothesis, we know that

$$[p,q] \subseteq \bigcup_{x \in [p,q]} (x - \delta_x, x + \delta_x) \subseteq (a,b),$$

where $\delta_x > 0$. Since $[p,q]$ is compact, we follow from Definition 4.12 (Compact Sets) that there are $x_1, x_2, \ldots, x_n \in [p,q]$ such that

$$[p,q] \subseteq \bigcup_{k=1}^{n} (x_k - \delta_{x_k}, x_k + \delta_{x_k}). \tag{7.54}$$

If $n = 1$, then we have $f(p) \geq f(q)$ which is a contradiction. Therefore, without loss of generality, we may assume that $n \geq 2$ and $p \leq x_1 < x_2 < \cdots < x_n \leq q$. In addition, we may assume further that there are no $i, j \in \{1, 2 \ldots, n\}$ such that

$$(x_i - \delta_{x_i}, x_i + \delta_{x_i}) \subseteq (x_j - \delta_{x_j}, x_j + \delta_{x_j}). \tag{7.55}$$

Otherwise, we may ignore the neighborhood $(x_i - \delta_{x_i}, x_i + \delta_{x_i})$ in the set relation (7.54). By the hypothesis, $p = x_1$ and $q = x_n$ so that

$$f(p) = f(x_1) \quad \text{and} \quad f(q) = f(x_n) \tag{7.56}$$

respectively. For each $k = 1, 2, \ldots, n-1$, it is clear that

$$(x_k - \delta_{x_k}, x_k + \delta_{x_k}) \cap (x_{k+1} - \delta_{x_{k+1}}, x_{k+1} + \delta_{x_{k+1}}) \neq \varnothing.$$

Otherwise, there exists a $m \in \{1, 2, \ldots, n-1\}$ such that

$$(x_m - \delta_{x_m}, x_m + \delta_{x_m}) \cap (x_{m+1} - \delta_{x_{m+1}}, x_{m+1} + \delta_{x_{m+1}}) = \varnothing.$$

By the relation (7.54) again, we can find a $p \in \{1, 2, \ldots, n-1\} \setminus \{m, m+1\}$ such that

$$(x_m + \delta_{x_m}, x_{m+1} + \delta_{x_{m+1}}) \subseteq (x_p - \delta_{x_p}, x_p + \delta_{x_p}). \tag{7.57}$$

If $x_p < x_m$, then we gain from the set relation (7.57) that $x_p + \delta_{x_p} > x_m + \delta_{x_m}$ and then it implies that

$$\delta_{x_p} > x_m - x_p + \delta_{x_m} > \delta_{x_m}.$$

In this case, we have
$$(x_m - \delta_{x_m}, x_m + \delta_{x_m}) \subseteq (x_p - \delta_{x_p}, x_p + \delta_{x_p})$$
which contradicts our assumption (7.55). If $x_{m+1} < x_p$, then the relation (7.57) evidently implies that $x_p - \delta_{x_p} < x_{m+1} - \delta_{x_{m+1}}$ but this shows that
$$\delta_{x_{m+1}} < \delta_{x_p} + x_{m+1} - x_p < \delta_{x_p}.$$
Therefore, we have
$$(x_{m+1} - \delta_{x_{m+1}}, x_{m+1} + \delta_{x_{m+1}}) \subseteq (x_p - \delta_{x_p}, x_p + \delta_{x_p}),$$
a contradiction again.

Now we let
$$y_k \in (x_k, x_k + \delta_{x_k}) \cap (x_{k+1} - \delta_{x_{k+1}}, x_{k+1})$$
for $k = 1, 2 \ldots, n-1$. Since f is decreasing in $(x_k - \delta_{x_k}, x_k + \delta_{x_k})$ and $(x_{k+1} - \delta_{x_{k+1}}, x_{k+1} + \delta_{x_{k+1}})$, we must have
$$f(x_k) \geq f(y_k) \geq f(x_{k+1}) \tag{7.58}$$
for every $k = 1, 2, \ldots, n-1$. Thus, by combining the equalities (7.56) and the inequalities (7.58), we conclude that
$$f(p) \geq f(x_1) \geq f(x_2) \geq \cdots \geq f(x_n) \geq f(q)$$
which is a contradiction. Hence f is decreasing in (a, b) and this completes the proof of the problem. ∎

Problem 7.35

(★)(★) Suppose that $f : [a, b] \to \mathbb{R}$ is monotonically decreasing and $a = p_0 < p_1 < \cdots < p_n = b$. Prove that
$$\sum_{k=1}^{n-1} [f(p_k-) - f(p_k+)] \leq f(a+) - f(b-).$$

Proof. By Theorem 7.18 and Remark 7.3, we see that both $f(p_k+)$ and $f(p_k-)$ exist and
$$f(p_k-) \geq f(p_k) \geq f(p_k+) \tag{7.59}$$
for $k = 1, 2, \ldots, n$. Let $q_1, q_2, \ldots, q_n \in [a, b]$ be points such that
$$a = p_0 < q_1 < p_1 < q_2 < \cdots < q_k < p_k < q_{k+1} < \cdots < p_{n-1} < q_n < p_n = b.$$
Particularly, we have
$$q_k < p_k < q_{k+1}, \tag{7.60}$$
where $k = 1, 2, \ldots, n-1$. Since f is monotonically decreasing on $[a, b]$, it follows from the inequalities (7.59) and (7.60) that
$$f(q_k) \geq f(p_k-) \geq f(p_k) \geq f(p_k+) \geq f(q_{k+1}) \tag{7.61}$$
which means that
$$f(q_k) - f(q_{k+1}) \geq f(p_k-) - f(p_k+)$$

7.6. Monotonic Functions

for each $k = 1, 2, \ldots, n-1$. Thus we deduce from the inequalities (7.61) and then using the decreasing property of f again to obtain that

$$\sum_{k=1}^{n-1}[f(p_k-) - f(p_k+)] \leq \sum_{k=1}^{n-1}[f(q_k) - f(q_{k+1})]$$
$$= [f(q_1) - f(q_2)] + [f(q_2) - f(q_3)] + \cdots + [f(q_{n-1}) - f(q_n)]$$
$$= f(q_1) - f(q_n)$$
$$\leq f(a+) - f(b-).$$

We end the proof of the problem. ∎

CHAPTER 8

Differentiation

In this section, we briefly review the definitions, notations and properties of derivatives of real functions defined on intervals (a,b) or $[a,b]$. The main references for this part are [2, Chap. 4], [3, Chap. 5], [5, Chap. 6] and [23, Chap. 5].

8.1 Fundamental Concepts

8.1.1 Definitions and Notations

Definition 8.1. Suppose that $f : [a,b] \to \mathbb{R}$ and $x \in [a,b]$. Define the function $\phi : [a,b]\setminus\{x\} \to \mathbb{R}$ by

$$\phi(t) = \frac{f(t) - f(x)}{t - x}$$

and define the limit (if it exists)

$$f'(x) = \lim_{t \to x} \phi(t) = \lim_{t \to x} \frac{f(t) - f(x)}{t - x}. \tag{8.1}$$

Equivalently, the number $f'(x)$ given in (8.1) can also be defined as

$$f'(x) = \lim_{h \to 0} \frac{f(x+h) - f(x)}{h}. \tag{8.2}$$

The new function f' is called the **derivative** of f whose domain is the subset $E \subseteq (a,b)$ at which the limit (8.1) exists and the process of obtaining f' from f is called **differentiation**. If f' is defined at x, then it is said that f is **differentiable** at x. Similarly, we say f is **differentiable** in E. At the end-points a and b, we consider the **right-hand derivative** $f'(a+)$ and the **left-hand derivative** $f'(b-)$ respectively.[a]

---

[a]They are

$$f'(a+) = \lim_{\substack{t \to a \\ t > a}} \frac{f(t) - f(a)}{t - a} \quad \text{and} \quad f'(b-) = \lim_{\substack{t \to b \\ t < b}} \frac{f(t) - f(b)}{t - b}.$$

> **Remark 8.1**
>
> We assume the reader is familiar with the derivatives of some well-known functions such as constant functions, linear functions, power functions, polynomials, rational functions, trigonometric functions, logarithmic/exponential functions and etc.

8.1.2 Elementary Properties of Derivatives

Theorem 8.2. *Suppose that $f : [a,b] \to \mathbb{R}$. If f is differentiable at $p \in [a,b]$, then f is continuous at p.*

The next result describes the usual operations for derivatives of the sum, difference, product and quotient of two differentiable functions.

Theorem 8.3 (Operations of Differentiable Functions)**.** *Suppose that $f, g : [a,b] \to \mathbb{R}$ and they are differentiable at $p \in [a,b]$. Then we have the following formulas:*

(a) $(f \pm g)'(p) = f'(p) \pm g'(p)$;

(b) $(f \cdot g)'(p) = f(p)g'(p) + f'(p)g(p)$;

(c) $\left(\dfrac{f}{g}\right)'(p) = \dfrac{g(p)f'(p) - g'(p)f(p)}{g^2(p)}$.

For composition of differentiable functions, we have the famous **Chain Rule**:

Theorem 8.4 (Chain Rule)**.** *Suppose that $f : [a,b] \to \mathbb{R}$ is continuous and $f'(p)$ exists at some point $p \in [a,b]$. Furthermore, if $g : I \to \mathbb{R}$, $f([a,b]) \subseteq I$ and g is differentiable at $f(p)$. If $h : [a,b] \to \mathbb{R}$ is defined by*
$$h(t) = g(f(t)),$$
then h is differentiable at p and its derivative is given by
$$h'(p) = g'(f(p)) \times f'(p).$$

8.1.3 Local Maxima/Minima and Zero Derivatives

Definition 8.5. *Let X be a metric space with metric d and $f : X \to \mathbb{R}$ be a function. Then f is said to have a **local maximum** at $p \in X$ if there is a $\delta > 0$ such that*
$$f(p) \geq f(x)$$
for all $x \in X$ with $d(x, p) < \delta$.

Local minima can be defined similarly. A point p is called a **local extreme** of f if it is either a local maximum or a local minimum of f.

8.1. Fundamental Concepts

> **Remark 8.2**
>
> It is clear that if f has an absolute maximum/minimum at p, then p is also a local maximum/minimum. However, the converse is false.

Theorem 8.6 (Fermat's Theorem). *Suppose that $f : [a,b] \to \mathbb{R}$ has a local maximum/minimum at a point $p \in (a,b)$ and $f'(p)$ exists. Then we have*

$$f'(p) = 0.$$

8.1.4 The Mean Value Theorem and the Intermediate Value Theorem for Derivatives

Theorem 8.7 (Rolle's Theorem). *Suppose that $f : [a,b] \to \mathbb{R}$ is continuous on $[a,b]$ and differentiable in (a,b). If $f(a) = f(b)$, then there exists a point $p \in (a,b)$ such that*

$$f'(p) = 0.$$

Theorem 8.8 (The Generalized Mean Value Theorem). *Suppose that $f, g : [a,b] \to \mathbb{R}$ are continuous on $[a,b]$ and differentiable in (a,b). Then there exists a point $p \in (a,b)$ such that*

$$[f(b) - f(a)]g'(p) = [g(b) - g(a)]f'(p).$$

We note that the above theorem is also called the **Cauchy Mean Value Theorem**.

The Mean Value Theorem for Derivatives. *If $f : [a,b] \to \mathbb{R}$ is continuous on $[a,b]$ and differentiable in (a,b), then there exists a point $p \in (a,b)$ such that*

$$f(b) - f(a) = (b-a)f'(p).$$

Geometrically, the above theorem states that a sufficiently "smooth" curve joining two points A and B in the plane has *at least* one tangent line with the **same** slope as the chord AB. The following result is an immediate consequence of the Mean Value Theorem for Derivatives.

Theorem 8.9. *Suppose that $f : [a,b] \to \mathbb{R}$ is continuous on $[a,b]$ and differentiable in (a,b).*

(a) *If $f'(x) \geq 0$ for all $x \in (a,b)$, then f is **monotonically increasing** on $[a,b]$.*

(b) *If $f'(x) \leq 0$ for all $x \in (a,b)$, then f is **monotonically decreasing** on $[a,b]$.*

(c) *If $f'(x) = 0$ for all $x \in (a,b)$, then $f \equiv c$ on $[a,b]$ for some constant c on $[a,b]$.*

> **Remark 8.3**
>
> If the inequalities in Theorem 8.9(a) and (b) are strict, then f is called **strictly increasing** on $[a,b]$ or **strictly decreasing** on $[a,b]$ respectively.

Similar to the Intermediate Value Theorem, we have an analogous result for the derivative of a function f.[b]

---

[b]This is also called **Darboux's Theorem**.

The Intermediate Value Theorem for Derivatives. *Suppose that $f : [a,b] \to \mathbb{R}$ is differentiable in $[a,b]$. If λ is a number between $f'(a)$ and $f'(b)$, then there exists a $p \in (a,b)$ such that*
$$f'(p) = \lambda.$$

8.1.5 L'Hôspital's Rule

Sometimes it is difficult to evaluate limits in the form
$$\lim_{x \to p} \frac{f(x)}{g(x)}$$
because $f(p) \to 0$ and $g(p) \to 0$ as $x \to p$. In this case, the limit of the quotient $\frac{f(x)}{g(x)}$ is said to be an **indeterminate form**. Symbolically, we use $\frac{0}{0}$ to denote this situation. Other indeterminate forms consist of the cases
$$\frac{\infty}{\infty}, \quad 0 \cdot \infty, \quad 0^0, \quad 1^\infty, \quad \infty^0 \quad \text{and} \quad \infty - \infty.$$

To compute limits in one of the above indeterminate forms, we may need the help of the so-called **L'Hôspital's Rule** and it is stated as follows:

Theorem 8.10 (L'Hôspital's Rule). *Suppose that $-\infty \le a < b < +\infty$, $f, g : (a,b) \to \mathbb{R}$ are differentiable in (a,b), $g'(x) \ne 0$ on (a,b) and*
$$\lim_{\substack{x \to a \\ x > a}} f(x) = \lim_{\substack{x \to a \\ x > a}} g(x) = 0.$$

Then we must have
$$\lim_{\substack{x \to a \\ x > a}} \frac{f(x)}{g(x)} = \lim_{\substack{x \to a \\ x > a}} \frac{f'(x)}{g'(x)} = L,$$

where $L \in [-\infty, +\infty]$.

> **Remark 8.4**
>
> In the preceding result, only the right-hand limit of $\frac{f(x)}{g(x)}$ at a is presented. In fact, similar results for the left-hand limit of $\frac{f(x)}{g(x)}$ at a or two-sided limits of a can be stated and treated.

The following result is very similar to Theorem 8.10 (L'Hôspital's Rule), but the main difference is that it considers the case where the denominator tends to ∞ from the right-hand side of a.

Theorem 8.11. *Suppose that $-\infty \le a < b < \infty$, $f, g : (a,b) \to \mathbb{R}$ are differentiable on (a,b), $g'(x) \ne 0$ on (a,b) and*
$$\lim_{\substack{x \to a \\ x > a}} g(x) = \pm\infty.$$

Then we must have
$$\lim_{\substack{x \to a \\ x > a}} \frac{f(x)}{g(x)} = \lim_{\substack{x \to a \\ x > a}} \frac{f'(x)}{g'(x)} = L,$$

where $L \in [-\infty, +\infty]$.

8.1.6 Higher Order Derivatives and Taylor's Theorem

Definition 8.12. *Suppose that f is a differentiable function in (a,b). If f', in turn, is defined on an open interval and if f' is itself differentiable, then we denote the first derivative of f' by f'', called the **second derivative** of f. Similarly, the **nth derivative** of f,[c] denoted by $f^{(n)}$, can be defined to be the first derivative of the function $f^{(n-1)}$ in an open interval. The function f itself can be written as $f^{(0)}$.*

We notice that if $f^{(n)}$ exists at a point $x \in (a,b)$, then the function $f^{(n-1)}(t)$ must exist for all $t \in (x - \delta, x + \delta) \subseteq (a,b)$ for some $\delta > 0$. Therefore, $f^{(n-1)}$ is also differentiable at x. Since $f^{(n-1)}$ exists on the open interval $(x - \delta, x + \delta)$, the function $f^{(n-2)}$ must be differentiable in $(x - \delta, x + \delta)$.

The following result is known as **Leibniz's rule** which generalizes the product rule of two differentiable functions.

Leibniz's Rule. *Suppose that f and g have nth derivatives at x. Then we have*

$$(fg)^{(n)}(x) = \sum_{k=0}^{n} C_k^n f^{(n-k)}(x) g^{(k)}(x),$$

where C_k^n is the binomial coefficient.

Furthermore, with the help of higher order derivatives, we can talk about the approximation of a function f by a particular polynomial at a point p. This is the main ingredient of the famous Taylor's Theorem.

Taylor's Theorem. *Let $n \in \mathbb{N}$. Suppose that $f : [a,b] \to \mathbb{R}$ satisfies the conditions:*

- *$f^{(n-1)}$ is continuous on $[a,b]$ and*
- *$f^{(n)}$ exists in (a,b).*

Then for every distinct points $x, p \in [a,b]$, there exists a point ξ between x and p such that[d]

$$f(x) = \underbrace{\sum_{k=0}^{n-1} \frac{f^{(k)}(p)}{k!}(x-p)^k}_{\text{Taylor's polynomial}} + \underbrace{\frac{f^{(n)}(\xi)}{n!}(x-p)^n}_{\text{Error term}}.$$

> **Remark 8.5**
>
> We notice that the domain $[a,b]$ in Taylor's Theorem can be replaced by any open interval containing the point p.

---

[c] Or the derivative of order n of f.
[d] That is either $\xi \in (p,x)$ or $\xi \in (x,p)$.

8.1.7 Convexity and Derivatives

The concepts of convexity and convex functions play an important role in different areas of mathematics. Particularly, they are useful and important in the study of optimization problems.

Definition 8.13 (Convex Functions). *A function $f : (a, b) \to \mathbb{R}$ is called **convex** on (a,b) if for every t satisfying $0 \leq t \leq 1$, we have*

$$f(tx_1 + (1-t)x_2) \leq tf(x_1) + (1-t)f(x_2) \tag{8.3}$$

whenever $x_1, x_2 \in (a, b)$.[e]

Geometrically, if f is convex on (a, b), then, for every $x_1, x_2 \in (a, b)$, the chord joining the two points $(x_1, f(x_1))$ and $(x_2, f(x_2))$ lies above the **graph** of f. One of the important properties of convex functions is the following result:

Theorem 8.14. *If $f : [a, b] \to \mathbb{R}$ is convex, then f is continuous on (a, b).*

However, a convex function f needs not be differentiable at a point, but if f has the first derivative or the second derivative, then we can establish some connections between f and its derivatives.

Theorem 8.15. *Suppose that f is a real differentiable function defined in (a, b). Then f is convex on (a, b) if and only if f' is **monotonically increasing** on (a, b).*

Theorem 8.16. *Suppose that f is a twice differentiable real function defined in (a, b). Then f is convex on (a, b) if and only if $f''(x) \geq 0$ for all $x \in (a, b)$.*

> **Remark 8.6**
>
> A function $f : (a, b) \to \mathbb{R}$ is called **concave** on (a, b) if the sign "$\leq$" in the inequality (8.3) is replaced by "$\geq$". In this case, the chord joining the two points $(x_1, f(x_1))$ and $(x_2, f(x_2))$ lies below the graph of f.

8.2 Properties of Derivatives

> **Problem 8.1**
>
> (★) *Suppose that $f : \mathbb{R} \to \mathbb{R}$ is continuous at p. Define $g : \mathbb{R} \to \mathbb{R}$ by*
>
> $$g(x) = (x - p)f(x)$$
>
> *for all $x \in \mathbb{R}$. Prove that $g'(p)$ exists.*

Proof. By the definition (8.2), we have

$$g'(p) = \lim_{h \to 0} \frac{g(p+h) - g(p)}{h} = \lim_{h \to 0} \frac{hf(p+h) - 0}{h} = \lim_{h \to 0} f(p+h).$$

---
[e]If the inequality is strict, then f is said to be **strictly convex** on (a, b).

8.2. Properties of Derivatives

Since f is continuous at p, it follows from Theorem 7.5 that

$$\lim_{h \to 0} f(p+h) = f(p)$$

so that

$$g'(p) = f(p).$$

This ends the proof of the problem. ∎

Problem 8.2

(⋆) Let $a > 1$ and $f : \mathbb{R} \to \mathbb{R}$ be a function defined by

$$f(x) = \begin{cases} x^a \sin \frac{1}{x}, & \text{if } x \neq 0; \\ 0, & \text{if } x = 0. \end{cases}$$

Prove that f is differentiable at 0.

Proof. By Definition 8.1, we have

$$\lim_{t \to 0} \frac{f(t) - f(0)}{t - 0} = \lim_{t \to 0} t^{a-1} \sin \frac{1}{t}.$$

Since $a > 1$ and $|\sin \frac{1}{x}| \leq 1$ for every $x \in \mathbb{R}$, we know that

$$0 \leq \left| t^{a-1} \sin \frac{1}{t} \right| \leq |t|^{a-1} \tag{8.4}$$

Apply Theorem 7.3 (Squeeze Theorem for Limits of Functions) to the inequalities (8.4) to get

$$\lim_{t \to 0} t^{a-1} \sin \frac{1}{t} = 0.$$

Thus f is differentiable at 0 and $f'(0) = 0$. This finishes the proof of the problem. ∎

Problem 8.3

(⋆) Let $m, n \in \mathbb{Z}$ and $f : \mathbb{R} \to \mathbb{R}$ be a function defined by

$$f(x) = \begin{cases} x^n \sin \frac{1}{x^m}, & \text{if } x \neq 0; \\ 0, & \text{if } x = 0. \end{cases}$$

Suppose that $n \geq m + 1$ and $n \geq 1$. Prove that there exists a $\delta > 0$ such that $f'(x)$ is bounded on the segment $(-\delta, \delta)$.

Proof. By applying Theorem 8.3 (Operations of Differentiable Functions) repeatedly, we have

$$f'(x) = nx^{n-1} \sin \frac{1}{x^m} - mx^{n-m-1} \cos \frac{1}{x^m} \tag{8.5}$$

if $x \in (-\delta, \delta)$ and $x \neq 0$. Since $n \geq 1$ and $n \geq m+1$, we have x^{n-1} and x^{n-m-1} are bounded on $(-1, 1)$. It is well-known that $\sin x$ and $\cos x$ are bounded on $\mathbb{R}$. Combining these facts and the derivative (8.5), we know that $f'(x)$ is bounded on $(-1, 1) \setminus \{0\}$.

To check the remaining case $f'(0)$, we notice from the definition (8.2) and an application of Theorem 7.3 (Squeeze Theorem for Limits of Functions) that

$$f'(0) = \lim_{h \to 0} \frac{f(0+h) - f(0)}{h} = \lim_{h \to 0} h^{n-1} \sin \frac{1}{h^m} = 0.$$

Hence we conclude that $f'(x)$ is bounded on $(-1, 1)$, completing the proof of the problem. ∎

Problem 8.4

(★) Prove that the function $f : \mathbb{R} \to \mathbb{R}$ defined by

$$f(x) = \begin{cases} x^3, & \text{if } x \in \mathbb{Q}; \\ 0, & \text{if } x \in \mathbb{R} \setminus \mathbb{Q} \end{cases}$$

is differentiable *only* at 0.

Proof. We note that

$$\frac{f(0+h) - f(0)}{h} = \frac{f(h)}{h} = \begin{cases} h^2, & \text{if } h \in \mathbb{Q}; \\ 0, & \text{if } h \in \mathbb{R} \setminus \mathbb{Q}. \end{cases}$$

By the definition (8.2), we have

$$f'(0) = \lim_{h \to 0} \frac{f(0+h) - f(0)}{h}$$
$$= \begin{cases} \lim_{h \to 0} \frac{h^2}{h}, & \text{if } h \in \mathbb{Q}; \\ \lim_{h \to 0} \frac{0}{h}, & \text{if } h \in \mathbb{R} \setminus \mathbb{Q} \end{cases}$$
$$= 0.$$

In other words, f is differentiable at 0.

Suppose that $x \in \mathbb{Q} \setminus \{0\}$. Then we deduce from Remark 2.1 that we can find a sequence $\{x_n\} \subseteq \mathbb{R} \setminus (\mathbb{Q} \cup \{x\})$ such that $x_n \to x$ as $n \to \infty$. By this sequence, we have

$$\lim_{n \to \infty} \frac{f(x_n) - f(x)}{x_n - x} = \lim_{n \to \infty} \frac{-x^3}{x_n - x}. \tag{8.6}$$

Since $x \neq 0$ and $x_n \to x$ as $n \to \infty$, the limit (8.6) is either $-\infty$ or $+\infty$ and this implies that f is not differentiable at every $x \in \mathbb{Q} \setminus \{0\}$. Similarly, if $x \in \mathbb{R} \setminus \mathbb{Q}$ and $x \neq 0$, then it follows from Theorem 2.2 (Density of Rationals) that there exists a sequence $\{y_n\} \subseteq \mathbb{Q} \setminus \{x, 0\}$ such that $y_n \to x$ as $n \to \infty$. By this sequence, we have

$$\lim_{n \to \infty} \frac{f(y_n) - f(x)}{y_n - x} = \lim_{n \to \infty} \frac{y_n^3}{y_n - x}. \tag{8.7}$$

8.2. Properties of Derivatives

Since $y_n \neq 0$ for all $n \in \mathbb{N}$, the limit (8.7) is either $-\infty$ or $+\infty$. Hence we conclude that f is not differentiable at every $x \in \mathbb{R} \setminus \{0\}$. We have completed the proof of the problem. ∎

Problem 8.5

(⋆) Prove that the function $f : \mathbb{R} \to \mathbb{R}$ defined by

$$f(x) = x|x|$$

is differentiable in $\mathbb{R}$.

Proof. It is clear that

$$f(x) = \begin{cases} x^2, & \text{if } x \geq 0; \\ -x^2, & \text{if } x < 0. \end{cases}$$

Therefore, f is differentiable at every $x \in \mathbb{R} \setminus \{0\}$. We have to check the differentiability of f at the point 0. By the definition (8.2), we have

$$f'(0+) = \lim_{\substack{h \to 0 \\ h > 0}} \frac{f(0+h) - f(0)}{h} = \lim_{\substack{h \to 0 \\ h > 0}} h = 0$$

and

$$f'(0-) = \lim_{\substack{h \to 0 \\ h < 0}} \frac{f(0+h) - f(0)}{h} = \lim_{\substack{h \to 0 \\ h < 0}} -h = 0.$$

By Definition 8.1, f is differentiable at 0. Hence f is differentiable in $\mathbb{R}$, completing the proof of the problem. ∎

Problem 8.6

(⋆)(⋆) Suppose that $f : \mathbb{R} \to \mathbb{R}$ is differentiable at p. Let $\{x_n\}$ and $\{y_n\}$ be increasing and decreasing sequences in $\mathbb{R} \setminus \{p\}$ respectively, both of them converge to p. Prove that

$$f'(p) = \lim_{n \to \infty} \frac{f(x_n) - f(y_n)}{x_n - y_n}.$$

Proof. Let

$$g(x) = \frac{f(x) - f(p)}{x - p} - f'(p). \tag{8.8}$$

Since $f'(p)$ exists, it follows from Definition 8.1 and Theorem 8.3 (Operations of Differentiable Functions) that

$$\lim_{x \to p} g(x) = f'(p) - f'(p) = 0. \tag{8.9}$$

Rewrite the expression (8.8) as

$$f(x) = f(p) + [f'(p) + g(x)](x - p)$$

which implies that

$$\frac{f(x_n) - f(y_n)}{x_n - y_n} = \frac{[f'(p) + g(x_n)](x_n - p) - [f'(p) + g(y_n)](y_n - p)}{x_n - y_n}$$

$$= \frac{f'(p)(x_n - y_n) + g(x_n)(x_n - p) - g(y_n)(y_n - p)}{x_n - y_n}$$

$$= f'(p) + g(x_n)\left(\frac{x_n - p}{x_n - y_n}\right) - g(y_n)\left(\frac{y_n - p}{x_n - y_n}\right)$$

for all $n \in \mathbb{N}$. Thus we have

$$\left|\frac{f(x_n) - f(y_n)}{x_n - y_n} - f'(p)\right| = \left|g(x_n)\left(\frac{x_n - p}{x_n - y_n}\right) - g(y_n)\left(\frac{y_n - p}{x_n - y_n}\right)\right|$$

$$\leq |g(x_n)| \times \left|\frac{x_n - p}{x_n - y_n}\right| + |g(y_n)| \times \left|\frac{y_n - p}{x_n - y_n}\right| \quad (8.10)$$

for all $n \in \mathbb{N}$. Since $\{x_n\}$ is increasing in $\mathbb{R} \setminus \{p\}$ and converging to p, we have $x_n < p$ for all $n \in \mathbb{N}$. Similarly, we have $p < y_n$ for all $n \in \mathbb{N}$. Therefore, we have $x_n < p < y_n$ for all $n \in \mathbb{N}$ and this implies that

$$\left|\frac{x_n - p}{x_n - y_n}\right| \leq 1 \quad \text{and} \quad \left|\frac{y_n - p}{x_n - y_n}\right| \leq 1. \quad (8.11)$$

By the inequalities (8.10) and (8.11), we obtain

$$\left|\frac{f(x_n) - f(y_n)}{x_n - y_n} - f'(p)\right| \leq |g(x_n)| + |g(y_n)| \quad (8.12)$$

for all $n \in \mathbb{N}$. By the limit (8.9), Theorems 5.6 (Squeeze Theorem for Convergent Sequences) and 7.1, we see that

$$\lim_{n \to \infty} |g(x_n)| = \lim_{n \to \infty} |g(y_n)| = 0.$$

Hence we deduce from the inequality (8.12) that

$$\lim_{n \to \infty} \left|\frac{f(x_n) - f(y_n)}{x_n - y_n} - f'(p)\right| = 0.$$

By definition (see §2.1.2), we have $-|a| \leq a \leq |a|$ and so

$$\lim_{n \to \infty} \frac{f(x_n) - f(y_n)}{x_n - y_n} = f'(p)$$

as required. This completes the proof of the problem. ∎

Problem 8.7

(★) Suppose that $f : \mathbb{R} \to \mathbb{R}$ is differentiable at p. Prove that

$$\lim_{x \to p} \frac{p^n f(x) - x^n f(p)}{x - p} = p^n f'(p) - nf(p)p^{n-1}.$$

8.2. Properties of Derivatives

Proof. We notice that

$$\frac{p^n f(x) - x^n f(p)}{x-p} = \frac{p^n f(x) - p^n f(p) + p^n f(p) - x^n f(p)}{x-p}$$
$$= p^n \cdot \frac{f(x) - f(p)}{x-p} - f(p) \cdot \frac{x^n - p^n}{x-p}. \quad (8.13)$$

Since

$$\lim_{x \to p} \frac{f(x) - f(p)}{x-p} = f'(p) \quad \text{and} \quad \lim_{x \to p} \frac{x^n - p^n}{x-p} = np^{n-1},$$

we deduce from the expression (8.13) that

$$\lim_{x \to p} \frac{p^n f(x) - x^n f(p)}{x-p} = p^n \lim_{x \to p} \frac{f(x) - f(p)}{x-p} - f(p) \lim_{x \to p} \frac{x^n - p^n}{x-p}$$
$$= p^n f'(p) - nf(p)p^{n-1}$$

which is our desired result. This finishes the proof of the problem. ∎

Problem 8.8

(⋆) *Is the converse of Theorem 8.6 (Fermat's Theorem) true? Furthermore, construct a function f having a local minimum at 0 but $f'(0) \neq 0$.*

Proof. For example, we consider the function $f(x) = x^3$ whose derivative is $f'(x) = 3x^2$ so that $f'(0) = 0$ if and only if $x = 0$. However, for every $\epsilon > 0$, we can always find $p, q \in (-\epsilon, \epsilon)$ such that

$$p^3 < 0 < q^3$$

which means that 0 is *not* a local maximum or a local minimum of f.

Furthermore, note that the function $f : \mathbb{R} \to \mathbb{R}$ defined by $f(x) = |x|$ has the absolute (and hence local) minimum at $x = 0$, but it is *not* differentiable at 0 because

$$f'(0+) = \lim_{\substack{t \to 0 \\ t > 0}} \frac{f(t) - f(0)}{t - 0} = 1 \quad \text{and} \quad f'(0-) = \lim_{\substack{t \to 0 \\ t < 0}} \frac{f(t) - f(0)}{t - 0} = -1.$$

Thus we have completed the proof of the problem. ∎

Problem 8.9

(⋆)(⋆) *Suppose that $f : [0,1] \to \mathbb{R}$ is differentiable in $[0,1]$ and there is no $p \in [0,1]$ such that $f(p) = f'(p) = 0$. Prove that the set*

$$Z = \{x \in [0,1] \,|\, f(x) = 0\}$$

is finite.

Proof. Assume that there was an infinite sequence $\{p_n\} \subseteq [0,1]$ such that $f(p_n) = 0$. Since $\{p_n\}$ is bounded, Problem 5.25 (The Bolzano-Weierstrass Theorem) shows that $\{p_n\}$ has a convergent subsequence $\{p_{n_k}\}$. Without loss of generality, we may assume that $\{p_n\}$ converges to p. Since $[0,1]$ is compact (or closed), we have $p \in [0,1]$. Since f is differentiable in $[0,1]$, it is continuous on $[0,1]$ by Theorem 8.2. Therefore, it follows from Theorem 7.5 that

$$f(p) = f\left(\lim_{n \to \infty} p_n\right) = \lim_{n \to \infty} f(p_n) = 0. \tag{8.14}$$

Recall that $f(p_n) = 0$ for all $n = 1, 2, \ldots$, we derive by using Theorem 7.5 and the value (8.14) that

$$f'(p) = \lim_{x \to p} \frac{f(x) - f(p)}{x - p} = \lim_{n \to \infty} \frac{f(p_n) - f(p)}{p_n - p} = 0$$

which contradicts the hypothesis. This ends the proof of the problem. ∎

Problem 8.10

(⋆) *Suppose that $f : \mathbb{R} \to \mathbb{R}$ is differentiable in $\mathbb{R}$. If the following formula*

$$f(x + h) = f(x) + hf'(x) \tag{8.15}$$

holds for every $x, h \in \mathbb{R}$, prove that there exist constants A and B such that

$$f(x) = Ax + B$$

for every $x \in \mathbb{R}$.

Proof. Take $h = y - 1$ and $x = 1$ in the formula (8.15), where $y \in \mathbb{R}$. Then it deduces that

$$f(1 + (y - 1)) = f(1) + (y - 1)f'(1)$$
$$f(y) = f'(1)y + [f(1) - f'(1)].$$

Thus we have

$$A = f'(1) \quad \text{and} \quad B = f(1) - f'(1).$$

Hence we have completed the proof of the problem. ∎

8.3 The Mean Value Theorem for Derivatives

Problem 8.11

(⋆) (⋆) *Suppose that $f : [0,1] \to \mathbb{R}$ is continuous and $f(0) = 0$. Furthermore, $f(x)$ is differentiable for every $x \in (0,1)$ and*

$$0 \le f'(x) \le 2f(x) \tag{8.16}$$

on $(0,1)$. Prove that $f \equiv 0$ on $[0,1]$.

8.3. The Mean Value Theorem for Derivatives

Proof. Define $F : [0,1] \to \mathbb{R}$ by
$$F(x) = e^{-2x} f(x)$$
for every $x \in [0,1]$. By Theorem 8.3 (Operations of Differentiable Functions), we see that
$$F'(x) = e^{-2x}[f'(x) - 2f(x)]$$
for every $x \in (0,1)$. By the hypothesis (8.16), we have
$$F'(x) \leq 0$$
for all $x \in (0,1)$. By Theorem 8.9(b), we derive that F is monotonically decreasing on $[0,1]$. Since $F(0) = e^0 f(0) = 0$, we conclude that
$$F(x) \leq 0 \tag{8.17}$$
on $[0,1]$. By the hypothesis (8.16) again, we have $f(x) \geq 0$ for every $(0,1)$. Thus it follows from this and the continuity of f that
$$f(x) \geq 0$$
on $[0,1]$. Therefore, we have
$$F(x) \geq 0 \tag{8.18}$$
on $[0,1]$. Combining the inequalities (8.17) and (8.18), we obtain $F(x) \equiv 0$ on $[0,1]$ which is equivalent to
$$f(x) \equiv 0$$
on $[0,1]$. This completes the proof of the problem. ∎

Problem 8.12

(★)(★) *Suppose that $f : [a,b] \to \mathbb{R}$ is continuous on $[a,b]$ and differentiable in (a,b). Furthermore, the derivative $f'(x)$ is bounded on (a,b). Prove that f is uniformly continuous on (a,b).*

Proof. Since $f'(x)$ is bounded on (a,b), there exists a $M > 0$ such that
$$|f'(x)| \leq M \tag{8.19}$$
for all $x \in (a,b)$. By the Mean Value Theorem for Derivatives, we have
$$f(x) - f(y) = f'(p)(x - y) \tag{8.20}$$
for some $p \in (x,y) \subseteq (a,b)$. For each $\epsilon > 0$, we take $\delta = \frac{\epsilon}{M}$. Therefore, if $x, y \in (a,b)$ and $|x - y| < \delta$, then we deduce from the inequality (8.19) and the expression (8.20) that
$$|f(x) - f(y)| = |f'(p)||x - y| \leq M|x - y| < M\delta < \epsilon.$$

Hence f is uniformly continuous on (a,b) by Definition 7.8 (Uniform Continuity). This finishes the proof of the problem. ∎

Problem 8.13

(★)(★) Suppose that $f : (a,b) \to \mathbb{R}$ is differentiable and its derivative f' is uniformly continuous on (a,b). Prove that

$$\lim_{n \to \infty} n\left[f(x) - f\left(x - \frac{1}{n}\right)\right] = f'(x)$$

for every $x \in (a,b)$.

Proof. Since f' is uniformly continuous on (a,b), for every $\epsilon > 0$, there exists a $\delta > 0$ such that

$$|f'(x) - f'(y)| < \epsilon \tag{8.21}$$

for every $x, y \in (a,b)$ with $|x - y| < \delta$. By Theorem 2.1 (The Archimedean Property), there exists a positive integer N such that $N\delta > 1$. Then for $x \in (a,b)$, we choose $n \geq N$ such that $x - \frac{1}{n} > a$. Therefore, we have

$$\left(x - \frac{1}{n}, x\right) \subseteq (a,b) \quad \text{and} \quad \left|x - \frac{1}{n} - x\right| = \frac{1}{n} < \delta. \tag{8.22}$$

By this, we obtain from the inequality (8.21) that

$$|f'(p) - f'(x)| < \epsilon \tag{8.23}$$

for all $p \in (x - \frac{1}{n}, x) \subseteq (a,b)$.

Since f is differentiable in (a,b), it is continuous on $[x - \frac{1}{n}, x] \subset (a,b)$. Applying the Mean Value Theorem for Derivatives, we conclude that

$$\left|n\left[f(x) - f\left(x - \frac{1}{n}\right)\right] - f'(x)\right| = \left|\frac{f(x) - f(x - \frac{1}{n})}{x - (x - \frac{1}{n})} - f'(x)\right| = |f'(p_n) - f'(x)| \tag{8.24}$$

for some $p_n \in (x - \frac{1}{n}, x)$ for those n satisfying the requirements (8.22). By the inequalities (8.23) and (8.24), we have

$$\left|n\left[f(x) - f\left(x - \frac{1}{n}\right)\right] - f'(x)\right| < \epsilon$$

for some $p_n \in (x - \frac{1}{n}, x)$ for those n satisfying the requirements (8.22). By Definition 8.1, we get the desired result that

$$\lim_{n \to \infty} n\left[f(x) - f\left(x - \frac{1}{n}\right)\right] = f'(x)$$

for every $x \in (a,b)$. This ends the proof of the problem. ∎

Problem 8.14

(★)(★) Prove that for $x, y > 0$ and $0 < \alpha < \beta$, we have

$$(x^\alpha + y^\alpha)^{\frac{1}{\alpha}} > (x^\beta + y^\beta)^{\frac{1}{\beta}}. \tag{8.25}$$

8.3. The Mean Value Theorem for Derivatives

Proof. If $x = y$, then we have
$$(x^\alpha + y^\alpha)^{\frac{1}{\alpha}} = 2^{\frac{1}{\alpha}} x \quad \text{and} \quad (x^\beta + y^\beta)^{\frac{1}{\beta}} = 2^{\frac{1}{\beta}} x.$$

Since $2^{\frac{1}{\alpha}} > 2^{\frac{1}{\beta}}$, the inequality holds in this case.

Without loss of generality, we may assume that $x \neq y$. Define the function $f : (0, \infty) \to \mathbb{R}$ by
$$f(t) = (x^t + y^t)^{\frac{1}{t}}.$$

By Theorem 8.4 (Chain Rule), we have
$$f'(t) = \frac{f(t)}{t^2(x^t + y^t)} \Big[x^t \ln x^t + y^t \ln y^t - (x^t + y^t) \ln(x^t + y^t) \Big]. \tag{8.26}$$

Let $X = x^t$ and $Y = y^t$. Then the brackets in the expression (8.26) can be written as
$$x^t \ln x^t + y^t \ln y^t - (x^t + y^t) \ln(x^t + y^t) = \ln(X^X Y^Y) - \ln(X + Y)^{X+Y}. \tag{8.27}$$

Since $x, y > 0$ and $t > 0$, $X > 0$ and $Y > 0$. By these facts, we have
$$X^X < (X+Y)^X \quad \text{and} \quad Y^Y < (X+Y)^Y$$

and thus
$$X^X Y^Y < (X+Y)^X \times (X+Y)^Y = (X+Y)^{X+Y}. \tag{8.28}$$

Now by substituting the inequality (8.28) and the expression (8.27) into the right-hand side of the derivative (8.26), we obtain
$$f'(t) < 0$$
for all $t \in (0, \infty)$. By Theorem 8.9(b) and Remark 8.3, we see that $f(t)$ is strictly decreasing on $(0, \infty)$. Hence, if $\beta > \alpha > 0$, then we have
$$f(\alpha) > f(\beta)$$
which is exactly the inequality (8.25). ∎

Problem 8.15

(⋆) *Prove that $|\cos x - \cos y| \leq |x - y|$ for every $x, y \in \mathbb{R}$.*

Proof. It is clear that the equality holds when $x = y$. Without loss of generality, we may assume that $x < y$. Since the function $f(t) = \cos t$ is continuous on $[x, y]$ and differentiable in (x, y), we obtain from the Mean Value Theorem for Derivatives that there exists a $p \in (x, y)$ such that
$$\cos y - \cos x = -(y - x) \sin p. \tag{8.29}$$

Since $|\sin t| \leq 1$ for all $t \in \mathbb{R}$, we know from this and the expression (8.29) that
$$|\cos x - \cos y| = |\sin p||x - y| \leq |x - y|.$$

This finishes the proof of the problem. ∎

> **Remark 8.7**
>
> We say that a function $f : [a,b] \to \mathbb{R}$ satisfies a **Lipschitz condition** if there exists a positive constant $K > 0$ such that
> $$|f(x) - f(y)| \leq K|x - y|$$
> for all $x, y \in [a,b]$. Here the K is called a **Lipschitz constant**. Thus Problem 8.15 says that the function $f(x) = \cos x$ is Lipschitz. Furthermore, we can see from the proof of Problem 8.12 that a function whose derivative is bounded on (a,b) is also Lipschitz.

> **Problem 8.16**
>
> (★)(★) *Suppose that $f : (a,b) \to \mathbb{R}$ is differentiable in (a,b) with unbounded derivative $f'(x)$. If f' is continuous on (a,b), prove that f is not Lipschitz on (a,b).*

Proof. Assume that f was Lipschitz on (a,b) with a Lipschitz constant K, i.e.,
$$|f(x) - f(y)| \leq K|x - y| \tag{8.30}$$
for all $x, y \in (a,b)$. Since f' is unbounded in (a,b), there exists a $p \in (a,b)$ such that
$$|f'(p)| > 2K.$$
Since $|f'| - 2K$ is continuous on (a,b) and (a,b) is open in $\mathbb{R}$, it yields from Problem 7.15 (The Sign-preserving Property) that there exists a $\delta > 0$ such that
$$|f'(x)| > 2K \tag{8.31}$$
for all $x \in (p-\delta, p+\delta) \subseteq (a,b)$. Pick $q \in (p-\delta, p)$. By the Mean Value Theorem for Derivatives, we know that
$$f(p) - f(q) = f'(\xi)(p - q) \tag{8.32}$$
for some $\xi \in (q,p)$. Combining the inequality (8.31), the expression (8.32) and then the Lipschitz condition (8.30), we have
$$2K|p - q| < |f'(\xi)| \times |p - q| = |f'(\xi)(p - q)| = |f(p) - f(q)| \leq K|p - q|$$
which implies $2K < K$, a contradiction. Hence f is not Lipschitz on (a,b) and we complete the proof of the problem. ∎

> **Problem 8.17**
>
> (★)(★) *Suppose that $f : (1, +\infty) \to \mathbb{R}$ is differentiable in $(1, +\infty)$ and*
> $$\lim_{x \to +\infty} f'(x) = 0.$$
> *Prove that*
> $$\lim_{x \to +\infty} \frac{f(x)}{x} = 0$$
> *holds.*

8.3. The Mean Value Theorem for Derivatives

Proof. By Definition 7.20 (Limits at Infinity), given $\epsilon > 0$, there exists a $M > 1$ such that $x > M$ implies that
$$|f'(x)| < \frac{\epsilon}{2}. \tag{8.33}$$
Pick $a > M$ and let $x > a$. Since f is differentiable in $(1, +\infty)$, it is also differentiable in (a, x) and continuous on $[a, x]$. By the Mean Value Theorem for Derivatives, we know that
$$f(x) - f(a) = f'(\xi)(x - a) \tag{8.34}$$
for some $\xi \in (a, x)$. Now we take absolute value to both sides of the expression (8.34) and then using the inequality (8.33), we get
$$|f(x) - f(a)| < \frac{\epsilon}{2}|x - a|.$$
Since $|f(x)| - |f(a)| \leq |f(x) - f(a)|$, we have
$$|f(x)| \leq |f(a)| + \frac{\epsilon}{2}|x - a|. \tag{8.35}$$
Since $f(a)$ is fixed, we may take $K > a$ large enough such that
$$\frac{|f(a)|}{K} < \frac{\epsilon}{2}. \tag{8.36}$$
Now if $x > K > a > M > 1$, then we follow from the inequalities (8.35) and (8.36) that
$$\left|\frac{f(x)}{x}\right| \leq \left|\frac{f(a)}{x}\right| + \frac{\epsilon}{2} \cdot \frac{|x-a|}{|x|} < \frac{|f(a)|}{K} + \frac{\epsilon}{2} < \frac{\epsilon}{2} + \frac{\epsilon}{2} = \epsilon$$
which means that
$$\lim_{x \to +\infty} \frac{f(x)}{x} = 0$$
holds. We end the proof of the problem. ∎

Problem 8.18

(⋆) *Find the number of solutions of the equation* $e^x = 1 - x$ *in* $\mathbb{R}$.

Proof. Let $f(x) = e^x + x - 1$. It is clear that
$$f(0) = e^0 + 0 - 1 = 1 + 0 - 1 = 0.$$
Thus 0 is a root of the equation $f(x) = 0$. Let $p \neq 0$ be another root of $f(x) = 0$. Without loss of generality, we may assume further that $p > 0$. Since f is differentiable in $(0, p)$, we apply the Mean Value Theorem for Derivatives to get
$$f(p) - f(0) = f'(\xi)(p - 0)$$
for some $\xi \in (0, p)$. Since $f(p) = 0$, this implies that
$$f'(\xi) = 0.$$
However, since $f'(x) = e^x + 1$, we have
$$e^\xi = -1$$
which is a contradiction. Hence we have finished the proof of the problem. ∎

Problem 8.19

(⋆) (⋆) Suppose that $a, b, c \in \mathbb{R}$ with $a < b < c$. Suppose further that $f : [a, c] \to \mathbb{R}$ is continuous on $[a, c]$ and differentiable in (a, c). Prove that there exist $\lambda \in (a, b)$ and $\theta \in (a, c)$ such that

$$f(b) - f(a) = (b - a)f'(\lambda) \quad \text{and} \quad f(c) - f(a) = (c - a)f'(\theta),$$

where $\lambda < \theta$.

Proof. Since f is continuous on $[a, c]$ and differentiable in (a, c), it is also continuous on $[a, b]$ and $[b, c]$ as well as differentiable in (a, b) and (b, c). Applying the Mean Value Theorem for Derivatives twice, we know that there exist $\lambda \in (a, b)$ and $\xi \in (b, c)$ such that

$$f(b) - f(a) = (b - a)f'(\lambda) \quad \text{and} \quad f(c) - f(b) = (c - b)f'(\xi),$$

By these facts, it is obvious that

$$\begin{aligned}
\frac{f(c) - f(a)}{c - a} &= \frac{f(c) - f(b) + f(b) - f(a)}{c - a} \\
&= \frac{f(c) - f(b)}{c - b} \cdot \frac{c - b}{c - a} + \frac{f(b) - f(a)}{b - a} \cdot \frac{b - a}{c - a} \\
&= f'(\xi) \cdot \frac{c - b}{c - a} + f'(\lambda) \cdot \frac{b - a}{c - a}.
\end{aligned} \quad (8.37)$$

If we let $x = \frac{c-b}{c-a}$, then $1 - x = 1 - \frac{c-b}{c-a} = \frac{b-a}{c-a}$ so that the expression (8.37) can be rewritten as

$$\frac{f(c) - f(a)}{c - a} = xf'(\xi) + (1 - x)f'(\lambda). \quad (8.38)$$

Since $a < b < c$, we have $0 < x < 1$ and this implies that the value

$$xf'(\xi) + (1 - x)f'(\lambda)$$

lies between $f'(\xi)$ and $f'(\lambda)$. By the Intermediate Value Theorem for Derivatives and then using the expression (8.38), there exists a $\theta \in (\lambda, \xi)$ such that

$$f'(\theta) = xf'(\xi) + (1 - x)f'(\lambda) = \frac{f(c) - f(a)}{c - a},$$

i.e.,

$$f(c) - f(a) = (c - a)f'(\theta).$$

This completes the proof of the problem. ∎

Problem 8.20

(⋆)(⋆) Let $p \in (a, b)$. Suppose that $f : [a, b] \to \mathbb{R}$ is continuous on $[a, b]$ and differentiable in $(a, b) \setminus \{p\}$. If $\lim\limits_{x \to p} f'(x)$ exists, prove that f is differentiable at p.

8.3. The Mean Value Theorem for Derivatives

Proof. We let
$$\lim_{x \to p} f'(x) = L$$

We want to show that
$$\lim_{x \to p} \frac{f(x) - f(p)}{x - p} = L \tag{8.39}$$

or equivalently, by Theorem 7.14,
$$F(p+) = F(p-) = L,$$

where
$$F(x) = \frac{f(x) - f(p)}{x - p}.$$

To this end, we pick a sequence $\{x_n\} \subseteq (p, b)$ converging to p. By Theorem 7.1, we have
$$\lim_{x \to p} f'(x) = \lim_{n \to \infty} f'(x_n) = L. \tag{8.40}$$

Now we consider the restrictions $f_{[p,x_n]} : [p, x_n] \to \mathbb{R}$. It is obvious that they are continuous on $[p, x_n]$ and differentiable in (p, x_n). By the Mean Value Theorem for Derivatives, there exist $\xi_n \in (p, x_n)$ such that
$$f(x_n) - f(p) = f'(\xi_n)(x_n - p). \tag{8.41}$$

Now $\xi_n \to p$ if and only if $x_n \to p$ and $x_n > p$. Therefore, we apply this fact and the limit (8.40) to get
$$L = \lim_{x \to p} f'(x) = \lim_{n \to \infty} f'(x_n) = \lim_{n \to \infty} f'(\xi_n). \tag{8.42}$$

By the expression (8.41) and the limit (8.42), we obtain
$$F(p+) = \lim_{n \to \infty} \frac{f(x_n) - f(p)}{x_n - p} = L.$$

Similarly, we can show that $F(p-) = L$. Hence we obtain the desired limit (8.39) so that f is differentiable at p, as required. This finishes the proof of the problem. ∎

Problem 8.21

(⋆)(⋆) *Suppose that $f : [a, b] \to \mathbb{R}$ is continuous on $[a, b]$ and differentiable in (a, b). Furthermore, let $f(a) = f(b) = 0$. Prove that for every $K \in \mathbb{R}$, there exists a $p \in (a, b)$ such that*
$$f'(p) = Kf(p).$$

Proof. Consider the function $F : [a, b] \to \mathbb{R}$ defined by
$$F(x) = e^{-Kx} f(x)$$

for every $x \in [a, b]$. It is clear that $F(a) = F(b) = 0$. By Theorem 8.7 (Rolle's Theorem), we see that there exists a point $p \in (a, b)$ such that
$$F'(p) = 0. \tag{8.43}$$

Since $F'(x) = e^{-Kx}[-Kf(x) + f'(x)]$ and $e^{-Kx} \neq 0$ on $[a,b]$, we follow from these and the result (8.43) that
$$f'(p) = Kf(p),$$
completing the proof of the problem. ∎

Problem 8.22

(★)(★) *Suppose that $f : [a,b] \to \mathbb{R}$ is continuous on $[a,b]$ and differentiable in (a,b). Prove that there exist $p, q \in (a,b)$ such that*
$$\frac{f'(p)}{a+b} = \frac{f'(q)}{2q}. \tag{8.44}$$

Proof. We apply the Mean Value Theorem for Derivatives to f, there exists a point $p \in (a,b)$ such that
$$f(b) - f(a) = (b-a)f'(p). \tag{8.45}$$
Next, we apply Theorem 8.8 (The Generalized Mean Value Theorem) to the functions $f(x)$ and $g(x) = x^2$ to conclude that there exists a $q \in (a,b)$ such that
$$2q[f(b) - f(a)] = (b^2 - a^2)f'(q) = (b-a)(b+a)f'(q). \tag{8.46}$$
Combining the expressions (8.45) and (8.46), we are able to show that
$$2q(b-a)f'(p) = (b-a)(b+a)f'(q)$$
which imply the formula (8.44) because $b - a \neq 0$. We complete the proof of the problem. ∎

8.4 L'Hôspital's Rule

Problem 8.23

(★)(★) *Reprove Problem 8.20 by using Theorem 8.10 (L'Hôspital's Rule).*

Proof. By Definition 8.1, we have to show that
$$\lim_{\substack{h \to 0 \\ h > 0}} \frac{f(p+h) - f(p)}{h} \quad \text{and} \quad \lim_{\substack{h \to 0 \\ h < 0}} \frac{f(p+h) - f(p)}{h}$$
exist and equal. Since f is continuous at p, we have
$$\lim_{\substack{h \to 0 \\ h > 0}} f(p+h) = \lim_{\substack{h \to 0 \\ h < 0}} f(p+h) = f(p)$$
by Theorem 7.14. By Theorem 8.10 (L'Hôspital's Rule), we follow that
$$\lim_{\substack{h \to 0 \\ h > 0}} \frac{f(p+h) - f(p)}{h} = \lim_{\substack{h \to 0 \\ h > 0}} \frac{f'(p+h)}{1} = \lim_{\substack{h \to 0 \\ h > 0}} f'(p+h) = f'(p+) \tag{8.47}$$

8.4. L'Hôspital's Rule

and
$$\lim_{\substack{h\to 0 \\ h<0}} \frac{f(p+h)-f(p)}{h} = \lim_{\substack{h\to 0 \\ h<0}} \frac{f'(p+h)}{1} = \lim_{\substack{h\to 0 \\ h<0}} f'(p+h) = f'(p-). \qquad (8.48)$$

Since $\lim_{x\to p} f'(x)$ exists, we apply Theorem 7.14 to the limits (8.47) and (8.48) to conclude that
$$\lim_{\substack{h\to 0 \\ h>0}} \frac{f(p+h)-f(p)}{h} = f'(p+) = f'(p-) = \lim_{\substack{h\to 0 \\ h<0}} \frac{f(p+h)-f(p)}{h},$$
as desired. We end the proof of the problem. ∎

Problem 8.24

(★) Prove, by using Theorem 8.10 (L'Hôspital's Rule), that
$$\lim_{x\to 0} \frac{\sin x}{x} = 1.$$

Proof. If we let $f(x) = \sin x$ and $g(x) = x$, then it is easy to check that they satisfy the hypotheses of Theorem 8.10 (L'Hôspital's Rule). Since $f'(x) = \cos x$ and $g'(x) = 1$, we have
$$\lim_{x\to 0} \frac{\cos x}{1} = 1$$
which implies that our desired result, completing the proof of the problem. ∎

Problem 8.25

(★)(★) Suppose that $f : (0, +\infty) \to \mathbb{R}$ is differentiable in $(0, +\infty)$ and $\lim_{x\to +\infty} [f(x) + f'(x)] = L$. Prove that
$$\lim_{x\to +\infty} f(x) = L.$$

Proof. By the facts
$$f(x) = \frac{f(x)\mathrm{e}^x}{\mathrm{e}^x}$$
on $(0, +\infty)$, $(\mathrm{e}^x)' \neq 0$ on $(0, +\infty)$ and $\mathrm{e}^x \to +\infty$ as $x \to +\infty$, it follows from Theorem 8.11 that
$$\lim_{x\to +\infty} f(x) = \lim_{x\to +\infty} \frac{f(x)\mathrm{e}^x}{\mathrm{e}^x} = \lim_{x\to +\infty} \frac{[f(x)+f'(x)]\mathrm{e}^x}{\mathrm{e}^x} = \lim_{x\to +\infty} [f(x)+f'(x)] = L,$$
as desired. This finishes the proof of the problem. ∎

Problem 8.26

(★)(★) Suppose that $f : (0, +\infty) \to \mathbb{R}$ is differentiable in $(0, +\infty)$ and $f'(x) + xf(x)$ is bounded on $(0, +\infty)$. Prove that
$$\lim_{x\to +\infty} f(x) = 0.$$

Proof. Since $f'(x) + xf(x)$ is bounded on $(0, +\infty)$, there exists a positive constant M such that
$$|f'(x) + xf(x)| \le M \tag{8.49}$$
for all $x \in (0, +\infty)$. By the facts
$$f(x) = \frac{f(x)e^{\frac{x^2}{2}}}{e^{\frac{x^2}{2}}}$$
on $(0, +\infty)$, $(e^{\frac{x^2}{2}})' = xe^{\frac{x^2}{2}} \ne 0$ on $(0, +\infty)$ and $e^{\frac{x^2}{2}} \to +\infty$ as $x \to +\infty$, it follows from Theorem 8.11 that
$$\lim_{x \to +\infty} f(x) = \lim_{x \to +\infty} \frac{f(x)e^{\frac{x^2}{2}}}{e^{\frac{x^2}{2}}} = \lim_{x \to +\infty} \frac{[f'(x) + xf(x)]e^{\frac{x^2}{2}}}{xe^{\frac{x^2}{2}}} = \lim_{x \to +\infty} \frac{f'(x) + xf(x)}{x}.$$

By using the bound (8.49) and Theorem 7.3 (Squeeze Theorem for Limits of Functions), we know that
$$\lim_{x \to +\infty} \frac{f'(x) + xf(x)}{x} = 0$$
which then implies our desired result that
$$\lim_{x \to +\infty} f(x) = 0,$$
completing the proof of the problem. ∎

Problem 8.27

(⋆) Construct an example of functions f, g having a finite derivatives in $(0, 1)$, $g'(x) \ne 0$ on $(0, 1)$ and
$$\lim_{\substack{x \to 0 \\ x > 0}} \frac{f(x)}{g(x)} = 0,$$
but $\lim\limits_{\substack{x \to 0 \\ x > 0}} \dfrac{f'(x)}{g'(x)}$ does not exist.

Proof. Consider the functions $f(x) = \sin \frac{1}{x}$ and $g(x) = \frac{1}{x}$. It is clear that $g'(x) = -\frac{1}{x^2} \ne 0$ on $(0, 1)$. By direct computation, we have
$$\lim_{\substack{x \to 0 \\ x > 0}} \frac{f(x)}{g(x)} = \lim_{\substack{x \to 0 \\ x > 0}} \frac{\sin \frac{1}{x}}{\frac{1}{x}} = \lim_{\substack{x \to 0 \\ x > 0}} x \sin \frac{1}{x} = 0,$$
but
$$\lim_{\substack{x \to 0 \\ x > 0}} \frac{f'(x)}{g'(x)} = \lim_{\substack{x \to 0 \\ x > 0}} \frac{-x^{-2} \cos \frac{1}{x}}{-x^{-2}} = \lim_{\substack{x \to 0 \\ x > 0}} \cos \frac{1}{x}$$
which does not exist. We end the proof of the problem. ∎

8.5 Higher Order Derivatives and Taylor's Theorem

> **Remark 8.8**
>
> Problem 8.27 shows that the conditions
> $$\lim_{\substack{x\to a\\ x>a}} f(x) = \lim_{\substack{x\to a\\ x>a}} g(x) = 0 \quad \text{and} \quad \lim_{\substack{x\to a\\ x>a}} g(x) = \pm\infty$$
> in Theorem 8.10 (L'Hôspital's Rule) and Theorem 8.11 cannot be omitted respectively.

8.5 Higher Order Derivatives and Taylor's Theorem

> **Problem 8.28**
>
> (⋆) Suppose that $f : \mathbb{R} \to \mathbb{R}$ is twice differentiable in $\mathbb{R}$, $f(0) = 0$, $f(\frac{1}{2}) = \frac{1}{2}$ and $f'(0) = 0$. Prove that
> $$f''(\xi) = 4$$
> for some $\xi \in (0, \frac{1}{2})$.

Proof. Take $n = 2$, $x = \frac{1}{2}$ and $p = 0$ in Taylor's Theorem and Remark 8.5, we have
$$f\left(\frac{1}{2}\right) = f(0) + \frac{f'(0)}{2} + \frac{f''(\xi)}{8} \tag{8.50}$$
for some $\xi \in (0, \frac{1}{2})$. By the hypotheses, we obtain from the expression (8.50) that
$$\frac{1}{2} = \frac{f''(\xi)}{8}$$
which means that $f''(\xi) = 4$, completing the proof of the problem. ∎

> **Problem 8.29**
>
> (⋆) Prove Leibniz's Rule.

Proof. When $n = 1$, the formula is clear because
$$(fg)' = f'g + g'f.$$
Assume that the statement is true for $n = m$ for some positive integer m, i.e.,
$$(fg)^{(m)} = \sum_{k=0}^{m} C_k^m f^{(m-k)} g^{(k)}.$$

When $n = k+1$, we have

$$(fg)^{(m+1)} = \frac{d}{dx}\left(\sum_{k=0}^{m} C_k^m f^{(m-k)} g^{(k)}\right)$$

$$= \sum_{k=0}^{m} C_k^m \frac{d}{dx}\left(f^{(m-k)} g^{(k)}\right)$$

$$= \sum_{k=0}^{m} C_k^m [f^{(m-k+1)} g^{(k)} + f^{(m-k)} g^{(k+1)}]$$

$$= \sum_{k=0}^{m} C_k^m f^{(m-k+1)} g^{(k)} + \sum_{k=0}^{m} C_k^m f^{(m-k)} g^{(k+1)}$$

$$= \sum_{k=1}^{m} C_k^m f^{(m-k+1)} g^{(k)} + f^{(m+1)} g^{(0)} + \sum_{k=0}^{m-1} C_k^m f^{(m-k)} g^{(k+1)} + f^{(0)} g^{(m+1)}$$

$$= \sum_{k=0}^{m-1} C_{k+1}^m f^{(m-k)} g^{(k+1)} + f^{(m+1)} g^{(0)} + \sum_{k=0}^{m-1} C_k^m f^{(m-k)} g^{(k+1)} + f^{(0)} g^{(m+1)}$$

$$= \sum_{k=0}^{m-1} [C_{k+1}^m + C_k^m] f^{(m-k)} g^{(k+1)} + f^{(m+1)} g^{(0)} + f^{(0)} g^{(m+1)}$$

$$= \sum_{k=0}^{m-1} C_{k+1}^{m+1} f^{(m-k)} g^{(k+1)} + f^{(m+1)} g^{(0)} + f^{(0)} g^{(m+1)}$$

$$= \sum_{k=1}^{m} C_k^{m+1} f^{(m-k+1)} g^{(k)} + f^{(m+1)} g^{(0)} + f^{(0)} g^{(m+1)}$$

$$= \sum_{k=0}^{m+1} C_k^{m+1} f^{(m-k+1)} g^{(k)}.$$

Thus the statement is true for $n = m+1$ if it is also true for $n = m$. Hence we follow from induction that Leibniz's Rule holds for all positive integers n. ∎

Problem 8.30

(★)(★) Suppose that $f : [0,1] \to \mathbb{R}$ is a function such that f' is continuous on $[0,1]$ and f'' exists in $(0,1)$. Suppose that $f(0) = f(1)$ and there exists a $M > 0$ such that $|f''(x)| \leq M$ on $(0,1)$. Prove that

$$|f'(x)| \leq \frac{M}{2}$$

for all $x \in (0,1)$.

Proof. Let $p \in (0,1)$. By Taylor's Theorem, there exists a ξ between x and p such that

$$f(x) = f(p) + f'(p)(x-p) + \frac{f''(\xi)}{2}(x-p)^2. \qquad (8.51)$$

Putting $x = 0$ and $x = 1$ into the formula (8.51), we obtain

$$f(0) = f(p) - f'(p)p + \frac{f''(\xi_0)}{2} p^2 \qquad (8.52)$$

8.5. Higher Order Derivatives and Taylor's Theorem

and
$$f(1) = f(p) + f'(p)(1-p) + \frac{f''(\xi_1)}{2}(1-p)^2 \tag{8.53}$$

respectively, where $\xi_0 \in (0, p)$ and $\xi_1 \in (p, 1)$. By the hypothesis $f(0) = f(1)$ and considering the substraction of the formulas (8.52) and (8.53), we yield that

$$0 = f'(p) + \frac{f''(\xi_1)}{2}(1-p)^2 - \frac{f''(\xi_0)}{2}p^2$$

and then
$$f'(p) = \frac{f''(\xi_0)}{2}p^2 - \frac{f''(\xi_1)}{2}(1-p)^2. \tag{8.54}$$

Since $|f''(x)| \le M$ on $(0,1)$, we get from the expression (8.54) that

$$|f'(p)| \le \frac{M}{2}[p^2 + (1-p)^2]. \tag{8.55}$$

It is obvious that
$$p^2 + (1-p)^2 = 2\left(p - \frac{1}{2}\right)^2 + \frac{1}{2} \le 1$$

for every $p \in (0, 1)$, we deduce from the inequality (8.55) that

$$|f'(p)| \le \frac{M}{2}$$

for all $p \in (0, 1)$. This completes the proof of the problem. ∎

Problem 8.31

(⋆) Prove that for $x > 0$, we have

$$\left| \ln(1+x) - \left(x - \frac{x^2}{2} + \frac{x^3}{3}\right) \right| \le \frac{x^4}{4}. \tag{8.56}$$

Proof. Let $f(x) = \ln x$. Note that

$$f'(x) = \frac{1}{x}, \quad f''(x) = -\frac{1}{x^2}, \quad f'''(x) = \frac{2}{x^3} \quad \text{and} \quad f^{(4)}(x) = -\frac{6}{x^4}, \tag{8.57}$$

where $x > 0$. Applying Taylor's Theorem with $n = 4$, x replaced by $1 + x$ and $p = 1$, we have

$$f(1+x) = f(1) + \frac{f'(1)}{1!}x + \frac{f''(1)}{2!}x^2 + \frac{f'''(1)}{3!}x^3 + \frac{f^{(4)}(\xi)}{4!}x^4 \tag{8.58}$$

for some $\xi \in (1, 1+x)$. Using the expressions (8.57), we deduce from the formula (8.58) that

$$\ln(1+x) = 0 + x - \frac{x^2}{2} + \frac{x^3}{3} - \frac{x^4}{4\xi^4}$$

so that
$$\left| \ln(1+x) - \left(x - \frac{x^2}{2} + \frac{x^3}{3}\right) \right| = \frac{x^4}{4\xi^4}. \tag{8.59}$$

Since $\xi \ge 1$, our desired inequality (8.56) follows immediately from the expression (8.59). This completes the proof of the problem. ∎

> **Problem 8.32**
>
> (★)(★) Suppose that $f : [a,b] \to \mathbb{R}$ is twice differentiable and $f'(a) = f'(b) = 0$. Prove that there exist $\xi_1, \xi_2 \in (a,b)$ such that
> $$f(b) - f(a) = \frac{1}{2} \times \left(\frac{b-a}{2}\right)^2 [f''(\xi_1) - f''(\xi_2)].$$

Proof. By Taylor's Theorem with $n = 2$, $x = \frac{a+b}{2}$ and $p = a$, we have

$$f\left(\frac{a+b}{2}\right) = f(a) + f'(a)\left(\frac{b-a}{2}\right) + \frac{f''(\xi_1)}{2}\left(\frac{b-a}{2}\right)^2 \tag{8.60}$$

for some $\xi_1 \in (a, \frac{a+b}{2})$. Similarly, we apply Taylor's Theorem with $n = 2$, $x = \frac{a+b}{2}$ and $p = b$ to get

$$f\left(\frac{a+b}{2}\right) = f(b) - f'(b)\left(\frac{b-a}{2}\right) + \frac{f''(\xi_2)}{2}\left(\frac{b-a}{2}\right)^2 \tag{8.61}$$

for some $\xi_2 \in (\frac{a+b}{2}, b)$. Since $f'(a) = f'(b) = 0$, it follows from the subtraction of the expressions (8.60) and (8.61) that

$$f(b) - f(a) = \frac{1}{2} \times \left(\frac{b-a}{2}\right)^2 [f''(\xi_1) - f''(\xi_2)],$$

finishing the proof of the problem. ∎

> **Problem 8.33**
>
> (★)(★)(★) Suppose that $f : \mathbb{R} \to \mathbb{R}$ is a twice differentiable function such that $f''(x) \le 0$ on $\mathbb{R}$. Prove that f is a constant function if f is bounded.

Proof. Assume that $f'(p) > 0$ for some $p \in \mathbb{R}$. Take $x > p$. Then Taylor's Theorem shows that there exists a $\xi \in (p, x)$ such that

$$f(x) = f(p) + f'(p)(x-p) + \frac{(x-p)^2}{2} f''(\xi). \tag{8.62}$$

Since $f''(x) \le 0$ on $\mathbb{R}$, we know from the expression (8.62) that

$$f(x) \ge f(p) + f'(p)(x - p). \tag{8.63}$$

Since $f'(p) > 0$, it follows from the inequality (8.63) that

$$\lim_{x \to +\infty} f(x) = +\infty,$$

a contradiction. Thus we must have
$$f'(x) \le 0$$
for all $x \in \mathbb{R}$.

8.6. Convexity and Derivatives 153

Next, we assume that $f'(p) < 0$ for some $p \in \mathbb{R}$. Take $x < p$ and we apply Taylor's Theorem again to ensure the existence of a $\theta \in (x, p)$ such that

$$f(x) = f(p) + f'(p)(x-p) + \frac{(x-p)^2}{2}f''(\theta). \tag{8.64}$$

Since $f''(x) \leq 0$ on $\mathbb{R}$, we derive from the expression (8.64) that

$$f(x) \geq f(p) + f'(p)(x-p). \tag{8.65}$$

We note that $x - p < 0$ so that the inequality (8.65) implies that

$$\lim_{x \to -\infty} f(x) = +\infty,$$

a contradiction again. Thus we have $f'(x) \geq 0$ for all $x \in \mathbb{R}$ and hence

$$f'(x) = 0$$

on $\mathbb{R}$. By Theorem 8.9(c), we conclude that f is a constant function, completing the proof of the problem. ∎

8.6 Convexity and Derivatives

Problem 8.34

(★)(★) Suppose that f is a real function defined in (a, b) and

$$\begin{vmatrix} 1 & 1 & 1 \\ x & y & z \\ f(x) & f(y) & f(z) \end{vmatrix} \geq 0$$

for every $x, y, z \in (a, b)$ and $x < y < z$. Prove that f is convex on (a, b).

Proof. Fix $x, y \in (a, b)$ and take $t \in (0, 1)$. Since $x < tx + (1-t)y < y$, we have

$$\begin{vmatrix} 1 & 1 & 1 \\ x & tx + (1-t)y & y \\ f(x) & f(tx + (1-t)y) & f(y) \end{vmatrix} \geq 0. \tag{8.66}$$

By using properties of determinants, the determinant on the left-hand side in (8.66) becomes

$$\begin{vmatrix} 1 & 0 & 1 \\ x & 0 & y \\ f(x) & f(tx+(1-t)y) - tf(x) - (1-t)f(y) & f(y) \end{vmatrix}$$

$$= -[f(tx+(1-t)y) - tf(x) - (1-t)f(y)] \begin{vmatrix} 1 & 1 \\ x & y \end{vmatrix}$$

$$= -(y-x)[f(tx+(1-t)y) - tf(x) - (1-t)f(y)]. \tag{8.67}$$

By combining the inequality (8.66), the expression (8.67) and using the fact that $x < y$, we obtain
$$f(tx + (1-t)y) - tf(x) - (1-t)f(y) \le 0$$
and this is equivalent to
$$f(tx + (1-t)y) \le tf(x) + (1-t)f(y) \tag{8.68}$$
for all $t \in (0,1)$. Since the equality in (8.68) holds trivially when $t = 0$ or $t = 1$, by Definition 8.13 (Convex Functions), f is convex on (a,b). This ends the proof of the problem. ∎

Problem 8.35

(★) Prove that the function $f : (0, \pi) \to \mathbb{R}$ defined by
$$f(x) = \frac{1}{\sin \frac{x}{2}}$$
is convex on $(0, \pi)$.

Proof. By direct differentiation, we have
$$f'(x) = -\frac{\cos \frac{x}{2}}{2\sin^2 \frac{x}{2}} \quad \text{and} \quad f''(x) = \frac{1 + \cos^2 \frac{x}{2}}{4\sin^3 \frac{x}{2}}.$$
For every $x \in (0, \pi)$, $\sin \frac{x}{2} > 0$ and $\cos \frac{x}{2} > 0$. Thus we have
$$f''(x) > 0$$
for all $x \in (0, \pi)$. By Theorem 8.16, f is (strictly) convex on $(0, \pi)$. This ends the proof of the problem. ∎

Problem 8.36

(★)(★) Suppose that $f : (a,b) \to \mathbb{R}$ is a convex function. Let $x_1, x_2, \ldots, x_n \in (a,b)$ and $\alpha_1, \alpha_2, \ldots, \alpha_n$ are nonnegative constants such that $\alpha_1 + \alpha_2 + \cdots + \alpha_n = 1$. Prove that
$$f(\alpha_1 x_1 + \cdots + \alpha_n x_n) \le \alpha_1 f(x_1) + \cdots + \alpha_n f(x_n). \tag{8.69}$$

Proof. Since $\alpha_1, \alpha_2, \ldots, \alpha_n$ are nonnegative constants such that $\alpha_1 + \alpha_2 + \cdots + \alpha_n = 1$, we have
$$\alpha_1 x_1 + \cdots + \alpha_n x_n \in (a,b).$$
It is clear that the inequality (8.69) is true for $n = 1$. Assume that the statement is true for $n = m$ for some positive integer m, i.e.,
$$f(\alpha_1 x_1 + \cdots + \alpha_m x_m) \le \alpha_1 f(x_1) + \cdots + \alpha_m f(x_m). \tag{8.70}$$

8.6. Convexity and Derivatives

For $n = m+1$, suppose that $x_1, \ldots, x_m, x_{m+1} \in (a,b)$ and $\alpha_1, \ldots, \alpha_m, \alpha_{m+1}$ are nonnegative constants such that $\alpha_1 + \cdots + \alpha_m + \alpha_{m+1} = 1$. Without loss of generality, we may assume that $0 < \alpha_{m+1} < 1$ and
$$\beta = \alpha_2 + \cdots + \alpha_{m+1} > 0.$$
Now it is evident that
$$\frac{\alpha_2}{\beta} + \cdots + \frac{\alpha_{m+1}}{\beta} = 1, \quad \alpha_1 + \beta = 1 \quad \text{and} \quad \frac{\alpha_2}{\beta}x_2 + \cdots + \frac{\alpha_{m+1}}{\beta}x_{m+1} \in (a,b).$$
Since f is convex on (a,b), we have
$$f(\alpha_1 x_1 + \cdots + \alpha_m x_m + \alpha_{m+1} x_{m+1}) = f\left(\alpha_1 x_1 + \beta\left(\frac{\alpha_2}{\beta}x_2 + \cdots + \frac{\alpha_{m+1}}{\beta}x_{m+1}\right)\right)$$
$$\leq \alpha_1 f(x_1) + \beta f\left(\frac{\alpha_2}{\beta}x_2 + \cdots + \frac{\alpha_{m+1}}{\beta}x_{m+1}\right). \quad (8.71)$$

Applying the assumption (8.70) to the second term of the right-hand side of the inequality (8.71), we obtain
$$f\left(\frac{\alpha_2}{\beta}x_2 + \cdots + \frac{\alpha_{m+1}}{\beta}x_{m+1}\right) \leq \frac{\alpha_2}{\beta}f(x_2) + \cdots + \frac{\alpha_{m+1}}{\beta}f(x_{m+1}). \quad (8.72)$$

By putting the inequality (8.72) back into the inequality (8.71), we conclude that
$$f(\alpha_1 x_1 + \cdots + \alpha_m x_m + \alpha_{m+1} x_{m+1}) \leq \alpha_1 f(x_1) + \beta\left[\frac{\alpha_2}{\beta}f(x_2) + \cdots + \frac{\alpha_{m+1}}{\beta}f(x_{m+1})\right]$$
$$= \alpha_1 f(x_1) + \cdots + \alpha_m f(x_m) + \alpha_{m+1} f(x_{m+1}).$$

Thus the statement is true for $n = m+1$ if it is true for $n = m$. Hence it follows from induction that the inequality (8.69) holds for all positive integers n. We have completed the proof of the problem. ∎

Remark 8.9

The inequality (8.69) is called **Jensen's inequality**.

Problem 8.37

(★)(★) Suppose that $f(x) = -\ln x$. Prove that
$$x_1^{\alpha_1} \cdots x_n^{\alpha_n} \leq \alpha_1 x_1 + \cdots + \alpha_n x_n$$
for all $x_1, \ldots, x_n \geq 0$ and $\alpha_1, \ldots, \alpha_n \geq 0$ with $\alpha_1 + \cdots + \alpha_n = 1$.

Proof. The inequality holds trivially when one of $x_1, \ldots, x_n$ is zero. Without loss of generality, we may assume that $x_1, \ldots, x_n > 0$. Since $f''(x) = \frac{1}{x^2} > 0$ for all $x > 0$, we deduce from Theorem 8.16 that f is convex on $(0, +\infty)$. By Problem 8.36, we have
$$-\ln(\alpha_1 x_1 + \cdots + \alpha_n x_n) \leq -\alpha_1 \ln x_1 - \cdots - \alpha_n \ln x_n$$
which implies that
$$x_1^{\alpha_1} \cdots x_n^{\alpha_n} \leq \alpha_1 x_1 + \cdots + \alpha_n x_n$$
as desired. This completes the proof of the problem. ∎

CHAPTER 9

The Riemann-Stieltjes Integral

There are two components in single variable calculus. One is differentiation which is reviewed in the previous chapter. The other one is **integration** which is the emphasis in this chapter. More precisely, we study properties of **definite integrals** (the integral of a real-valued function on a bounded interval) and the connection between derivatives and integrals. The main references for this part are [2, Chap. 1, 2 & 5], [3, Chap. 7], [5, Chap. 7], [6, Chap. 6], [23, Chap. 6] and [27, Chap. 11].

9.1 Fundamental Concepts

9.1.1 Definitions and Notations

A **partition** P of $[a,b]$ is a *finite* set of points, namely

$$P = \{x_0, x_1, \ldots, x_n\},$$

where $a = x_0 < x_1 < \cdots < x_{n-1} < x_n = b$. A partition P^* is called a **refinement** of P if $P \subseteq P^*$, i.e., every point of P is a point of P^*. If $\alpha : [a,b] \to \mathbb{R}$ is monotonically increasing, then we define, for each $k = 1, 2, \ldots, n$,

$$\Delta\alpha_k = \alpha(x_k) - \alpha(x_{k-1}) \qquad (9.1)$$

so that

$$\sum_{k=1}^{n} \Delta\alpha_k = \alpha(b) - \alpha(a).$$

Since α is monotonically increasing, it is trivial from the definition (9.1) that $\Delta\alpha_k \geq 0$ for each $k = 1, 2, \ldots, n$.

Let $f : [a,b] \to \mathbb{R}$ be bounded and $I_k = [x_{k-1}, x_k]$, where $k = 1, 2, \ldots, n$. Now for each $k = 1, 2, \ldots, n$, we define

$$M_k = \sup_{x \in I_k} f(x) \quad \text{and} \quad m_k = \inf_{x \in I_k} f(x). \qquad (9.2)$$

Definition 9.1 (The Riemann-Stieltjes Integral). *Suppose that $f : [a,b] \to \mathbb{R}$ is bounded and P is a partition of $[a,b]$. Furthermore, suppose that*

$$U(P,f,\alpha) = \sum_{k=1}^{n} M_k \Delta \alpha_k \quad \text{and} \quad L(P,f,\alpha) = \sum_{k=1}^{n} m_k \Delta \alpha_k,$$

where M_k and m_k are defined in (9.2). Finally, we define the two numbers

$$\overline{\int_a^b} f \,\mathrm{d}\alpha = \inf_P U(P,f,\alpha) \quad \text{and} \quad \underline{\int_a^b} f \,\mathrm{d}\alpha = \sup_P L(P,f,\alpha), \tag{9.3}$$

where the sup and inf in the numbers (9.3) take over all partitions P of $[a,b]$. If we have

$$\overline{\int_a^b} f \,\mathrm{d}\alpha = \underline{\int_a^b} f \,\mathrm{d}\alpha,$$

then we simply write the number as

$$\int_a^b f \,\mathrm{d}\alpha \quad \text{or} \quad \int_a^b f(x) \,\mathrm{d}\alpha(x) \tag{9.4}$$

*and we say that f is **integrable with respect to α in the Riemann sense** and write "$f \in \mathscr{R}(\alpha)$ on $[a,b]$". In this case, we call the number (9.4) the **Riemann-Stieltjes integral** of f with respect to α on $[a,b]$.*

Particularly, if we take $\alpha(x) = x$, then we write $U(P,f)$, $L(P,f)$ and $f \in \mathscr{R}$ instead of $U(P,f,\alpha)$, $L(P,f,\alpha)$ and $f \in \mathscr{R}(\alpha)$ respectively. In this case, the numbers (9.3)

$$\overline{\int_a^b} f \,\mathrm{d}x = \inf_P U(P,f) \quad \text{and} \quad \underline{\int_a^b} f \,\mathrm{d}x = \sup_P L(P,f)$$

are said to be the **upper Riemann integral** and the **lower Riemann integral** of f on $[a,b]$ respectively. Furthermore, the integral (9.4) is then called the **Riemann integral** and is denoted by

$$\int_a^b f(x) \,\mathrm{d}x.$$

Remark 9.1

The number (9.4) depends **only** on a, b, f and α, but *not* the variable of integration x. In fact, the symbol x is a "dummy variable" and may be replaced by other symbols.

Two particular examples should be mentioned. It can be easily seen from Definition 9.1 (The Riemann-Stieltjes Integral) that

$$\int_a^b 1 \,\mathrm{d}\alpha = \alpha(b) - \alpha(a) \quad \text{and} \quad \int_a^b 0 \,\mathrm{d}\alpha = 0. \tag{9.5}$$

9.1.2 Criteria for Integrability of Real Functions and their Properties

Theorem 9.2 (The Riemann Integrability Condition). *We have $f \in \mathscr{R}(\alpha)$ on $[a,b]$ if and only if for every $\epsilon > 0$, there exists a partition P such that*

$$U(P,f,\alpha) - L(P,f,\alpha) < \epsilon. \tag{9.6}$$

In addition, the inequality (9.6) holds for every refinement of P.

Although the above result provides us a way to test whether f is integrable with respect to α on $[a,b]$ or not by simply checking the inequality (9.5), it is desirable to know some sufficient conditions for $f \in \mathscr{R}(\alpha)$ on $[a,b]$ and the following result serves this purpose.

Theorem 9.3. *Suppose that $f : [a,b] \to \mathbb{R}$ is bounded and $\alpha : [a,b] \to \mathbb{R}$ is monotonically increasing.*

(a) *If f is continuous on $[a,b]$, then we have $f \in \mathscr{R}(\alpha)$ on $[a,b]$.*

(b) *If f is monotonic on $[a,b]$ and α is continuous on $[a,b]$, then $f \in \mathscr{R}(\alpha)$ on $[a,b]$.*

(c) *If f has finitely many points of discontinuity on $[a,b]$ and α is continuous at every point at which f is discontinuous, then we have $f \in \mathscr{R}(\alpha)$ on $[a,b]$.*

Recall from Theorem 7.19 (Froda's Theorem) that the set of discontinuities of a monotonic function is **at most countable**. Thus Theorem 9.3 suggests a connection of the integrability of f with respect to α and the cardinalities of the discontinuities of f and α exists. In fact, if we employ a concept of Lebesgue's measure, then we can obtain something stronger for Riemann integrals.

Theorem 9.4 (The Lebesgue's Integrability Condition). *We have $f \in \mathscr{R}$ on $[a,b]$ if and only if the set of discontinuous points of f in $[a,b]$ is of **measure zero**.*

For the definition of a set of measure zero, please refer to [23, §11.4, 11.11, pp. 302, 303, 309]. In particular, any countable set is of measure zero, so it follows from Theorem 9.4 (The Lebesgue's Integrability Condition) that if $f : [a,b] \to \mathbb{R}$ is a bounded function whose points of discontinuity form a countable set, then $f \in \mathscr{R}$ on $[a,b]$.

Theorem 9.5 (Composition Theorem). *Suppose that $m \leq f(x) \leq M$ on $[a,b]$ for some constants M and m, $f \in \mathscr{R}(\alpha)$ on $[a,b]$ and $g : [m,M] \to \mathbb{R}$ is continuous. Then we have*

$$h = g \circ f \in \mathscr{R}(\alpha)$$

on $[a,b]$.

Theorem 9.6 (Operations of Integrable Functions).

(a) *If $f, g \in \mathscr{R}(\alpha)$ on $[a,b]$ and $A, B \in \mathbb{R}$, then $Af + Bg \in \mathscr{R}(\alpha)$ on $[a,b]$ and*

$$\int_a^b (Af + Bg) \, d\alpha = A \int_a^b f \, d\alpha + B \int_a^b g \, d\alpha.$$

(b) If $f, g \in \mathscr{R}(\alpha)$ on $[a,b]$ and $f(x) \leq g(x)$ on $[a,b]$, then we have
$$\int_a^b f\,d\alpha \leq \int_a^b g\,d\alpha.$$

(c) If $f \in \mathscr{R}(\alpha)$ on $[a,b]$ and $a < c < b$, then we have $f \in \mathscr{R}(\alpha)$ on $[a,c]$ and $[c,b]$. Besides, we have
$$\int_a^b f\,d\alpha = \int_a^c f\,d\alpha + \int_c^b f\,d\alpha.$$

(d) If $f \in \mathscr{R}(\alpha)$ and $f \in \mathscr{R}(\beta)$ on $[a,b]$ and $A, B \in \mathbb{R}^+$, then we have $f \in \mathscr{R}(A\alpha + B\beta)$ and
$$\int_a^b f\,d(A\alpha + B\beta) = A\int_a^b f\,d\alpha + B\int_a^b f\,d\beta.$$

(e) If $f, g \in \mathscr{R}(\alpha)$ on $[a,b]$, then $fg \in \mathscr{R}(\alpha)$ on $[a,b]$.

(f) If $f \in \mathscr{R}(\alpha)$ on $[a,b]$, then $|f| \in \mathscr{R}(\alpha)$ on $[a,b]$ and
$$\left|\int_a^b f\,d\alpha\right| \leq \int_a^b |f|\,d\alpha.$$

Apart from showing $f \in \mathscr{R}(\alpha)$ on $[a,b]$, people are also interested in evaluating the exact value of the number (9.4). To this end, we need the concept of the **unit step function**[a] I whose definition is given by
$$I(x) = \begin{cases} 0, & \text{if } x \leq 0; \\ 1, & \text{if } x > 0. \end{cases}$$

Then we have the following result:

Theorem 9.7. Suppose that $\{c_n\}$ is a sequence of nonnegative numbers and $\{s_n\}$ is a sequence of **distinct** points in (a,b). If $\sum_{n=1}^{\infty} c_n$ converges, $f : [a,b] \to \mathbb{R}$ is continuous and
$$\alpha(x) = \sum_{n=1}^{\infty} c_n I(x - s_n),$$
then we have
$$\int_a^b f\,d\alpha = \sum_{n=1}^{\infty} c_n f(s_n).$$

9.1.3 The Substitution Theorem and the Change of Variables Theorem

The Substitution Theorem. Suppose that $\alpha : [a,b] \to \mathbb{R}$ is monotonically increasing and $\alpha' \in \mathscr{R}$ on $[a,b]$. Then $f \in \mathscr{R}(\alpha)$ on $[a,b]$ if and only if $f\alpha' \in \mathscr{R}$ on $[a,b]$. Furthermore, we have
$$\int_a^b f\,d\alpha = \int_a^b f(x)\alpha'(x)\,dx.$$

---
[a] Or called Heaviside step function.

9.1. Fundamental Concepts

The Change of Variables Theorem. *Suppose that $\varphi : [A, B] \to [a, b]$ is a strictly increasing continuous onto function. Furthermore, $\alpha : [a, b] \to \mathbb{R}$ is monotonically increasing on $[a, b]$ and $f \in \mathscr{R}(\alpha)$ on $[a, b]$. We define $\beta, g : [A, B] \to \mathbb{R}$ by*

$$\beta = \alpha \circ \varphi \quad \text{and} \quad g = f \circ \varphi.$$

Then we have $g \in \mathscr{R}(\beta)$ and

$$\int_a^b f \, d\alpha = \int_A^B g \, d\beta.$$

> **Remark 9.2**
>
> Both the Substitution Theorem and the Change of Variables Theorem provide us some convenient methods to *evaluate* the integral
>
> $$\int_a^b f \, d\alpha.$$
>
> Furthermore, we note from Remark 7.7 that φ is actually one-to-one.

9.1.4 The Fundamental Theorem of Calculus

There is a close connection between the concepts of differentiation and integration. In fact, there are two important and useful results related to this connection and they are classically combined and called the **First Fundamental Theorem of Calculus** and the **Second Fundamental Theorem of Calculus**.

The First Fundamental Theorem of Calculus. *Let $f \in \mathscr{R}$ on $[a, b]$ and $a \le x \le b$. Define*

$$F(x) = \int_a^x f(t) \, dt.$$

Then $F : [a, b] \to \mathbb{R}$ is continuous. In addition, if f is continuous at $p \in [a, b]$, then F is differentiable at p and

$$F'(p) = f(p)$$

holds.

The Second Fundamental Theorem of Calculus. *Let $f \in \mathscr{R}$ on $[a, b]$. If $F : [a, b] \to \mathbb{R}$ is a differentiable function such that $F' = f$, then we have*

$$\int_a^b f(x) \, dx = F(b) - F(a).$$

> **Remark 9.3**
>
> An **antiderivative** or a **primitive function** of a function f is a differentiable function F such that $F' = f$. Then the First Fundamental Theorem of Calculus implies the existence of antiderivatives for continuous functions and the Second Fundamental Theorem of Calculus comes up with a practical way of evaluating the integral by using a antiderivative F of f explicitly.

As an immediate application of the Second Fundamental Theorem of Calculus, we have the following important and practical skill in integral calculus: the **Integration by Parts**.

The Integration by Parts. *Suppose that $F, G : [a, b] \to \mathbb{R}$ are differentiable functions. Furthermore, we suppose that $F', G' \in \mathscr{R}$ on $[a, b]$. Then the following formula holds*

$$\int_a^b F(x)G'(x)\,dx = F(b)G(b) - F(a)G(a) - \int_a^b F'(x)G(x)\,dx.$$

9.1.5 The Mean Value Theorems for Integrals

In §8.1.4, we discuss the Mean Value Theorem for Derivatives. In integral calculus, we also have mean value theorems and the key message of one of them is to guarantee the *existence* of a rectangle with the same area and width.

The First Mean Value Theorem for Integrals. *Suppose that $f : [a, b] \to \mathbb{R}$ is a continuous function. Then there exists a $p \in (a, b)$ such that*

$$\int_a^b f(x)\,dx = f(p)(b - a).$$

The Second Mean Value Theorem for Integrals. *Suppose that $f : [a, b] \to \mathbb{R}$ is monotonic increasing. Let $A \leq f(a+)$ and $B \geq f(b-)$. If $g : [a, b] \to \mathbb{R}$ is continuous on $[a, b]$, then there exists a $p \in [a, b]$ such that*

$$\int_a^b f(x)g(x)\,dx = A\int_a^p g(x)\,dx + B\int_p^b g(x)\,dx.$$

In particular, if $f(x) \geq 0$ on $[a, b]$, then we have

$$\int_a^b f(x)g(x)\,dx = B\int_p^b g(x)\,dx$$

for some $p \in [a, b]$.

> **Remark 9.4**
>
> The particular case of the Second Mean Value Theorem for Integrals is also known as **Bonnet's Theorem**.

9.2 Integrability of Real Functions

> **Problem 9.1**
>
> (★) If $\alpha \equiv 0$ on $[a, b]$ and $f \in \mathscr{R}(\alpha)$ on $[a, b]$, prove that
>
> $$\int_a^b f\,d\alpha = 0.$$

9.2. Integrability of Real Functions

Proof. Since $\alpha \equiv 0$ on $[a,b]$, we have $\Delta \alpha_k = 0$ by the expression (9.1). By Definition 9.1 (The Riemann-Stieltjes Integral), we get

$$U(P, f, \alpha) = L(P, f, \alpha) = 0$$

so that

$$\overline{\int_a^b} f \, d\alpha = \underline{\int_a^b} f \, d\alpha = 0.$$

Hence, it follows from the expression (9.4) that

$$\int_a^b f \, d\alpha = 0.$$

This completes the proof of the problem. ∎

Problem 9.2

(⋆) *Suppose that $\theta \in \mathbb{R}$ and $f_\theta : [a,b] \to \mathbb{R}$ is defined by*

$$f_\theta(x) = \begin{cases} 1, & \text{if } x \in (a,b]; \\ \theta, & \text{if } x = a. \end{cases}$$

Prove that $f_\theta \in \mathscr{R}$ on $[a,b]$ for all $\theta \in \mathbb{R}$.

Proof. If $\theta = 1$, then we have $f_1 \equiv 1$ on $[a,b]$. By Theorem 9.3(a), it is clear that $f_1 \in \mathscr{R}$ on $[a,b]$. If $\theta \neq 1$, then f has *only* one point of discontinuity at $x = a$. By Theorem 9.3(c), we see that $f_\theta \in \mathscr{R}$ on $[a,b]$ in this case. Hence we have $f_\theta \in \mathscr{R}$ on $[a,b]$ for all $\theta \in \mathbb{R}$, completing the proof of the problem. ∎

Problem 9.3

(⋆) *Suppose that for every monotonic decreasing function $f : [a,b] \to \mathbb{R}$, we have*

$$\int_a^b f \, d\alpha = 0.$$

Prove that α is a constant function on $[a,b]$.

Proof. Suppose that $p \in (a,b)$ and define

$$f(x) = \begin{cases} 1, & \text{if } a \leq x \leq p; \\ 0, & \text{if } p < x \leq b. \end{cases}$$

Since f is monotonic decreasing on $[0,1]$, it follows from the hypothesis that

$$\int_a^b f \, d\alpha = 0. \tag{9.7}$$

By Theorem 9.6(c) (Operations of Integrable Functions) and the examples (9.5), the equation (9.7) implies that

$$\int_a^p f\,d\alpha + \int_p^b f\,d\alpha = 0$$
$$\alpha(p) - \alpha(a) + 0 = 0$$
$$\alpha(p) = \alpha(a) \tag{9.8}$$

for every $p \in (a,b)$. Besides, if we take $f(x) = 1$ for all $x \in [a,b]$, then our hypothesis and the examples (9.5) again show that

$$\alpha(b) - \alpha(a) = \int_a^b d\alpha = 0$$

or equivalently

$$\alpha(b) = \alpha(a). \tag{9.9}$$

Combining the expressions (9.8) and (9.9), we conclude that α is a constant function on $[a,b]$, finishing the proof of the problem. ∎

Problem 9.4

(★) Suppose that $\alpha : [a,b] \to \mathbb{R}$ is a monotonically increasing function and $\mathscr{R}_\alpha[a,b]$ is the set of all bounded functions which are Riemann-Stieltjes integrable with respect to α on $[a,b]$. Prove that $\mathscr{R}_\alpha[a,b]$ is a vector space.

Proof. Recall that a set V is a **vector space** if

$$x + y \in V \quad \text{and} \quad cx \in V$$

for every $x, y \in V$ and scalar c. Then our desired result follows immediately from Theorem 9.6 (Operations of Integrable Functions), completing the proof of the problem. ∎

Problem 9.5

(★)(★) Let $f : [a,b] \to \mathbb{R}$ be a function and $p \in (a,b)$. Suppose that $\alpha : [a,b] \to \mathbb{R}$ is monotonically increasing. Suppose, further that, there exists a $\epsilon > 0$ such that for every $\delta > 0$, we have $x, y \in (p, p+\delta)$ such that

$$|f(x) - f(p)| \geq \epsilon \quad \text{and} \quad \alpha(y) - \alpha(p) \geq \epsilon. \tag{9.10}$$

Prove that $f \notin \mathscr{R}(\alpha)$ on $[a,b]$.

Proof. Without loss of generality, we may assume that $\epsilon = 1$ in the inequalities (9.10). Suppose that $P = \{x_0, x_1, \ldots, x_n\}$ is a partition of $[a,b]$ such that $p = x_{j-1}$ for some $j = 2, 3, \ldots, n$. Consider the difference

$$U(P, f, \alpha) - L(P, f, \alpha) = \sum_{k=1}^n (M_k - m_k)\Delta\alpha_k. \tag{9.11}$$

9.2. Integrability of Real Functions

Since we always have $M_k - m_k \geq 0$ for all $k = 1, 2, \ldots, n$ and $p = x_{j-1}$, we deduce from the expression (9.11) that

$$U(P, f, \alpha) - L(P, f, \alpha) \geq (M_j - m_j)\Delta\alpha_j$$
$$= (M_j - m_j)[\alpha(x_j) - \alpha(p)]. \tag{9.12}$$

By the assumption (9.10), we may take $x_j = y$ so that

$$\alpha(x_j) - \alpha(p) \geq 1.$$

It is clear from the other assumption (9.10) that

$$M_j - m_j \geq |f(x) - f(p)| \geq 1$$

for every $x \in (p, x_j]$. In fact, the inequality $M_j - m_j \geq 1$ also holds on $[p, x_j]$, so we obtain from the inequality (9.12) that

$$U(P, f, \alpha) - L(P, f, \alpha) \geq 1. \tag{9.13}$$

Assume that $f \in \mathscr{R}(\alpha)$ on $[a, b]$. By Theorem 9.2 (The Riemann Integrability Condition), we must have

$$U(P, f, \alpha) - L(P, f, \alpha) < 1 \tag{9.14}$$

for some partition P of $[a, b]$. Since the inequality (9.14) also holds for any refinement P^* of P, we may assume that $p \in P$. However, the inequality (9.14) will contradict the inequality (9.13) in this case. Hence we have shown that $f \notin \mathscr{R}(\alpha)$ on $[a, b]$ which completes the proof of the problem. ∎

Remark 9.5

The conditions (9.10) mean that both f and α are discontinuous from the right at p. Similarly, we can show that $f \notin \mathscr{R}(\alpha)$ on $[a, b]$ if they are discontinuous from the left at p.

Problem 9.6

(★) Recall the Dirichlet function $D(x)$ is given by

$$D(x) = \begin{cases} 1, & \text{if } [a,b] \cap \mathbb{Q}; \\ 0, & \text{otherwise.} \end{cases} \tag{9.15}$$

Prove that $D(x) \notin \mathscr{R}$ on $[a, b]$.

Proof. Let $P = \{x_0, x_1, \ldots, x_n\}$ be a partition of $[a, b]$. By the definition (9.15), we know that

$$M_k = \sup_{x \in I_k} D(x) = 1 \quad \text{and} \quad m_k = \inf_{x \in I_k} D(x) = 0,$$

where $I_k = [x_{k-1}, x_k]$ and $k = 1, 2, \ldots, n$. Thus we obtain from Definition 9.1 (The Riemann-Stieltjes Integral) that

$$U(P, D(x)) = b - a \quad \text{and} \quad L(P, D(x)) = 0$$

which imply that
$$\overline{\int_a^b} D(x)\,\mathrm{d}x = b - a \ne 0 = \underline{\int_a^b} D(x)\,\mathrm{d}x.$$
Hence we obtain $D(x) \notin \mathscr{R}$ on $[a,b]$ and we finish the proof of the problem. ∎

Problem 9.7

(⋆) Construct a function $f : [a,b] \to \mathbb{R}$ such that $|f| \in \mathscr{R}$ on $[a,b]$, but $f \notin \mathscr{R}$ on $[a,b]$.

Proof. Consider the function $f : [a,b] \to \mathbb{R}$ defined by
$$f(x) = \begin{cases} 1, & \text{if } x \in [a,b] \cap \mathbb{Q}; \\ -1, & \text{otherwise.} \end{cases}$$
Since $|f| \equiv 1$ on $[a,b]$, we must have $|f| \in \mathscr{R}$ on $[a,b]$. Assume that $f \in \mathscr{R}$ on $[a,b]$. Since
$$D(x) = \frac{1}{2}[f(x) + 1],$$
Theorem 9.6(a) (Operations of Integrable Functions) shows that $D(x) \in \mathscr{R}$ on $[a,b]$ which contradicts Problem 9.6. ∎

Problem 9.8

(⋆) Suppose that $f : [0,1] \to \mathbb{R}$ and $g : [0,1] \to [0,1]$ are functions such that $f, g \in \mathscr{R}$ on $[0,1]$. Prove or disprove $f \circ g \in \mathscr{R}$ on $[0,1]$.

Proof. Define $f : [0,1] \to \mathbb{R}$ by
$$f(x) = \begin{cases} 1, & \text{if } x \in (0,1]; \\ 0, & \text{if } x = 0. \end{cases}$$
By Theorem 9.3(c), we have $f \in \mathscr{R}$ on $[0,1]$. By [23, Exercise 18, p. 100], we know that the function $g : [0,1] \to [0,1]$ defined by[b]
$$g(x) = \begin{cases} \frac{1}{n}, & \text{if } x = \frac{m}{n}, m \in \mathbb{Z}, n \in \mathbb{N}, m \text{ and } n \text{ are coprime}, x \in [0,1]; \\ 0, & \text{otherwise} \end{cases} \quad (9.16)$$
is continuous at every irrational point of $[0,1]$ and discontinuous at every rational point of $[0,1]$. We have to show that $g \in \mathscr{R}$ on $[0,1]$. Since the set of all rational points of $[0,1]$ is countable, it is of measure zero. By Theorem 9.4 (The Lebesgue's Integrability Condition), we see that $g \in \mathscr{R}$ on $[0,1]$. However, it is easy to check that
$$D(x) = f(g(x))$$
on $[0,1]$, where D is the Dirichlet function. By Problem 9.6, we know that $D(x) \notin \mathscr{R}$ on $[0,1]$. This ends the proof of the problem. ∎

---

[b] We notice that 1 is the only positive integer which is coprime to 0.

9.2. Integrability of Real Functions

> **Remark 9.6**
>
> The function $g(x)$ defined in (9.16) is called the **Riemann function**, the **Thomae's function**, the **popcorn function** or the **ruler function**.

> **Problem 9.9**
>
> (★)(★) *Suppose that $f : [a,b] \to \mathbb{R}$ is bounded and there exists a sequence $\{P_n\}$ of partitions of $[a,b]$ such that*
> $$\lim_{n\to\infty}[U(P_n,f) - L(P_n,f)] = 0. \qquad (9.17)$$
> *Prove that $f \in \mathscr{R}$ on $[a,b]$ and*
> $$\int_a^b f(x)\,\mathrm{d}x = \lim_{n\to\infty} U(P_n,f) = \lim_{n\to\infty} L(P_n,f).$$

Proof. Given $\epsilon > 0$. By the hypothesis (9.17), there exists a positive integer N such that $n \geq N$ implies
$$U(P_n,f) - L(P_n,f) = |U(P_n,f) - L(P_n,f) - 0| < \epsilon. \qquad (9.18)$$

Thus it follows from Theorem 9.2 (The Riemann Integrability Condition) that $f \in \mathscr{R}$ on $[a,b]$. For the second assertion, we obtain from the paragraph following Definition 9.1 (The Riemann-Stieltjes Integral) that

$$L(P_n,f) \leq \sup_P L(P,f) = \int_a^b f(x)\,\mathrm{d}x = \inf_P U(P,f) \leq U(P_n,f).$$

Therefore, we establish from this and the inequality (9.18) that

$$\left| U(P_n,f) - \int_a^b f(x)\,\mathrm{d}x \right| \leq |U(P_n,f) - L(P_n,f)| < \epsilon$$

for every $n \geq N$. Thus we have

$$\int_a^b f(x)\,\mathrm{d}x = \lim_{n\to\infty} U(P_n,f). \qquad (9.19)$$

Similarly, we can show that the expression (9.19) also holds when $U(P_n,f)$ is replaced by $L(P_n,f)$. This completes the proof of the problem. ∎

> **Problem 9.10**
>
> (★)(★) *Suppose that $f,g \in \mathscr{R}(\alpha)$ on $[a,b]$. Prove the **Schwarz Inequality for Integral***
> $$\left| \int_a^b fg\,\mathrm{d}\alpha \right| \leq \left(\int_a^b f^2\,\mathrm{d}\alpha \right)^{\frac{1}{2}} \left(\int_a^b g^2\,\mathrm{d}\alpha \right)^{\frac{1}{2}}. \qquad (9.20)$$

Proof. By Theorem 9.6(e) (Operations of Integrable Functions), we know that $f^2, g^2, fg \in \mathscr{R}(\alpha)$ on $[a, b]$. We consider the case that

$$\int_a^b f^2 \, d\alpha > 0 \quad \text{and} \quad \int_a^b g^2 \, d\alpha > 0.$$

Let

$$F(x) = \frac{f(x)}{\left(\int_a^b f^2 \, d\alpha\right)^{\frac{1}{2}}} \quad \text{and} \quad G(x) = \frac{g(x)}{\left(\int_a^b g^2 \, d\alpha\right)^{\frac{1}{2}}}. \tag{9.21}$$

Since $f^2, g^2, fg \in \mathscr{R}(\alpha)$ on $[a, b]$, it is obvious from Theorem 9.6 (Operations of Integrable Functions) that $F^2, G^2, FG \in \mathscr{R}(\alpha)$ on $[a, b]$ and

$$\int_a^b F^2 \, d\alpha = \int_a^b G^2 \, d\alpha = 1.$$

By applying the A.M. $\geq$ G.M. to F^2 and G^2, we see that

$$\int_a^b FG \, d\alpha \leq \int_a^b \left(\frac{F^2 + G^2}{2}\right) d\alpha = \frac{1}{2} \int_a^b F^2 \, d\alpha + \frac{1}{2} \int_a^b G^2 \, d\alpha = 1. \tag{9.22}$$

Thus, after putting the two expressions (9.21) into the inequality (9.22), we get

$$\int_a^b fg \, d\alpha \leq \left(\int_a^b f^2 \, d\alpha\right)^{\frac{1}{2}} \left(\int_a^b g^2 \, d\alpha\right)^{\frac{1}{2}}. \tag{9.23}$$

Next, we suppose that

$$\int_a^b f^2 \, d\alpha = 0.$$

Given $\epsilon > 0$. Then the A.M. $\geq$ G.M. implies that

$$fg = (\epsilon^{-1} f)(\epsilon g) \leq \frac{(\epsilon^{-1} f)^2 + (\epsilon g)^2}{2}.$$

By Theorem 9.6 (Operations of Integrable Functions), we see that

$$\int_a^b fg \, d\alpha \leq \frac{\epsilon^{-2}}{2} \int_a^b f^2 \, d\alpha + \frac{\epsilon^2}{2} \int_a^b g^2 \, d\alpha = \frac{\epsilon^2}{2} \int_a^b g^2 \, d\alpha. \tag{9.24}$$

Since ϵ is arbitrary, it follows from the inequality (9.24) that

$$\int_a^b fg \, d\alpha \leq 0. \tag{9.25}$$

By a similar argument, we can show that the inequality (9.25) also holds when

$$\int_a^b g^2 \, d\alpha = 0.$$

Hence what we have shown is that the inequality (9.23) holds *for any* $f, g \in \mathscr{R}$ on $[a, b]$.

9.2. Integrability of Real Functions

Finally, if we replace f by $-f$ in the inequality (9.23), then we achieve

$$-\int_a^b fg\,d\alpha \le \Big(\int_a^b (-f)^2\,d\alpha\Big)^{\frac{1}{2}} \Big(\int_a^b g^2\,d\alpha\Big)^{\frac{1}{2}}$$
$$= \Big(\int_a^b f^2\,d\alpha\Big)^{\frac{1}{2}} \Big(\int_a^b g^2\,d\alpha\Big)^{\frac{1}{2}}. \qquad (9.26)$$

Hence the expected inequality (9.20) follows immediately from combining the inequalities (9.23) and (9.26). This completes the proof of the problem. ∎

Problem 9.11

(★) Given that $\int_0^1 x^2\,dx = \frac{1}{3}$. Suppose that $f \in \mathscr{R}$ on $[0,1]$ and

$$\int_0^1 f(x)\,dx = \int_0^1 xf(x)\,dx = 1.$$

Prove that

$$\int_0^1 f^2(x)\,dx \ge 3. \qquad (9.27)$$

Proof. By using Problem 9.10 directly with $g(x) = x$, we gain

$$\int_0^1 f^2(x)\,dx \times \int_0^1 x^2\,dx \ge \Big(\int_0^1 xf(x)\,dx\Big)^2 = 1. \qquad (9.28)$$

By the given hypothesis

$$\int_0^1 x^2\,dx = \frac{1}{3},$$

the desired result (9.27) follows immediately from the inequality (9.28). Hence we have completed the proof of the problem. ∎

Problem 9.12

(★)(★) Suppose that $\alpha : [a,b] \to \mathbb{R}$ is a monotonically increasing function and $f : [a,b] \to \mathbb{R}$ is a bounded function such that $f \in \mathscr{R}(\alpha)$ on $[a,b]$. Recall from Problem 7.12 that

$$\varphi(x) = \max(f(x), 0) \quad \text{and} \quad \psi(x) = \min(f(x), 0)$$

Prove that $\varphi, \psi \in \mathscr{R}(\alpha)$ on $[a,b]$.

Proof. By the proof of Problem 7.12, we see that

$$\varphi(x) = \frac{1}{2}[f(x) + |f(x)|] \quad \text{and} \quad \psi(x) = \frac{1}{2}[f(x) - |f(x)|].$$

By Theorem 9.6(f) and then (a) (Operations of Integrable Functions), we are able to conclude that $\varphi, \psi \in \mathscr{R}(\alpha)$ on $[a,b]$. We have completed the proof of the problem. ∎

9.3 Applications of Integration Theorems

> **Problem 9.13**
>
> (★) Suppose that $f : [a, b] \to \mathbb{R}$ is bounded and $f \in \mathscr{R}$ on $[a, b]$ and $a \leq x \leq b$. Define
> $$F(x) = \int_a^x f(t)\, dt.$$
> Prove that there exists a positive constant M such that
> $$|F(x) - F(y)| \leq M|x - y| \qquad (9.29)$$
> for all $x, y \in [a, b]$.

Proof. Since f is bounded on $[a, b]$, there exists a positive constant M such that
$$|f(x)| \leq M \qquad (9.30)$$
for all $x \in [a, b]$. Let $x \geq y$. Then we deduce from Theorem 9.6(c), (f) (Operations of Integrable Functions) and the bound (9.30) that
$$\begin{aligned}
|F(x) - F(y)| &= \left| \int_a^x f(t)\, dt - \int_a^y f(t)\, dt \right| \\
&= \left| \int_a^y f(t)\, dt + \int_y^x f(t)\, dt - \int_a^y f(t)\, dt \right| \\
&= \left| \int_y^x f(t)\, dt \right| \\
&\leq \left| \int_y^x |f(t)|\, dt \right| \\
&\leq \left| \int_y^x M\, dt \right| \\
&= M|x - y|.
\end{aligned}$$
The case for $x < y$ is similar. Hence the inequality (9.29) holds for all $x, y \in [a, b]$ and we have finished the proof of the problem. ∎

> **Problem 9.14**
>
> (★)(★) Suppose that $\alpha : [0, 4] \to \mathbb{R}$ is a function defined by
> $$\alpha(x) = \begin{cases} x^2, & \text{if } 0 \leq x \leq 2; \\ x^4, & \text{if } 2 < x \leq 4. \end{cases}$$
> Evaluate the integral
> $$\int_0^4 x\, d\alpha. \qquad (9.31)$$

9.3. Applications of Integration Theorems

Proof. It is clear that α is a monotonically increasing function and it has a discontinuity of the first kind at 2 (see Definition 7.15 (Types of Discontinuity)). By Theorem 9.3(a), the integral (9.31) is well-defined. Now we write $\alpha = \beta + \gamma$, where $\beta, \gamma : [0,4] \to \mathbb{R}$ are functions defined by

$$\beta(x) = \begin{cases} 0, & \text{if } 0 \leq x \leq 2; \\ 12, & \text{if } 2 < x \leq 4 \end{cases} \quad \text{and} \quad \gamma(x) = \begin{cases} x^2, & \text{if } 0 \leq x \leq 2; \\ x^4 - 12, & \text{if } 2 < x \leq 4. \end{cases}$$

Then it is obvious that both β and γ are monotonically increasing functions on $[0,4]$. By Theorem 9.6(d) and then (c)(Operations of Integrable Functions), we obtain

$$\int_0^4 x\, d\alpha = \int_0^4 x\, d\beta + \int_0^4 x\, d\gamma = \int_0^4 x\, d\beta + \int_0^2 x\, d\gamma + \int_2^4 x\, d\gamma. \tag{9.32}$$

We note that $(x^2)' = 2x \in \mathscr{R}$ on $[0,2]$ and $(x^4 - 12)' = 4x^3 \in \mathscr{R}$ on $[2,4]$, so the Substitution Theorem yields

$$\int_0^2 x\, d\gamma = \int_0^2 2x^2\, dx = \frac{16}{3}$$

and

$$\int_2^4 x\, d\gamma = \int_2^4 4x^4\, dx = \left[\frac{4x^5}{5}\right]_2^4 = \frac{3968}{5},$$

so the integral (9.32) reduces to

$$\int_0^4 x\, d\alpha = \int_0^4 x\, d\beta + \frac{16}{3} + \frac{3968}{5}. \tag{9.33}$$

Since $\beta(x) = 12I(x-2)$, we follow from Theorem 9.7 that

$$\int_0^4 x\, d\beta = 12 \times 2 = 24. \tag{9.34}$$

Hence we obtain from the integrals (9.33) and (9.34) that

$$\int_0^4 x\, d\alpha = 24 + \frac{16}{3} + \frac{3968}{5} = \frac{12344}{15},$$

completing the proof of the problem. ∎

Problem 9.15

(★)(★)(★) Suppose that $f \in \mathscr{R}$ on $[a,b]$ and $g : [a,b] \to \mathbb{R}$ is a function such that the set

$$E = \{x \in [a,b] \mid f(x) \neq g(x)\}$$

is finite. Prove that $g \in \mathscr{R}$ on $[a,b]$ and

$$\int_a^b f(x)\, dx = \int_a^b g(x)\, dx$$

without using Theorem 9.4 (The Lebesgue's Integrability Condition).

Proof. Given $\epsilon > 0$. Let $h : [a,b] \to \mathbb{R}$ be defined by

$$h(x) = f(x) - g(x)$$

and $E = \{x_1, \ldots, x_n\}$ for some positive integer n. There are three cases for consideration.

- **Case (1):** $a < x_1 < \cdots < x_n < b$. Let

$$M = \max(|h(x_1)|, |h(x_2)|, \ldots, |h(x_n)|) > 0.$$

By Theorem 2.2 (Density of Rationals), there exists a rational $\delta > 0$ such that

$$a < x_1 - \delta, \quad x_n + \delta < b \quad \text{and} \quad x_k + \delta < x_{k+1} - \delta,$$

where $k = 1, 2, \ldots, n-1$. Furthermore, we may assume that

$$\delta < \frac{\epsilon}{2Mn}. \tag{9.35}$$

We consider the partition

$$P = \{a, x_1 - \delta, x_1 + \delta, x_2 - \delta, x_2 + \delta, \ldots, x_n - \delta, x_n + \delta, b\}.$$

For each $k = 1, 2, \ldots, n$, in the interval $I_k = [x_k - \delta, x_k + \delta]$, since $|h(x)| \leq M$ for all $x \in [a,b]$, we have[c]

$$-M \leq M_k = \sup_{I_k} h(x) \leq M \quad \text{and} \quad -M \leq m_k = \inf_{I_k} h(x) \leq M. \tag{9.36}$$

In the interval $I = [a, x_1 - \delta]$ or $I' = [x_n + \delta, b]$, we have $h(x) = 0$ so that

$$\begin{aligned} M_I &= \sup_I h(x) = 0, \quad m_I = \inf_I h(x) = 0, \\ M_{I'} &= \sup_{I'} h(x) = 0, \quad m_{I'} = \inf_{I'} h(x) = 0. \end{aligned} \tag{9.37}$$

By definition, the sup and the inf given in (9.37), we have

$$U(P, h) = M_I(x_j - \delta - a) + \sum_{k=1}^{n} M_k(2\delta) + M_{I'}(b - x_j - \delta) = 2\delta \sum_{k=1}^{n} M_k$$

and

$$L(P, h) = m_I(x_j - \delta - a) + \sum_{k=1}^{n} m_k(2\delta) + m_{I'}(b - x_j - \delta) = 2\delta \sum_{k=1}^{n} m_k.$$

Thus we deduce from the inequality (9.35), the sup and the inf in (9.36) that

$$-\epsilon < -2nM\delta \leq U(P, h) \leq 2nM\delta < \epsilon \tag{9.38}$$

and

$$-\epsilon < -2nM\delta \leq L(P, h) \leq 2nM\delta < \epsilon. \tag{9.39}$$

---

[c] It may happen that $h(x_k) < 0$ for some k.

9.3. Applications of Integration Theorems

Since ϵ is arbitrary, we get from the estimates (9.38) and (9.39) that
$$U(P, h) = L(P, h) = 0.$$

Now we are able to conclude from the paragraph following Definition 9.1 (The Riemann-Stieltjes Integral) that
$$\int_a^b h(x)\,dx = \overline{\int_a^b} h(x)\,dx = \underline{\int_a^b} h(x)\,dx = 0. \tag{9.40}$$

Hence we apply Theorem 9.6(a) (Operations of Integrable Functions) to establish that
$$\int_a^b f(x)\,dx = \int_a^b f(x)\,dx - \int_a^b h(x)\,dx = \int_a^b [f(x) - h(x)]\,dx = \int_a^b g(x)\,dx. \tag{9.41}$$

- **Case (2):** $a = x_1$. In this case, we still have
$$M = \max(|h(a)|, |h(x_2)|, \ldots, |h(x_n)|) > 0$$
and a $\delta > 0$ such that
$$a + \delta < x_2 - \delta, \quad x_n + \delta < b \quad \text{and} \quad x_k + \delta < x_{k+1} - \delta,$$
where $k = 2, 3, \ldots, n-1$. By this setting, it is easy to check that the inequalities (9.36) also hold for $k = 2, 3, \ldots, n$ and
$$M_{I'} = m_{I'} = 0, \tag{9.42}$$
where $I' = [x_n + \delta, b]$. Now we have to check the remaining interval $I_1 = [a, a + \delta]$. Since $h(a) \neq 0$, it is evident that
$$-M \leq M_1 = \sup_{I_1} h(x) \leq M \quad \text{and} \quad -M \leq m_1 = \inf_{I_1} h(x) \leq M. \tag{9.43}$$

Thus we combine the inequalities (9.36), (9.43) and the values (9.42) to get the inequalities (9.38) and (9.39). Hence, by a similar argument as in the proof of **Case (1)**, we see that the integral (9.40) and then the integral equation (9.41) also hold in this case.

- **Case (3):** $b = x_n$. Since the argument of this part is very similar to that proven in **Case (2)**, we omit the details here.

We have completed the proof of the problem. ∎

Problem 9.16

(⋆) (⋆) (⋆) *Suppose that $\mathscr{C}([0,1])$ denotes the set of all continuous real functions on $[0,1]$. For every $f, g \in \mathscr{C}([0,1])$, we define*
$$d(f, g) = \int_0^1 \frac{|f(x) - g(x)|}{1 + |f(x) - g(x)|}\,dx. \tag{9.44}$$

Prove that d is a metric in $\mathscr{C}([0,1])$. Is $\mathscr{C}([0,1])$ a complete metric space with respect to this metric?

Proof. Since f and g are continuous functions on $[0,1]$, $f,g \in \mathscr{R}$ on $[0,1]$ and Theorem 9.6 (Operations of Integrable Functions) guarantees that the integral (9.44) is well-defined.

Next, we recall from Problem 2.7 that

$$\frac{|c|}{1+|c|} \leq \frac{|a|}{1+|a|} + \frac{|b|}{1+|b|}, \tag{9.45}$$

where $c = a + b$. If we put $a = f - g$ and $b = g - h$, then $c = f - h$ and the inequality (9.45) implies that

$$\frac{|f-h|}{1+|f-h|} \leq \frac{|f-g|}{1+|f-g|} + \frac{|g-h|}{1+|g-h|}. \tag{9.46}$$

Thus we obtain from the inequality (9.46) that

$$d(f,h) \leq d(f,g) + d(g,h)$$

which proves the triangle inequality is valid. Hence d is a metric in $\mathscr{C}([0,1])$.

We claim that $\mathscr{C}([0,1])$ is *not* a complete metric space. To this end, we consider the functions $f_n : [0,1] \to \mathbb{R}$ defined by

$$f_n(x) = \begin{cases} n^2 x, & \text{if } 0 \leq x < \frac{1}{n}; \\ \dfrac{1}{x}, & \text{if } \frac{1}{n} \leq x \leq 1. \end{cases} \tag{9.47}$$

It is clear that each f_n is continuous at every point on $[0,1]$ *except possibly* the point $\frac{1}{n}$. We check the continuity of f_n at $\frac{1}{n}$. Since $f_n(\frac{1}{n}+) = \frac{1}{\frac{1}{n}} = n$ and $f_n(\frac{1}{n}-) = n^2 \times \frac{1}{n} = n$, each f_n is continuous at $\frac{1}{n}$. In other words, we have $\{f_n\} \subseteq \mathscr{C}([0,1])$.

Given $\epsilon > 0$. Take N to be a positive integer such that $\frac{1}{N} < \epsilon$. Now if $x \geq \max(\frac{1}{m}, \frac{1}{n})$, then we have $f_n(x) = f_m(x) = \frac{1}{x}$ so that

$$\int_{\max(\frac{1}{m}, \frac{1}{n})}^{1} \frac{|f_n(x) - f_m(x)|}{1 + |f_n(x) - f_m(x)|} \, dx = 0. \tag{9.48}$$

Therefore, for $m, n \geq N$, we deduce from the definition (9.44) and the result (9.48) that

$$d(f_m, f_n) = \int_0^{\max(\frac{1}{m}, \frac{1}{n})} \frac{|f_n(x) - f_m(x)|}{1 + |f_n(x) - f_m(x)|} \, dx \leq \int_0^{\max(\frac{1}{m}, \frac{1}{n})} dx = \max\left(\frac{1}{m}, \frac{1}{n}\right) \leq \frac{1}{N} < \epsilon.$$

By Definition 5.12, $\{f_n\}$ is a Cauchy sequence.

Assume that $\mathscr{C}([0,1])$ was complete. It means that $\{f_n\}$ converges to a function $f \in \mathscr{C}([0,1])$, see the paragraph following Theorem 5.13. By the definition of f_n in (9.47), it is reasonable to *conjecture* that

$$f(x) = \frac{1}{x}$$

on $(0,1]$. In fact, if $f(p) \neq \frac{1}{p}$ for some $p \in (0,1]$, then the continuity of f ensures that there exist $\epsilon > 0$ and $\delta > 0$ such that

$$\left| f(x) - \frac{1}{x} \right| \geq \epsilon \tag{9.49}$$

9.3. Applications of Integration Theorems

for all $x \in [p - \delta, p]$. Without loss of generality, we may take $\epsilon = \delta$. Since the function $g(x) = f(x) - \frac{1}{x}$ is continuous on $[p - \epsilon, p]$, it follows from the Extreme Value Theorem that there exists a positive constant M such that

$$\left| f(x) - \frac{1}{x} \right| \leq M \tag{9.50}$$

on $[p - \delta, p]$. Now, for sufficiently large n, we have $[p - \epsilon, p] \subseteq (\frac{1}{n}, 1]$ so that $f_n(x) = \frac{1}{x}$. Thus we follow from this and the inequalities (9.49) and (9.50) that

$$d(f_n, f) = \int_0^1 \frac{|f(x) - f_n(x)|}{1 + |f(x) - f_n(x)|} \, dx \geq \int_{p-\epsilon}^p \frac{|f(x) - \frac{1}{x}|}{1 + |f(x) - \frac{1}{x}|} \, dx \geq \int_{p-\epsilon}^p \frac{\epsilon}{1 + M} \, dx > 0.$$

In other words, f_n *does not* converge to f and then

$$f(p) = \frac{1}{p}$$

for all $p \in (0, 1]$, but this contradicts the continuity of f on $[0, 1]$. Hence $\mathscr{C}([0, 1])$ is not complete with respect to the metric d and we have completed the proof of the problem. ∎

Problem 9.17

(⋆)(⋆) Suppose that $f : [a, b] \to \mathbb{R}$ is a function such that

$$\lim_{x \to p} f(x)$$

exists for every $p \in [a, b]$. Prove that $f \in \mathscr{R}$ on $[a, b]$.

Proof. By the hypothesis, it means that both $f(p+)$ and $f(p-)$ exist and equal for every $p \in [a, b]$. Thus any discontinuity of f must be **simple** (see Definition 7.15 (Types of Discontinuity)). Therefore, by Theorem 7.16 (Countability of Simple Discontinuities), the set of all simple discontinuities of f is at most countable. Denote this set to be E. By the paragraph following Theorem 9.4 (The Lebesgue's Integrability Condition), E is of measure zero and hence we obtain our desired result, completing the proof of the problem. ∎

Problem 9.18

(⋆)(⋆) Suppose that $f : [0, 1] \to \mathbb{R}$ is continuous and there exists a positive constant M such that

$$M \int_0^x f(t) \, dt \leq f(x) \tag{9.51}$$

for all $x \in [0, 1]$. Prove that $f(x) \geq 0$ on $[0, 1]$.

Proof. For every $x \in [0, 1]$, we define

$$F(x) = \int_0^x f(t) \, dt.$$

Since f is continuous on $[0,1]$, the First Fundamental Theorem of Calculus implies that

$$F'(x) = f(x) \geq MF(x) \tag{9.52}$$

on $[0,1]$. Next, we consider the function $G : [0,1] \to \mathbb{R}$ given by

$$G(x) = F(x)\mathrm{e}^{-Mx}.$$

By the inequality (9.52), we deduce that

$$G'(x) = \frac{\mathrm{d}}{\mathrm{d}x}[F(x)\mathrm{e}^{-Mx}] = \mathrm{e}^{-Mx}F'(x) - M\mathrm{e}^{-Mx}F(x) = \mathrm{e}^{-Mx}[F'(x) - MF(x)] \geq 0$$

for every $x \in [0,1]$. By Theorem 8.9(a), G is monotonically increasing on $[0,1]$ and thus

$$G(x) \geq G(0) \tag{9.53}$$

for all $x \in [0,1]$. Since $G(0) = F(0)\mathrm{e}^0 = 0$, we establish from the inequality (9.53) that

$$F(x)\mathrm{e}^{-Mx} \geq 0$$

and then $F(x) \geq 0$ on $[0,1]$. Now we conclude from this and the hypothesis (9.51) that

$$f(x) \geq MF(x) \geq 0$$

on $[0,1]$. This completes the proof of the problem. ∎

Problem 9.19

(★)(★) Suppose that $f : [0,1] \to \mathbb{R}$ is continuous on $[0,1]$. Let n be a positive integer. Prove that there exists a $\alpha \in [0,1]$ such that

$$\int_0^1 x^n f(x)\,\mathrm{d}x = \frac{1}{n+1}f(\alpha). \tag{9.54}$$

Proof. Since f is continuous on $[0,1]$, the Extreme Value Theorem tells us that there exist $p, q \in [0,1]$ such that

$$f(p) = \sup_{x \in [0,1]} f(x) \quad \text{and} \quad f(q) = \inf_{x \in [0,1]} f(x).$$

By Theorem 9.6(a) and (b) (Operations of Integrable Functions), we see that

$$f(q) \int_0^1 x^n\,\mathrm{d}x \leq \int_0^1 x^n f(x)\,\mathrm{d}x \leq f(p) \int_0^1 x^n\,\mathrm{d}x. \tag{9.55}$$

By the Second Fundamental Theorem of Calculus, since

$$\frac{\mathrm{d}}{\mathrm{d}x}\left(\frac{x^{n+1}}{n+1}\right) = x^n,$$

we have

$$\int_0^1 x^n\,\mathrm{d}x = \frac{x^{n+1}}{n+1}\bigg|_0^1 = \frac{1}{n+1}.$$

9.3. Applications of Integration Theorems

Thus we obtain from this and the inequalities (9.55) that
$$f(q) \leq (n+1)\int_0^1 x^n f(x)\,dx \leq f(p).$$

Since f is continuous on $[0,1]$, we follow from the Intermediate Value Theorem that there exists a $\alpha \in [0,1]$ such that
$$(n+1)\int_0^1 x^n f(x)\,dx = f(\alpha)$$
which implies the desired result (9.54). We end the proof of the problem. ∎

Problem 9.20

(★)(★) *Suppose that $\psi : [a,b] \to \mathbb{R}$ has second derivative in $[a,b]$ and*
$$\psi(a) = \psi(b) = \psi'(a) = \psi'(b) = 0. \tag{9.56}$$

Prove that there exists a positive constant M such that
$$\left|\int_a^b \cos(\theta x)\psi(x)\,dx\right| \leq \frac{(b-a)M}{\theta^2}$$
for all $\theta > 1$.

Proof. Since ψ and $\cos(\theta x)$ are differentiable in $[a,b]$, we follow from the Integration by Parts that
$$\int_a^b \cos(\theta x)\psi(x)\,dx = \int_a^b \psi(x)\frac{d}{dx}\left(\frac{\sin(\theta x)}{\theta}\right)dx$$
$$= \frac{\sin(\theta x)}{\theta}\psi(x)\Big|_a^b - \frac{1}{\theta}\int_a^b \sin(\theta x)\psi'(x)\,dx. \tag{9.57}$$

By the hypotheses (9.56), the expression (9.57) reduces to
$$\int_a^b \cos(\theta x)\psi(x)\,dx = -\frac{1}{\theta}\int_a^b \sin(\theta x)\psi'(x)\,dx. \tag{9.58}$$

Now we apply the Integration by Parts to the right-hand side of the expression (9.58) and then using the hypotheses (9.56) again, we derive that
$$-\frac{1}{\theta}\int_a^b \sin(\theta x)\psi'(x)\,dx = \frac{1}{\theta}\int_a^b \psi'(x)\left(\frac{\cos(\theta x)}{\theta}\right)'dx$$
$$= -\frac{1}{\theta^2}\int_a^b \cos(\theta x)\psi''(x)\,dx. \tag{9.59}$$

Combining the two expressions (9.58) (9.59) and then using Theorem 9.6(f) (Operations of Integrable Functions), we obtain
$$\left|\int_a^b \cos(\theta x)\psi(x)\,dx\right| = \left|\frac{1}{\theta^2}\int_a^b \cos(\theta x)\psi''(x)\,dx\right| \leq \frac{1}{\theta^2}\int_a^b |\cos(\theta x)||\psi''(x)|\,dx. \tag{9.60}$$

Since ψ'' is continuous on $[a,b]$, the Extreme Value Theorem ensures that there exists a positive constant M such that $|\psi''(x)| \leq M$ for all $x \in [a,b]$. Furthermore, since $|\cos(\theta x)| \leq 1$ for every $x \in [a,b]$ and $\theta > 1$, the inequality (9.60) can be reduced to

$$\left| \int_a^b \cos(\theta x)\psi(x)\,dx \right| \leq \frac{(b-a)M}{\theta^2}$$

which is our desired result, completing the proof of the problem. ∎

Problem 9.21

(★) Suppose that $f : [0,1] \to [0,+\infty)$ is differentiable in $[0,1]$ and $|f'(x)| \leq M$ for some positive constant M. Let

$$F(x) = \int_0^{f(x)} e^{-2t}\,dt,$$

where $x \geq 0$. Prove that

$$|F'(x)| \leq M.$$

Proof. Since f is differentiable in $[0,1]$ and $f(x) \geq 0$ on $[0,+\infty)$, it follows from the Chain Rule and the First Fundamental Theorem of Calculus that

$$F'(x) = \frac{d(f(x))}{dx} \cdot \frac{d}{d(f(x))} \int_0^{f(x)} e^{-2t}\,dt = f'(x)e^{-2f(x)}$$

so that

$$|F'(x)| \leq \frac{|f'(x)|}{e^0} \leq M$$

which is exactly the desired result. This completes the proof of the problem. ∎

Problem 9.22

(★)(★) Suppose that $f : [a,b] \to \mathbb{R}$ has nth continuous derivative in $[a,b]$. Then we have

$$f(x) = \sum_{k=0}^{n-1} \frac{f^{(k)}(p)}{k!}(x-p)^k + \frac{1}{(n-1)!}\int_p^x f^{(n)}(t)(x-t)^{n-1}\,dt, \qquad (9.61)$$

where $[p,x] \subseteq [a,b]$.

Proof. Since f' is continuous on $[a,b]$, $f' \in \mathscr{R}$ on $[a,b]$ by Theorem 9.3(a) and we follow from the Second Fundamental Theorem of Calculus that

$$f(x) - f(p) = \int_p^x f'(t)\,dt = -\int_p^x \underbrace{f'(t)}_{F(t)} \underbrace{\frac{d}{dt}(x-t)}_{G(t)}\,dt. \qquad (9.62)$$

Apply the Integration by Parts to the formula (9.62), we have

$$f(x) = f(p) - f'(t)(x-t)\Big|_p^x + \int_p^x f''(t)(x-t)\,dt$$

9.3. Applications of Integration Theorems

$$= f(p) + f'(p)(x-p) + \int_p^x f''(t)(x-t)\,dt. \tag{9.63}$$

Now we may express the formula (9.63) in the following form

$$f(x) = f(p) + f'(p)(x-p) - \frac{1}{2}\int_p^x \underbrace{f''(t)}_{F(t)} \underbrace{\frac{d}{dt}(x-t)^2}_{G(t)}\,dt$$

so that the Integration by Parts can be used again to obtain

$$f(x) = f(p) + f'(p)(x-p) + \frac{1}{2}f''(p)(x-p)^2 + \frac{1}{2}\int_p^x f'''(t)(x-t)^2\,dt.$$

Since f has continuous derivative in $[a,b]$ up to order nth, the above process can be continued $(n-3)$-steps to obtain

$$f(x) = \sum_{k=0}^{n-1} \frac{f^{(k)}(p)}{k!}(x-p)^k + \frac{1}{(n-1)!}\int_p^x f^{(n)}(t)(x-t)^{n-1}\,dt$$

which is our expected result (9.61). ∎

> **Remark 9.7**
>
> Problem 9.22 says that we can express the remainder of Taylor's Theorem in an **exact form**.

> **Problem 9.23**
>
> (★) Prove the First Mean Value Theorem for Integrals by using the First Fundamental Theorem of Calculus.

Proof. Define

$$F(x) = \int_a^x f(t)\,dt. \tag{9.64}$$

Since f is continuous on $[a,b]$, we have F is differentiable in $[a,b]$ by the First Fundamental Theorem of Calculus. By the Mean Value Theorem for Derivatives, there exists a $p \in (a,b)$ such that

$$F(b) - F(a) = F'(p)(b-a). \tag{9.65}$$

By the definition (9.64), we have

$$F(b) = \int_a^b f(t)\,dt \quad \text{and} \quad F(a) = \int_a^b f(t)\,dt = 0$$

so that

$$F(b) - F(a) = \int_a^b f(t)\,dt. \tag{9.66}$$

Thus, by substituting the result (9.66) and the fact $F'(p) = f(p)$ into the formula (9.65), we have

$$\int_a^b f(t)\,dt = f(p)(b-a)$$

which is our desired result, completing the proof of the problem. ∎

> **Problem 9.24**
>
> (★) (★) Suppose that $\varphi : [A, B] \to [a, b]$ is a strictly increasing continuous onto function. Furthermore, suppose that φ has a continuous derivative on $[A, B]$ and $f : [a, b] \to \mathbb{R}$ is continuous in $[a, b]$. Prove that
>
> $$\int_{\varphi(A)}^{\varphi(B)} f(x) \, dx = \int_A^B f(\varphi(y)) \varphi'(y) \, dy. \qquad (9.67)$$

Proof. Since f is continuous on $[a, b]$, we have $f \in \mathscr{R}(\alpha)$ on $[a, b]$. In the Change of Variables Theorem, if we take $\alpha(x) = x$ which is monotonically increasing on $[a, b]$, then we have $\beta = \varphi$ and

$$\int_a^b f(x) \, dx = \int_A^B f(\varphi) \, d\varphi. \qquad (9.68)$$

Since φ' is continuous on $[A, B]$, it follows from Theorem 9.3(a) that $\varphi' \in \mathscr{R}$ on $[A, B]$. Thus the Substitution Theorem implies that

$$\int_A^B f(\varphi) \, d\varphi = \int_A^B f(\varphi(y)) \varphi'(y) \, dy. \qquad (9.69)$$

Hence our desired result (9.67) follows by combining the two expressions (9.68) and (9.69) and using the facts that $\varphi(A) = a$ and $\varphi(B) = b$. This completes the proof of the problem. ∎

> **Problem 9.25**
>
> (★) (★) Suppose that $f : [a, b] \to \mathbb{R}$ is continuous on $[a, b]$. Let $[p, q] \subset [a, b]$ and x be a variable such that $[p + x, q + x] \subset [a, b]$. Prove that
>
> $$\frac{d}{dx} \int_p^q f(x + y) \, dy = f(q + x) - f(p + x).$$

Proof. Fix x, since $f(x + y) \in [a, b]$ for every $y \in [p, q]$, we have $f(x + y) \in \mathscr{R}$ on $[p, q]$. Let $\varphi : [p, q] \to [p + x, q + x]$ be defined by

$$\varphi(y) = x + y.$$

When $y = p$, $\varphi(p) = p + x$; when $y = q$, $\varphi(q) = q + x$. By Problem 9.24, we have

$$\frac{d}{dx} \int_p^q f(x + y) \, dy = \frac{d}{dx} \int_p^q f(\varphi(y)) \varphi'(y) \, dy$$

$$= \frac{d}{dx} \int_{\varphi(p)}^{\varphi(q)} f(t) \, dt$$

$$= \frac{d}{dx} \int_{p+x}^{q+x} f(t) \, dt$$

$$= \frac{d}{dx} \Big[\int_{p+x}^a f(t) \, dt + \int_a^{q+x} f(t) \, dt \Big]$$

9.3. Applications of Integration Theorems

$$= \frac{d}{dx}\left[-\int_a^{p+x} f(t)\,dt + \int_a^{q+x} f(t)\,dt\right]. \tag{9.70}$$

Next, we apply the Chain Rule and then the First Fundamental Theorem of Calculus to the expression (9.70), we obtain

$$\frac{d}{dx}\int_p^q f(x+y)\,dy = -f(p+x) + f(q+x) = f(q+x) - f(p+x),$$

as desired. We have completed the proof of the problem. ∎

Problem 9.26

(⋆)(⋆) *Suppose $f : [a,b] \to \mathbb{R}$ is continuous on $[a,b]$. Prove that the formula*

$$2\int_a^b \int_a^x f(x)f(y)\,dy\,dx = \left(\int_a^b f(x)\,dx\right)^2 \tag{9.71}$$

holds.

Proof. For $x \in [a,b]$, let

$$F(x) = \int_a^x f(y)\,dy.$$

Then we have

$$\int_a^b \int_a^x f(x)f(y)\,dy\,dx = \int_a^b f(x)\left[\int_a^x f(y)\,dy\right]dx = \int_a^b f(x)F(x)\,dx. \tag{9.72}$$

Since f is continuous on $[a,b]$, the First Fundamental Theorem of Calculus implies that

$$F'(x) = f(x)$$

on $[a,b]$. Thus the expression (9.72) can be further rewritten as

$$\int_a^b \int_a^x f(x)f(y)\,dy\,dx = \int_a^b F(x)F'(x)\,dx. \tag{9.73}$$

Apply the Integration by Parts to the right-hand side of the expression (9.73), we get

$$\int_a^b F(x)F'(x)\,dx = F(b)F(b) - F(a)F(a) - \int_a^b F'(x)F(x)\,dx$$

which means that

$$\int_a^b F(x)F'(x)\,dx = \frac{1}{2}[F^2(b) - F^2(a)].$$

Since

$$F(b) = \int_a^b f(x)\,dx \quad \text{and} \quad F(a) = 0,$$

we have

$$\int_a^b F(x)F'(x)\,dx = \frac{1}{2}\left(\int_a^b f(x)\,dx\right)^2. \tag{9.74}$$

Hence our desired result (9.71) follows immediately if we substitute the expression (9.74) back into the expression (9.73). This ends the proof of the problem. ∎

9.4 The Mean Value Theorems for Integrals

> **Problem 9.27**
>
> (★) Let $a \in \mathbb{R}$. Let $f : [a, a+1] \to \mathbb{R}$ be continuous and
> $$\int_a^{a+1} f(x)\,\mathrm{d}x = 1.$$
> Prove that $f(p) = 1$ for some $p \in (a, a+1)$.

Proof. A direct application of the First Mean Value Theorem for Integrals shows that there exists a $p \in (a, a+1)$ such that
$$1 = \int_a^{a+1} f(x)\,\mathrm{d}x = f(p)(a+1-a) = f(p).$$
This completes the proof of the problem. ∎

> **Problem 9.28**
>
> (★) Let $\alpha > 0$ and $n \in \mathbb{N}$, prove that
> $$\int_n^{n+\alpha} \frac{\cos x}{x^2}\,\mathrm{d}x = \frac{\alpha \cos p}{p^2}$$
> for some $p \in (n, n+\alpha)$.

Proof. Let $f : [n, n+\alpha] \to \mathbb{R}$ be defined by
$$f(x) = \frac{\cos x}{x^2}.$$
Since $n \geq 1$, it is clear that f is continuous on $[n, n+\alpha]$. By the First Mean Value Theorem for Integrals, *there exists a* $p \in (n, n+\alpha)$ such that
$$\int_n^{n+\alpha} \frac{\cos x}{x^2}\,\mathrm{d}x = \frac{\cos p}{p^2}(n+\alpha-n) = \frac{\alpha \cos p}{p^2}.$$
This completes the proof of the problem. ∎

> **Problem 9.29**
>
> (★)(★) Suppose that $\varphi : [a, b] \to \mathbb{R}$ is differentiable in $[a, b]$, $\varphi'(x) \geq \eta > 0$, φ' is monotonically decreasing and continuous on $[a, b]$. Prove that
> $$\left| \int_a^b \sin \varphi(x)\,\mathrm{d}x \right| \leq \frac{2}{\eta}.$$

9.4. The Mean Value Theorems for Integrals

Proof. Since $\varphi'(x) > 0$ and φ' is monotonically decreasing on $[a,b]$, $\frac{1}{\varphi'}$ is well-defined, monotonically increasing on $[a,b]$ and
$$0 < \frac{1}{\varphi'(x)} \le \frac{1}{\eta}$$
on $[a,b]$. Since φ' is continuous on $[a,b]$, the function $\varphi' \sin \varphi$ is also continuous on $[a,b]$. Thus by the special case of the Second Mean Value Theorem For Integrals, we can find a $p \in [a,b]$ such that
$$\int_a^b \sin \varphi(x)\,\mathrm{d}x = \int_a^b \underbrace{\frac{1}{\varphi'(x)}}_{f(x)} \times \underbrace{\varphi'(x) \sin \varphi(x)}_{g(x)}\,\mathrm{d}x$$
$$= \frac{1}{\eta} \int_p^b [\sin \varphi(x)] \times \varphi'(x)\,\mathrm{d}x. \tag{9.75}$$

Since $\varphi'(x) > 0$ on $[a,b]$, recall from Theorem 8.9(a) and Remark 8.3 that φ is strictly increasing on $[a,b]$. Since φ' is continuous on $[a,b]$, φ satisfies the conditions of Problem 9.24. It is clear that $f(x) = \sin x$ is continuous on $[\varphi(p), \varphi(b)]$, so we obtain from the formula (9.67) that
$$\int_p^b [\sin \varphi(x)] \times \varphi'(x)\,\mathrm{d}x = \int_{\varphi(p)}^{\varphi(b)} \sin x\,\mathrm{d}x = -\cos x \Big|_{\varphi(p)}^{\varphi(b)} = \cos \varphi(p) - \cos \varphi(b). \tag{9.76}$$

Hence we derive from the expressions (9.75) and (9.76) that
$$\left| \int_a^b \sin \varphi(x)\,\mathrm{d}x \right| = \left| \frac{1}{\eta} [\cos \varphi(p) - \cos \varphi(b)] \right| \le \frac{2}{\eta},$$
completing the proof of the problem. ∎

Problem 9.30

(★)(★) Suppose that $f, g : [0,1] \to \mathbb{R}$ are monotonically increasing continuous functions. Prove that
$$\int_0^1 f(x)\,\mathrm{d}x \times \int_0^1 g(x)\,\mathrm{d}x \le \int_0^1 f(x)g(x)\,\mathrm{d}x.$$

Proof. By Theorem 9.6(e) (Operations of Integrable Functions), we see that $fg \in \mathscr{R}$ on $[0,1]$. Define the function $\varphi : [0,1] \to \mathbb{R}$ by
$$\varphi(x) = g(x) - \int_0^1 g(t)\,\mathrm{d}t.$$

By the First Mean Value Theorem for Integrals, there exists a $p \in (0,1)$ such that
$$\int_0^1 g(t)\,\mathrm{d}t = g(p)$$
which gives
$$\varphi(x) = g(x) - g(p). \tag{9.77}$$

By Theorem 9.6 (Operations of Integrable Functions) again, we have $\varphi \in \mathscr{R}$ on $[0,1]$ which implies that $f\varphi \in \mathscr{R}$ on $[0,1]$. Furthermore, since g is monotonically increasing on $[0,1]$, we deduce from the result (9.77) that if $0 \leq x \leq p$, then $\varphi(x) \leq 0$ and if $p \leq x \leq 1$, then $\varphi(x) \geq 0$. By these and the fact that f is monotonically increasing on $[0,1]$, we conclude that

$$\int_0^1 f(x)\varphi(x)\,dx = \int_0^p f(x)\varphi(x)\,dx + \int_p^1 f(x)\varphi(x)\,dx$$
$$\geq f(p)\int_0^p \varphi(x)\,dx + f(p)\int_p^1 \varphi(x)\,dx$$
$$= f(p)\int_0^1 \varphi(x)\,dx. \qquad (9.78)$$

By the definition of φ, we must have

$$\int_0^1 \varphi(x)\,dx = \int_0^1 g(x)\,dx - \int_0^1 \left(\int_0^1 g(t)\,dt\right)dx = 0 \qquad (9.79)$$

Therefore, we know from the inequality (9.78) and the result (9.79) that

$$\int_0^1 f(x)\varphi(x)\,dx \geq 0$$

and this implies that

$$\int_0^1 f(x)g(x)\,dx \geq \int_0^1 f(x)\,dx \times \int_0^1 g(x)\,dx.$$

This completes the proof of the problem. ∎

Problem 9.31

(★) *Suppose that $f : [0, 2\pi] \to \mathbb{R}$ is strictly increasing on $[0, 2\pi]$ and $f(x) \geq 0$ on $[0, 2\pi]$. Prove that*
$$\int_0^{2\pi} f(x)\sin x\,dx < 0.$$

Proof. It is clear that the function f satisfies the hypotheses in the special case of the Second Mean Value Theorem for Integrals, so there exists a $p \in (0, 2\pi)$ such that

$$\int_0^{2\pi} f(x)\sin x\,dx = f(2\pi-)\int_p^{2\pi} \sin x\,dx$$
$$= f(2\pi-)\left[-\cos x\Big|_p^{2\pi}\right]$$
$$= -f(2\pi-)(1-\cos p). \qquad (9.80)$$

Since f is strictly increasing on $[0, 2\pi]$ and $f(0) \geq 0$, we must have $f(2\pi-) > 0$. Since $p \in (0, 2\pi)$, we have $1 - \cos p > 0$. Hence these two facts imply that the right-hand side of the expression (9.80) must be negative. We end the proof of the problem. ∎

9.4. The Mean Value Theorems for Integrals

Problem 9.32

$(\star)(\star)(\star)$ Suppose that $f, g : [a,b] \to \mathbb{R}$ has continuous derivative on $[a,b]$. Suppose, further, that g is convex on $[a,b]$, $f(a) = g(a)$, $f(b) = g(b)$ and $g(x) \geq f(x)$ on $[a,b]$. Prove that

$$\int_a^b \sqrt{1 + [g'(x)]^2}\, dx \leq \int_a^b \sqrt{1 + [f'(x)]^2}\, dx.$$

Proof. Define the function $\varphi : \mathbb{R} \to \mathbb{R}$ by

$$\varphi(t) = \sqrt{1 + t^2}.$$

Since $\varphi''(t) = \frac{1}{(1+t^2)^{\frac{3}{2}}} \geq 0$ for all $t \in \mathbb{R}$, φ is convex on $\mathbb{R}$ by Theorem 8.16. Let $a < s < u < t < b$, if $\lambda = \frac{t-u}{t-s}$, then we have $0 < \lambda < 1$ and $\lambda s + (1-\lambda)t = u$. Since φ is convex on (a,b), we have

$$\varphi(\lambda s + (1-\lambda)t) \leq \lambda \varphi(s) + (1-\lambda)\varphi(t)$$
$$\varphi(u) \leq \frac{t-u}{t-s}\varphi(s) + \left(1 - \frac{t-u}{t-s}\right)\varphi(t)$$
$$(t-s)\varphi(u) \leq (t-u)\varphi(s) + (u-s)\varphi(t)$$
$$\frac{\varphi(u) - \varphi(s)}{u-s} \leq \frac{\varphi(t) - \varphi(s)}{t-s}. \tag{9.81}$$

If we take $u \to s$ to both sides of the inequality (9.81), then we have

$$\varphi'(s) \leq \frac{\varphi(t) - \varphi(s)}{t-s}. \tag{9.82}$$

Since $f(x) \leq g(x)$ on $[a,b]$, we have $f'(x) \leq g'(x)$ on $[a,b]$. If we put $t = f'(x)$ and $s = g'(x)$ into the inequality (9.82), then we see that

$$\varphi(g'(x)) + [f'(x) - g'(x)]\varphi'(g'(x)) \leq \varphi(f'(x)). \tag{9.83}$$

Since f', g' and φ, φ' are continuous on $[a,b]$ and $\mathbb{R}$ respectively, they are Riemann integrable on $[a,b]$ and $\mathbb{R}$ respectively. By Theorem 9.5 (Composition Theorem), $\varphi(f'), \varphi(g')$ and $\varphi'(g')$ are Riemann integrable on $[a,b]$. Therefore, we apply Theorem 9.6(b) (Operations of Integrable Functions) to the inequality (9.83) to get

$$\int_a^b \varphi(g'(x))\, dx + \int_a^b [f'(x) - g'(x)]\varphi'(g'(x))\, dx \leq \int_a^b \varphi(f'(x))\, dx$$

which is equivalent to

$$\int_a^b \sqrt{1 + [g'(x)]^2}\, dx + \int_a^b [f'(x) - g'(x)]\varphi'(g'(x))\, dx \leq \int_a^b \sqrt{1 + [f'(x)]^2}\, dx. \tag{9.84}$$

Since g and φ are convex on (a,b) and $\mathbb{R}$ respectively, we follow from Theorem 8.15 that g' and φ' are monotonically increasing on (a,b) and $\mathbb{R}$ respectively. Thus $\varphi'(g')$ is also monotonically increasing on (a,b). In addition, since g' and φ' are continuous on $[a,b]$ and $\mathbb{R}$ respectively, the function $\varphi'(g')$ is continuous on $[a,b]$. Therefore, this implies that $\varphi'(g')$ is actually monotonically increasing on $[a,b]$. Hence it follows from the Second Mean Value Theorem for Integrals and then the Second Fundamental Theorem of Calculus that there exists a $p \in [a,b]$ such that

$$\int_a^b \underbrace{[f'(x)-g'(x)]}_{g(x)} \underbrace{\varphi'(g'(x))}_{f(x)}\,\mathrm{d}x = \varphi'(g'(a)) \int_a^p [f'(x)-g'(x)]\,\mathrm{d}x$$

$$+ \varphi'(g'(b)) \int_p^b [f'(x)-g'(x)]\,\mathrm{d}x$$
$$= \varphi'(g'(a))[f(p)-g(p)-f(a)+g(a)]$$
$$+ \varphi'(g'(b))[f(b)-f(p)-g(b)+g(p)]$$
$$= [\varphi'(g'(b))-\varphi'(g'(a))][g(p)-f(p)]. \qquad (9.85)$$

Recall that $g(x) \geq f(x)$ on $[a,b]$ and $\varphi'(g')$ is monotonically increasing on $[a,b]$, we are able to obtain from the expression (9.85) that

$$\int_a^b [f'(x)-g'(x)]\varphi'(g'(x))\,\mathrm{d}x \geq 0$$

so that the inequality (9.84) implies that

$$\int_a^b \sqrt{1+[g'(x)]^2}\,\mathrm{d}x \leq \int_a^b \sqrt{1+[f'(x)]^2}\,\mathrm{d}x.$$

This completes the proof of the problem. ∎

CHAPTER 10

Sequences and Series of Functions

10.1 Fundamental Concepts

In Chapters 5 and 6, we consider sequences and series of real numbers. In this chapter, we study sequences and series whose terms are **functions**. In fact, the representation of a function as the limit of a sequence or an infinite series of functions arises naturally in advanced analysis. The main references in this chapter are [3, Chap. 9], [5, Chap. 8], [27, Chap. 3] and [34, Chap. 16].

10.1.1 Pointwise and Uniform Convergence

Definition 10.1 (Pointwise Convergence). *Suppose that $\{f_n\}$ is a sequence of real-valued or complex-valued functions on a set $E \subseteq \mathbb{R}$. For every $x \in E$, if the sequence $\{f_n(x)\}$ converges, then the function f defined by the equation*
$$f(x) = \lim_{n \to \infty} f_n(x) \tag{10.1}$$
*is called the **limit function** of $\{f_n\}$ and we can say that $\{f_n\}$ **converges pointwise** to f on E. Similarly, if the series $\sum f_n(x)$ converges pointwise for every $x \in E$, then we can define the function*
$$f(x) = \sum_{n=1}^{\infty} f_n(x) \tag{10.2}$$
*on E and it is called the **sum of the series**.*

Definition 10.2 (Uniform Convergence). *A sequence $\{f_n\}$ of functions defined on $E \subseteq \mathbb{R}$ is said to **converge uniformly** to f on E if for every $\epsilon > 0$, there exists an $N(\epsilon) \in \mathbb{N}$ such that $n \geq N(\epsilon)$ implies that*
$$|f_n(x) - f(x)| < \epsilon \tag{10.3}$$
*for all $x \in E$. In this case, f is called the **uniform limit** of $\{f_n\}$ on E. Similarly, we say that the series $\sum f_n(x)$ **converges uniformly** to f on E if the sequence $\{s_n\}$ of its partial sums, where*
$$s_n(x) = \sum_{k=1}^{n} f_k(x),$$
converges uniformly to f on E.

Figure 10.1 shows the sequence of functions $\{f_n\}$ on $[0, 1)$, where $f_n(x) = x^n$. This sequence of functions converges pointwise but *not* uniformly to $f = 0$ because if we take $\epsilon = \frac{1}{2}$, then for each $n \in \mathbb{N}$, one can find a x_n such that $f_n(x_n) > \frac{1}{2}$, i.e., points above the red dotted line. See also Problem 10.4 below.

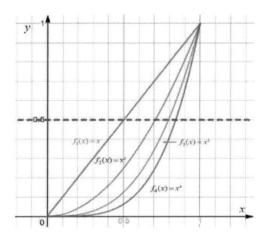

Figure 10.1: An example of pointwise convergence.

Now Figure 10.2 gives the sequence of functions $\{f_n\}$ on $[0, 1]$, where $f_n(x) = \frac{1}{n^2+x^2}$. In fact, the idea of the inequality (10.3) can be "seen" in the figure that the graphs of all f_n for $n \geq N$ lie "below" the line $y = \epsilon$.

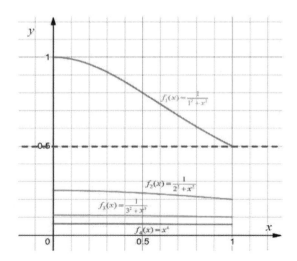

Figure 10.2: An example of uniform convergence.

10.1. Fundamental Concepts

> **Remark 10.1**
>
> (a) It is obvious that the uniform convergence of $\{f_n\}$ to f implies its pointwise convergence to f, but the converse is false.
>
> (b) Some books use $f_n \to f$ and $f_n \rightrightarrows f$ on E to denote pointwise convergence and uniform convergence respectively.

The core interest for the convergence problem here is that we want to determine what kinds of properties of functions f_n that will be "preserved" under the limiting processes (10.1) or (10.2). In fact, we discover that uniform convergence preserves continuity, differentiation and integrability of the functions f_n.

10.1.2 Criteria for Uniform Convergence

In the following, we state some methods of testing whether a sequence $\{f_n\}$ converges to its pointwise limit f uniformly.

Theorem 10.3 (Cauchy Criterion for Uniform Convergence). *Suppose that $\{f_n\}$ is a sequence of functions defined on $E \subseteq \mathbb{R}$.*

(a) *The sequence $\{f_n\}$ converges uniformly on E if and only if for every $\epsilon > 0$, there is an $N(\epsilon) \in \mathbb{N}$ such that $m, n \geq N(\epsilon)$ imply that*

$$|f_n(x) - f_m(x)| < \epsilon$$

for all $x \in E$.

(b) *The series $\sum f_n$ converges uniformly on E if and only if for every $\epsilon > 0$, there is an $N(\epsilon) \in \mathbb{N}$ such that $m > n \geq N(\epsilon)$ imply that*

$$\left|\sum_{k=n}^{m} f_k(x)\right| < \epsilon$$

for all $x \in E$.

Theorem 10.4. *The sequence $\{f_n\}$ converges uniformly to f on $E \subseteq \mathbb{R}$ if and only if*

$$M_n = \sup_{x \in E} |f_n(x) - f(x)| \to 0$$

as $n \to \infty$.

Theorem 10.5 (Weierstrass M-test). *Suppose that $\{f_n\}$ is a sequence of functions defined on $E \subseteq \mathbb{R}$ and $\{M_n\}$ is a sequence of nonnegative numbers such that $|f_n(x)| \leq M_n$ for all $n = 1, 2, 3, \ldots$. If $\sum M_n$ converges, then $\sum f_n$ converges uniformly on E.*

10.1.3 Preservation Theorems

Theorem 10.6 (Uniform Convergence and Continuity). *Suppose that $f_n \to f$ uniformly on $E \subseteq \mathbb{R}$. If every f_n is continuous at $p \in E$, then the uniform limit f is also continuous at p.*

Theorem 10.7 (Uniform Convergence and Riemann-Stieltjes Integration)**.** Let $a < b$. Suppose that α is monotonically increasing on $[a,b]$, all $f_n \in \mathscr{R}(\alpha)$ on $[a,b]$ and $f_n \to f$ uniformly on $[a,b]$. Then we have $f \in \mathscr{R}(\alpha)$ on $[a,b]$ and

$$\int_a^b f \, d\alpha = \int_a^b \lim_{n \to \infty} f_n \, d\alpha = \lim_{n \to \infty} \int_a^b f_n \, d\alpha.$$

In particular, if the series

$$\sum_{n=1}^\infty f_n \to f$$

uniformly to f on $[a,b]$, then we have

$$\int_a^b f \, d\alpha = \int_a^b \sum_{n=1}^\infty f_n \, d\alpha = \sum_{n=1}^\infty \int_a^b f_n \, d\alpha.$$

Theorem 10.8 (Uniform Convergence and Differentiability)**.** Suppose that $\{f_n\}$ is a sequence of differentiable functions on $[a,b]$, where $a < b$. We suppose that

(a) $\{f_n(p)\}$ converges for at least one point $p \in [a,b]$, and

(b) $\{f_n'\}$ converges uniformly on $[a,b]$.

Then there exists a function f defined on $[a,b]$ such that $f_n \to f$ uniformly on $[a,b]$ and

$$f'(x) = \lim_{n \to \infty} f_n'(x) \tag{10.4}$$

on $[a,b]$.

Furthermore, if the sequences $\{f_n(p)\}$ and $\{f_n'\}$ in parts (a) and (b) are replaced by the series $\sum f_n(p)$ and $\sum f_n'$ respectively, then we have the same conclusion and the equation (10.4) is replaced by

$$f'(x) = \sum_{n=1}^\infty f_n'(x).$$

10.1.4 Uniformly Boundedness and Equicontinuity

Now we want to find any similarity between sequences of numbers $\{x_n\}$ and sequences of functions $\{f_n\}$. In fact, we have two important and basic questions. The first question origins from the Bolzano-Weierstrass Theorem (Problem 5.25): Every bounded sequence of real (or complex in fact) sequence has a convergent subsequence. *Is there any similar result for sequences of functions?* To answer this question, we have to define two kinds of boundedness first.

Definition 10.9 (Pointwise Boundedness and Uniformly Boundedness)**.** Suppose that $\{f_n\}$ is a sequence of functions defined on $E \subseteq \mathbb{R}$.

(a) It is said that $\{f_n\}$ is **pointwise bounded** on E if for every $x \in E$, there is a *finite-valued* function $g : E \to \mathbb{R}$ such that
$$|f_n(x)| < g(x)$$
for all $x \in E$ and $n = 1, 2, \ldots$.

10.1. Fundamental Concepts

(b) It is said that $\{f_n\}$ is **uniformly bounded** on E if there is a positive constant M such that
$$|f_n(x)| < M$$
for all $x \in E$ and $n = 1, 2, \ldots$.

> **Remark 10.2**
>
> It is clear from the definitions that uniformly boundedness certainly implies pointwise boundedness.

It is well-known that if $\{f_n\}$ is **pointwise bounded** on a *countable* set E, then it has a subsequence $\{f_{n_k}\}$ such that $\{f_{n_k}(x)\}$ converges for every $x \in E$.[a] This answers the first question partially. However, the general situation *does not* hold even if E is compact and $\{f_n\}$ is a uniformly bounded sequence of continuous functions.

Next, the second question origins from the fact that a sequence $\{x_n\}$ converges if and only if every subsequence $\{x_{n_k}\}$ converges (Theorem 5.3). *Is every convergent sequence of functions contains a uniformly convergent subsequence?* The answer is negative even if we assume stronger conditions that $\{f_n\}$ is uniformly bounded on a compact set E.

Both failures of the above two questions are due to the lack of the concept of **equicontinuity** which is stated as follows:

Definition 10.10 (Equicontinuity). Let X be a metric space with metric d and $\mathscr{F}$ a family of functions defined on $E \subseteq X$. The family $\mathscr{F}$ is **equicontinuous** on E if for every $\epsilon > 0$, there exists a $\delta > 0$ such that
$$|f(x) - f(y)| < \epsilon$$
for all $f \in \mathscr{F}$ and all $x, y \in X$ with $d(x, y) < \delta$.

Theorem 10.11. *Suppose that K is a compact metric space. If $\{f_n\}$ is a sequence of continuous functions on K and it converges uniformly on K, then $\{f_n\}$ is equicontinuous on K.*

The Arzelà-Ascoli Theorem. *Suppose that K is a compact metric space and $\{f_n\}$ is a sequence of continuous functions on K. If $\{f_n\}$ is pointwise bounded and equicontinuous on K, then it is uniformly bounded on K. Furthermore, $\{f_n\}$ contains a uniformly convergent subsequence $\{f_{n_k}\}$ on K.*

10.1.5 The Space $\mathscr{C}(X)$ and the Approximation by Polynomials

Definition 10.12. Suppose that X is a metric space. Then $\mathscr{C}(X)$ denotes the set of all complex-valued bounded continuous functions on X. For each $f \in \mathscr{C}(X)$, we define
$$\|f\| = \sup_{x \in X} |f(x)|$$
which is called the **supremum norm**. Then $\mathscr{C}(X)$ is a complete metric space with this metric $\|\cdot\|$.

---

[a] See, for example, [23, Theorem 7.23, p. 156].

> **Remark 10.3**
>
> If $\mathscr{A}$ is closed in $\mathscr{C}(X)$, then $\mathscr{A}$ is call **uniformly closed** and its closure $\overline{\mathscr{A}}$ is called **uniform closure**.

By Theorem 10.6 (Uniform Convergence and Continuity) or Problem 10.15, we know that the uniform limit of a sequence of polynomials is a continuous function. It is natural to ask its converse: Can a continuous function be approximated *uniformly* by polynomials? The answer to this question is affirmative and in fact, we state it as follows:

The Weierstrass Approximation Theorem. *If $f : [a,b] \to \mathbb{C}$ is continuous, then there exists a sequence of polynomials $\{P_n\}$ such that $P_n \to f$ uniformly on $[a,b]$. If f is real, then we may take every P_n to be real.*

10.2 Uniform Convergence for Sequences of Functions

> **Problem 10.1**
>
> (⋆) For each $n \in \mathbb{N}$, we define
> $$f_n(x) = \begin{cases} \frac{x}{n}, & \text{if } n \text{ is odd;} \\ \frac{1}{n}, & \text{otherwise.} \end{cases}$$
> Prove that $\{f_n\}$ converges pointwise but not uniformly on $\mathbb{R}$.

Proof. For every $x \in \mathbb{R}$, since
$$\lim_{k \to \infty} f_{2k+1}(x) = \lim_{k \to \infty} \frac{x}{2k+1} = 0 \quad \text{and} \quad \lim_{k \to \infty} f_{2k}(x) = \lim_{k \to \infty} \frac{1}{2k} = 0,$$
we conclude that $\{f_n\}$ converges pointwise to $f = 0$ on $\mathbb{R}$. However, since
$$M_{2k+1} = \sup_{x \in \mathbb{R}} |f_{2k+1}(x) - 0| = \sup_{x \in \mathbb{R}} \left|\frac{x}{2k}\right| = +\infty,$$
Theorem 10.4 says that $\{f_n\}$ does not converge uniformly on $\mathbb{R}$. This completes the proof of the problem. ∎

> **Problem 10.2**
>
> (⋆) Verify that the sequence of functions $f_n(x) = \frac{x^2+nx}{n}$ converges uniformly on $[0,1]$, but not on $\mathbb{R}$.

10.2. Uniform Convergence for Sequences of Functions

Proof. Obviously, we have $f_n(x) \to x$ as $n \to \infty$ for every $x \in \mathbb{R}$, so its pointwise limit is $f(x) = x$. Now for each *fixed* $n \in \mathbb{N}$, we have

$$M_n = \sup_{x \in \mathbb{R}} |f_n(x) - f(x)| = \sup_{x \in \mathbb{R}} \frac{x^2}{n} = \infty,$$

so Theorem 10.4 implies that $\{f_n\}$ does not converge uniformly to $f(x) = x$ on $\mathbb{R}$. However, on the interval $[0,1]$, we note that

$$M_n = \frac{1}{n}$$

for every $n \in \mathbb{N}$ so that $M_n \to 0$ as $n \to \infty$. Hence it deduces from Theorem 10.4 again that $\{f_n\}$ converges uniformly to $f(x) = x$ on $[0,1]$. We have completed the proof of the problem. ∎

Problem 10.3 (Dini's Theorem)

(⋆) Suppose that $\{f_n\}$ is a sequence of continuous functions defined on $[a,b]$ such that $f_n(x) \geq f_{n+1}(x)$ for all $x \in [a,b]$ and $n \in \mathbb{N}$. If $\{f_n\}$ converges pointwise to a continuous function f on $[a,b]$, prove that $f_n \to f$ uniformly on $[a,b]$.

Proof. Without loss of generality, we assume that $f = 0$. Since each f_n is uniformly continuous on $[a,b]$, the number

$$M_n = \sup_{x \in [a,b]} |f_n(x)|$$

is well-defined. Since $f_n(x) \geq f_{n+1}(x)$ on $[a,b]$, $\{M_n\}$ is decreasing. If M_n does not converge to 0, then there exists a $\epsilon > 0$ such that $M_n > \epsilon$ for all $n \in \mathbb{N}$. This means that for each n, we can find a $x_n \in [a,b]$ with

$$f_n(x_n) > \epsilon. \tag{10.5}$$

Since $\{x_n\} \subseteq [a,b]$, it follows from the Bolzano-Weierstrass Theorem that $\{x_n\}$ has a convergent subsequence. Let the limit of the subsequence be p. By the hypothesis, we have $f_n(p) \to 0$ as $n \to \infty$. In other words, there is an $N \in \mathbb{N}$ such that

$$f_N(p) < \epsilon.$$

Since f_N is continuous at p, the Sign-preserving Property (see Problem 7.15) ensures that there is a $\delta > 0$ such that

$$f_N(x) < \epsilon \tag{10.6}$$

for all $x \in [a,b]$ with $|x - p| < \delta$. Now the property $f_n(x) \geq f_{n+1}(x)$ on $[a,b]$ implies that the inequality (10.6) also holds for all f_n with $n \geq N$. By the definition of p, we can choose x_n with $n \geq N$ and $|x_n - p| < \delta$ so that the inequality (10.6) gives

$$f_n(x_n) < \epsilon \tag{10.7}$$

for all $n \geq N$. Now it is obvious that the inequalities (10.5) and (10.7) are contrary. Hence we see that $M_n \to 0$ as $n \to \infty$, completing the proof of the problem. ∎

Problem 10.4

(⋆) Show that the hypothesis "f is continuous" in Problem 10.3 (Dini's Theorem) cannot be dropped.

Proof. Let $f_n(x) = x^n$ on $[0, 1]$. Then it is easy to see that $f_n(1) = 1$ for all $n \in \mathbb{N}$ and $f_n(x) \to 0$ as $n \to \infty$ for all $x \in [0, 1)$. Thus the limit function f of this sequence $\{f_n\}$ is given by

$$f(x) = \begin{cases} 1, & \text{if } x = 1; \\ 0, & \text{otherwise.} \end{cases} \tag{10.8}$$

Assume the convergence was uniform. Then Theorem 10.6 (Uniform Convergence and Continuity) shows that the function f must also be continuous on $[0, 1]$, but it contradicts the definition (10.8). Thus the condition "f is continuous" in Problem 10.3 (Dini's Theorem) cannot be dropped, completing the proof of the problem. ∎

Problem 10.5

(⋆) Construct a sequence of continuous functions $\{f_n\}$ defined on a compact set K with pointwise continuous limit function f but it does not converge uniformly to f on K.

Proof. Suppose that $K = [0, 1]$ and each $f_n : [0, 1] \to \mathbb{R}$ is defined by

$$f_n(x) = \begin{cases} nx, & \text{if } x \in [0, \frac{1}{n}]; \\ 2 - nx, & \text{if } x \in (\frac{1}{n}, \frac{2}{n}]; \\ 0, & \text{otherwise.} \end{cases}$$

Then it is clear that every f_n is continuous on $[0, 1]$. If $x = 0$, then $f_n(0) = 0$ for every $n \in \mathbb{N}$. If $x \in (0, 1]$, then there exists an $N \in \mathbb{N}$ such that $x > \frac{2}{N}$ so that $x \notin (0, \frac{2}{N}]$ and thus $f_n(x) = 0$ for every $n \geq N$. In other words, its limit function is

$$f = 0$$

which is continuous on $[0, 1]$. However, we note that for each $n \in \mathbb{N}$, we have

$$M_n = \sup_{x \in [0,1]} |f_n(x) - 0| = f_n\left(\frac{1}{n}\right) = 1$$

and we follow from Theorem 10.4 that $\{f_n\}$ does not converge to $f = 0$ uniformly. This is the end of the proof. ∎

Problem 10.6

(⋆) Suppose $f_n \to f$ uniformly on $E \subseteq \mathbb{R}$ and there is a positive constant M such that $|f_n(x)| \leq M$ on E and all $n \in \mathbb{N}$. Let g be continuous on $[-M, M]$. Define $h = g \circ f$ and $h_n = g \circ f_n$ for every $n \in \mathbb{N}$. Show that $h_n \to h$ uniformly on E.

Proof. Since g is continuous on $[-M, M]$, we know from Theorem 7.10 that g is uniformly continuous on $[-M, M]$. Given $\epsilon > 0$, there is a $\delta > 0$ such that

$$|g(x) - g(y)| < \epsilon \tag{10.9}$$

10.2. Uniform Convergence for Sequences of Functions

for all $x, y \in [-M, M]$ with $|x - y| < \delta$. Since $f_n \to f$ uniformly on E, there is an $N \in \mathbb{N}$ such that $n \geq N$ implies that
$$|f_n(x) - f(x)| < \delta \tag{10.10}$$
for every $x \in E$. Combining the inequalities (10.9) and (10.10), it can be shown that $n \geq N$ implies
$$|h_n(x) - h(x)| = |g(f_n(x)) - g(f(x))| < \epsilon$$
for every $x \in E$. By Definition 10.2 (Uniform Convergence), $\{h_n\}$ converges uniformly to h on E. We complete the proof of the problem. ∎

Problem 10.7

(★) Define $f_n : [0, 1] \to \mathbb{R}$ by
$$f_n(x) = \frac{1}{1 + x^n}.$$
Prove that $\{f_n\}$ converges uniformly to f on $[0, \alpha]$, but not on $[0, 1]$, where $0 < \alpha < 1$.

Proof. Let f be the pointwise limit function of $\{f_n\}$. If $x = 1$, then $f_n(1) = \frac{1}{2}$ for every $n \in \mathbb{N}$. Thus we have $f(1) = \frac{1}{2}$. Let $x \in [0, 1)$. Since
$$\lim_{n \to \infty} f_n(x) = \lim_{n \to \infty} \frac{1}{1 + x^n} = 1,$$
we obtain $f(x) = 1$ on $x \in [0, 1)$. In other words, we have
$$f(x) = \begin{cases} \frac{1}{2}, & \text{if } x = 1; \\ 1, & \text{otherwise.} \end{cases} \tag{10.11}$$

Next, we fix $\alpha \in (0, 1)$. For every $x \in [0, \alpha]$, we have $x^n \leq \alpha^n$ so that $x^n + x^n \alpha^n \leq \alpha^n + x^n \alpha^n$. By this and the definition (10.11), we know that
$$|f_n(x) - f(x)| = \left| \frac{1}{1 + x^n} - 1 \right| = \frac{x^n}{1 + x^n} \leq \frac{\alpha^n}{1 + \alpha^n}. \tag{10.12}$$

Since $0 < \alpha < 1$, we have
$$\lim_{n \to \infty} \frac{\alpha^n}{1 + \alpha^n} = 0.$$
Therefore, we conclude from this and the inequality (10.12) that $f_n \to f$ uniformly on $[0, \alpha]$.

However, given a positive integer n, we let $x \in (\sqrt[n]{\frac{1}{2}}, 1)$ so that $\frac{1}{2} < x^n < 1$. By this and the definition (10.11) again, it yields that
$$|f_n(x) - f(x)| = \left| \frac{1}{1 + x^n} - 1 \right| = \frac{x^n}{1 + x^n} > \frac{\frac{1}{2}}{1 + 1} = \frac{1}{4}$$
and then $\{f_n\}$ does not converge to f on $[0, 1]$ by Theorem 10.4. Hence we complete the proof of the problem. ∎

Problem 10.8

(★) For $n \in \mathbb{N}$, we suppose that $f_n = \frac{1}{n}\exp(-n^2x^2)$ on $\mathbb{R}$. Prove that $f_n \to 0$ uniformly on $\mathbb{R}$, $f_n' \to 0$ pointwise on $\mathbb{R}$, but not uniformly on $(-M, M)$ for every $M > 0$.

Proof. For every $x \in \mathbb{R}$, note that $e^{x^2} \geq 1$ and so
$$|f_n(x)| = \frac{1}{ne^{n^2x^2}} \leq \frac{1}{n}.$$

By Theorem 10.4, $f_n \to 0$ uniformly on $\mathbb{R}$. By direct differentiation, we have $f_n'(x) = -2nxe^{-n^2x^2}$, so it is easy to see that $f_n' \to 0$ pointwise on $\mathbb{R}$.

Assume that there was a $M > 0$ such that $f_n' \to 0$ uniformly on $(-M, M)$. Pick $\epsilon = e^{-1}$. By assumption, there exists an $N \in \mathbb{N}$ such that $n \geq N$ implies that
$$|f_n'(x)| = \frac{2n|x|}{e^{n^2x^2}} < e^{-1} \tag{10.13}$$

on $(-M, M)$. Now we may take N to be large enough so that $N > \frac{1}{M}$, i.e., $\frac{1}{N} \in (0, M)$. Then it is legal to put $x = \frac{1}{N}$ into the inequality (10.13) to get the contradiction that
$$2e^{-1} < e^{-1}.$$

Hence $\{f_n'\}$ does not converge uniformly to $f' = 0$ on $(-M, M)$, completing the proof of the problem. ∎

Problem 10.9

(★) Suppose that $\{f_n\}$ is a sequence of uniformly continuous functions on $\mathbb{R}$. If $f_n \to f$ uniformly on $\mathbb{R}$, prove that f is uniformly continuous on $\mathbb{R}$.

Proof. Given $\epsilon > 0$. By the hypotheses, there exists an $N \in \mathbb{N}$ such that $n \geq N$ implies that
$$|f_n(x) - f(x)| < \frac{\epsilon}{3} \tag{10.14}$$

on $\mathbb{R}$. Since f_N is uniformly continuous on $\mathbb{R}$, there exists a $\delta > 0$ such that
$$|f_N(x) - f_N(y)| < \frac{\epsilon}{3} \tag{10.15}$$

for every pair $x, y \in \mathbb{R}$ with $|x - y| < \delta$. Combining the inequalities (10.14) and (10.15), we see immediately that
$$|f(x) - f(y)| \leq |f(x) - f_N(x)| + |f_N(x) - f_N(y)| + |f_N(y) - f(y)| < \epsilon$$

for all $x, y \in \mathbb{R}$ with $|x - y| < \delta$. Hence f is uniformly continuous on $\mathbb{R}$ which completes the proof of the problem. ∎

10.2. Uniform Convergence for Sequences of Functions

Problem 10.10

(★) Let $\{f_n\}$ be a sequence of bounded functions on $E \subseteq \mathbb{R}$. If $f_n \to f$ uniformly on E, prove that f is bounded on E. Is it true for pointwise convergence on bounded E?

Proof. Given $\epsilon > 0$. Then one can find an $N \in \mathbb{N}$ such that $n \geq N$ implies that

$$|f_n(x) - f(x)| < \epsilon.$$

For this N, since f_N is bounded on E, there is a $M > 0$ such that $|f_N(x)| \leq M$ on E. Thus we obtain

$$|f(x)| \leq |f(x) - f_N(x)| + |f_N(x)| < \epsilon + M$$

on E, i.e., f is bounded on E.

The second assertion is negative. For example, we consider $E = (0, 1)$ and

$$f_n(x) = \min(x^{-2}, n).$$

Then it is trivial to check that $f_n(x) \to x^{-2}$ pointwise on $(0, 1)$. Since x^{-2} is unbounded on the bounded set $(0, 1)$, this counterexample completes the proof of the problem. ∎

Problem 10.11

(★) Prove that

$$\lim_{n \to \infty} \int_1^\pi e^{-nx^2} \, dx = 0.$$

Proof. Consider $f_n(x) = e^{-nx^2}$ defined on $[1, \pi]$. Clearly, we have $f_n \in \mathscr{R}$ on $[1, \pi]$ because f_n is continuous on $[1, \pi]$ and

$$M_n = \sup_{x \in [1, \pi]} |f_n(x)| \leq \frac{1}{e^n} \to 0$$

as $n \to \infty$. Thus $\{f_n\}$ converges uniformly to 0 on $[1, \pi]$ by Theorem 10.4. By Theorem 10.7 (Uniform Convergence and Riemann-Stieltjes Integration), we see easily that

$$\lim_{n \to \infty} \int_1^\pi e^{-nx^2} \, dx = \int_1^\pi \lim_{n \to \infty} e^{-nx^2} \, dx = 0$$

which completes the proof of the problem. ∎

Problem 10.12

(★) Let $f_n : \mathbb{R} \to \mathbb{R}$. Prove that if $\{f_n\}$ converges pointwise to f on $\mathbb{R}$, then it converges uniformly to f on any finite subset of $\mathbb{R}$.

Proof. Let $\{x_1, x_2, \ldots, x_m\}$ be a finite subset of $\mathbb{R}$. Given $\epsilon > 0$. Since $f_n \to f$ pointwise on $\mathbb{R}$, there exists an $N(\epsilon, x_k) \in \mathbb{N}$ such that $n \geq N(\epsilon, x_k)$ implies that

$$|f_n(x_k) - f(x_k)| < \epsilon, \qquad (10.16)$$

where $k = 1, 2, \ldots, m$. Take

$$N(\epsilon) = \max(N(\epsilon, x_1), N(\epsilon, x_2), \ldots, N(\epsilon, x_m)).$$

Then it is obvious that the inequality (10.16) holds for all $k = 1, 2, \ldots, m$ when $n \geq N(\epsilon)$. In other words, $\{f_n\}$ converges uniformly to f on $\{x_1, x_2, \ldots, x_m\}$ which ends the proof of the problem. ∎

Problem 10.13

(★)(★) Let $f_n : \mathbb{R} \to \mathbb{R}$. Prove that if $\{f_n\}$ converges uniformly to f on all countable subsets of $\mathbb{R}$, then $\{f_n\}$ converges uniformly to f on $\mathbb{R}$. Can the hypothesis "all countable subsets" be replaced by the condition "all Cauchy sequences"?

Proof. Assume that $\{f_n\}$ did not converge uniformly to f on $\mathbb{R}$. In other words, *for some $\epsilon > 0$ and for every $N \in \mathbb{N}$, there exists* a $x \in \mathbb{R}$ such that

$$|f_n(x) - f(x)| \geq \epsilon$$

for some $n \geq N$. Particularly, pick $N = 1$, then one can find a $x_1 \in \mathbb{R}$ such that

$$|f_n(x_1) - f(x_1)| \geq \epsilon$$

hold for some $n \geq 1$. In fact, we can find a countable subset $E = \{x_k\}$ of $\mathbb{R}$ such that

$$|f_n(x_k) - f(x_k)| \geq \epsilon \qquad (10.17)$$

hold for some $n \geq k$, where $k = 1, 2, \ldots$. However, the inequality (10.17) means that $\{f_n\}$ does *not* converge uniformly on the countable set E, a contradiction.

The second assertion is false. For example, we consider

$$f_n(x) = \begin{cases} 0, & \text{if } x \leq n - \frac{1}{2}; \\ x - n + \frac{1}{2}, & \text{if } n - \frac{1}{2} \leq x \leq n; \\ \frac{1}{2}, & \text{otherwise.} \end{cases} \qquad (10.18)$$

Let $\{p_k\}$ be a Cauchy sequence in $\mathbb{R}$. We claim that $f_n \to 0$ uniformly on $\{p_k\}$. We know from Theorem 5.2 that $\{p_k\}$ is bounded by a positive constant M. On the one hand, if $n > M + \frac{1}{2}$, then $p_k < n - \frac{1}{2}$ for every $k = 1, 2, \ldots$ and we obtain from the definition (10.18) that

$$f_n(p_k) = 0$$

for every $k = 1, 2, \ldots$. Consequently, $f_n \to 0$ uniformly on $\{p_k\}$. On the other hand, for every $n \in \mathbb{N}$, since

$$M_n = \sup_{x \in \mathbb{R}} |f_n(x) - 0| = \frac{1}{2},$$

Theorem 10.4 implies that $\{f_n\}$ does not converge uniformly on $\mathbb{R}$. This completes the proof of the problem. ∎

10.2. Uniform Convergence for Sequences of Functions

Problem 10.14

(★) (★) Suppose that $\{f_n\}$ is a sequence of continuous functions defined on a compact set $K \subset \mathbb{R}$, $f_n \to f$ pointwise on K and f is continuous. Prove that $f_n \to f$ uniformly on K if and only if for every $\epsilon > 0$, there is a $m \in \mathbb{N}$ and a $\delta > 0$ such that $n > m$ and $|f_k(x) - f(x)| < \delta$ imply that
$$|f_{k+n}(x) - f(x)| < \epsilon \tag{10.19}$$
for all $x \in K$ and $k \in \mathbb{N}$.

Proof. Suppose that $f_n \to f$ uniformly on K. Thus given $\epsilon > 0$, there exists an $N \in \mathbb{N}$ such that $n > N$ implies that
$$|f_n(x) - f(x)| < \epsilon$$
for all $x \in K$. If we take $m = N$ and $\delta = \epsilon$, then the inequality (10.19) holds trivially.

Conversely, given $\epsilon > 0$. By the hypotheses, there exists a $m \in \mathbb{N}$ and a $\delta > 0$ such that $n > m$ and $|f_k(x) - f(x)| < \delta$ imply the inequality (10.19) on K and all $k \in \mathbb{N}$. Fix $p \in K$. Since $f_n(p) \to f(p)$ as $n \to \infty$, there exists a $k \in \mathbb{N}$ such that
$$|f_k(p) - f(p)| < \frac{\delta}{3}. \tag{10.20}$$

Recall that both f_k and f are continuous at p, so there exists a neighborhood $B(p, r_p)$ of radius $r_p > 0$ such that $x \in B(p, r_p) \cap K$ implies
$$|f_k(x) - f_k(p)| < \frac{\delta}{3} \quad \text{and} \quad |f(x) - f(p)| < \frac{\delta}{3}. \tag{10.21}$$

Combining the inequalities (10.20) and (10.21), we see that
$$|f_k(x) - f(x)| \leq |f_k(x) - f_k(p)| + |f_k(p) - f(p)| + |f(p) - f(x)| < \delta \tag{10.22}$$
for all $x \in B(p, r_p) \cap K$. It is clear that $\{B(p, r_p) \,|\, p \in K\}$ is an open cover of K, so there exists a finite index $\{p_1, p_2, \ldots, p_s\}$ such that
$$K \subseteq B(p_1, r_{p_1}) \cup B(p_2, r_{p_2}) \cup \cdots \cup B(p_s, r_{p_s}).$$

Suppose that k_i is the corresponding positive integer satisfying the inequality (10.22) for the point p_i, where $i = 1, 2, \ldots, s$. Then for arbitrary $x \in K$, we have $x \in B(p_i, r_{p_i})$ for some i so that $x \in B(p_i, r_{p_i}) \cap K$ and thus
$$|f_{k_i}(x) - f(x)| < \delta.$$
By our hypotheses, if $n > m$, then we have
$$|f_{k_i + n}(x) - f(x)| < \epsilon \tag{10.23}$$
for all $x \in B(p_i, r_{p_i}) \cap K$. Put $N = \max(k_1, k_2, \ldots, k_s) + m$. If $n' > N \geq k_i + m$ for every $i = 1, 2, \ldots, s$, then we deduce from the inequality (10.23) that
$$|f_{n'}(x) - f(x)| < \epsilon. \tag{10.24}$$

Since x is arbitrary in K, we conclude from the inequality (10.24) that $f_n \to f$ uniformly on K and we complete the proof of the problem. ∎

> **Problem 10.15**
>
> (★)(★) Suppose that $\{P_n\}$ is a sequence of polynomials on $\mathbb{R}$ with complex coefficients and $P_n \to P$ uniformly on $\mathbb{R}$. Is P again a polynomial on $\mathbb{R}$?

Proof. We claim that there exists an $N \in \mathbb{N}$ such that
$$\deg P_n = \deg P_N$$
for all $n \geq N$. Assume that it was not the case. Since $P_n \to P$ uniformly on $\mathbb{R}$, Theorem 10.3 (Cauchy Criterion for Uniform Convergence) ensures that there exists a $k \in \mathbb{N}$ such that $m, n \geq k$ imply that
$$|P_m(x) - P_n(x)| < 1 \tag{10.25}$$
for all $x \in \mathbb{R}$. For this particular k, our assumption shows that there is an $n_k > k$ with the property
$$\deg P_{n_k} \neq \deg P_k$$
and this definitely shows that
$$\sup_{x \in \mathbb{R}} |P_{n_k}(x) - P_k(x)| = \infty.$$
which contradicts the inequality (10.25). Hence this proves the claim.

Now we let $\deg P_N = K$ and
$$P_n(x) = a_{n,K} x^K + a_{n,K-1} x^{K-1} + \cdots + a_{n,0}, \tag{10.26}$$
for all $n \geq N$, where $a_{n,K}, a_{n,K-1}, \ldots, a_{n,0} \in \mathbb{C}$. If $a_{n,i} \neq a_{N,i}$ for some $i \in \{1, 2, \ldots, K\}$, then we have
$$\sup_{x \in \mathbb{R}} |P_n(x) - P_N(x)| = \infty$$
which contradicts the inequality (10.25) again. In other words, we have $a_{n,i} = a_{N,i}$ for all $n \geq N$ and $i = 1, 2, \ldots, K$. For simplicity, we write $a_{N,i} = a_i$ and the polynomial (10.26) can be expressed as
$$P_n(x) = a_K x^K + a_{K-1} x^{K-1} + \cdots + a_1 x + a_{n,0},$$
for all $n \geq N$. If $a_{n,0} \to a_0$ as $n \to \infty$, then the function $P(x)$ must be in the form
$$P(x) = a_K x^K + a_{K-1} x^{K-1} + \cdots + a_1 x + a_0$$
which is a polynomial on $\mathbb{R}$ with complex coefficients. This completes the proof of the problem. ∎

> **Problem 10.16**
>
> (★)(★) Suppose that $f_n : \mathbb{R} \to \mathbb{R}$ and all f_n satisfy **Lipschitz condition** with the same **Lipschitz constant** K. If $f_n \to f$ uniformly on $\mathbb{R}$, prove that f also satisfies Lipschitz condition with Lipschitz constant K.

10.3. Uniform Convergence for Series of Functions

Proof. For the definitions of Lipschitz condition and Lipschitz constant, you are suggested to read Remark 8.7.

Given $\epsilon > 0$, there exists an $N \in \mathbb{N}$ such that $n \geq N$ implies that
$$|f_n(x) - f(x)| < \frac{\epsilon}{2} \tag{10.27}$$
for all $x \in \mathbb{R}$. Now for any pair $x, p \in \mathbb{R}$, we deduce from the inequality (10.27) that
$$|f(x) - f(p)| \leq |f(x) - f_N(x)| + |f_N(x) - f_N(p)| + |f_N(p) - f(p)| < \epsilon + |f_N(x) - f_N(p)|. \tag{10.28}$$
Since f_N satisfies the Lipschitz condition with Lipschitz constant K, it means that
$$|f_N(x) - f_N(p)| \leq K|x - p|$$
for all $x, p \in \mathbb{R}$. Thus the inequality (10.28) gives
$$|f(x) - f(p)| < \epsilon + K|x - p| \tag{10.29}$$
for all $x, p \in \mathbb{R}$. Since ϵ is arbitrary, the inequality (10.29) actually implies that
$$|f(x) - f(p)| \leq K|x - p|$$
for all $x, p \in \mathbb{R}$. Hence f also satisfies Lipschitz condition with Lipschitz constant K, completing the proof of the problem. ∎

10.3 Uniform Convergence for Series of Functions

Problem 10.17

(★) Let $p > 1$. Show that the series
$$\sum_{n=1}^{\infty} \left[\frac{\pi}{2} - \arctan[n^p(1 + x^2)] \right]$$
converges uniformly on $\mathbb{R}$.

Proof. Recall the fact that $\arctan x + \arctan \frac{1}{x} = \frac{\pi}{2}$ for $x > 0$, so
$$0 < \frac{\pi}{2} - \arctan[n^p(1 + x^2)] = \arctan \frac{1}{n^p(1 + x^2)} \tag{10.30}$$
for all $x \in \mathbb{R}$. Let $f(x) = x - \arctan x$ for $x > 0$. Since $f'(x) = \frac{x^2}{1+x^2} > 0$, f is strictly increasing on $(0, \infty)$ by Remark 8.3 and thus
$$x > \arctan x$$

if $x > 0$. Therefore, we further reduce the inequality (10.30) to

$$0 < \frac{\pi}{2} - \arctan[n^p(1+x^2)] < \frac{1}{n^p(1+x^2)} \leq \frac{1}{n^p}$$

for all $x \in \mathbb{R}$. By Theorem 6.10, the series $\sum_{n=1}^{\infty} \frac{1}{n^p}$ converges so that we follow from Theorem 10.5 (Weierstrass M-test) that the series

$$\sum_{n=1}^{\infty} \left[\frac{\pi}{2} - \arctan[n^p(1+x^2)]\right]$$

converges uniformly on $\mathbb{R}$. This completes the proof of the problem. ∎

Problem 10.18

(★) Let $p > \frac{1}{2}$. Prove that the series

$$\sum_{n=1}^{\infty} \frac{x}{n^p(1+nx^2)}$$

converges uniformly on $\mathbb{R}$.

Proof. For each $n \in \mathbb{N}$, let

$$f_n(x) = \frac{x}{n^p(1+nx^2)}$$

be defined on $\mathbb{R}$. Using the A.M. $\geq$ G.M., we know that

$$\frac{n^p(1+nx^2)}{2} = \frac{n^p + n^{p+1}x^2}{2} \geq \sqrt{n^{2p+1}x^2} = n^{p+\frac{1}{2}}|x|$$

which implies

$$\left|\frac{x}{n^p(1+nx^2)}\right| \leq \frac{1}{2n^{p+\frac{1}{2}}}$$

for all $x \in \mathbb{R}$. Since $p > \frac{1}{2}$, it yields from Theorem 6.10 that $\sum_{n=1}^{\infty} \frac{1}{n^{p+\frac{1}{2}}}$ converges and Theorem 10.5 (Weierstrass M-test) shows that the series

$$\sum_{n=1}^{\infty} \frac{x}{n^p(1+nx^2)}$$

converges uniformly on $\mathbb{R}$. We have completed the proof of the problem. ∎

10.3. Uniform Convergence for Series of Functions

> **Problem 10.19**
>
> (⋆) Suppose that each $f_n : [0, 1] \to (0, \infty)$ is continuous and
> $$f(x) = \sum_{n=1}^{\infty} f_n(x)$$
> is also continuous on $[0, 1]$. Prove that the series $\sum_{n=1}^{\infty} f_n(x)$ converges uniformly on $[0, 1]$.

Proof. Let $S_n(x) = \sum_{k=1}^{n} f_k(x)$ and
$$R_n(x) = \sum_{k=n+1}^{\infty} f_k(x) = f(x) - S_n(x),$$
where $n = 1, 2, \ldots$ and $x \in [0, 1]$. Since f and S_n are continuous on $[0, 1]$, each R_n is continuous on $[0, 1]$. By the definition, we also have
$$R_n(x) \geq R_{n+1}(x)$$
for all $x \in [0, 1]$ and all $n \in \mathbb{N}$. Furthermore, $R_n(x) \to 0$ pointwise on $[0, 1]$. Hence it follows from Problem 10.3 (Dini's Theorem) that $R_n(x) \to 0$ uniformly on $[0, 1]$, i.e.,
$$\sum_{k=1}^{n} f_k(x) \to f(x)$$
uniformly on $[0, 1]$. This ends the proof of the problem. ∎

> **Problem 10.20**
>
> (⋆) Prove that
> $$f(x) = \sum_{n=1}^{\infty} \frac{x^3 \sin nx}{n^2}$$
> is continuous on $\mathbb{R}$.

Proof. Let $N \in \mathbb{N}$. On $(-N, N)$, we have
$$\left| \frac{x^3 \sin nx}{n^2} \right| \leq \frac{N^3}{n^2}.$$
Since $\sum_{n=1}^{\infty} \frac{N^3}{n^2} < \infty$, Theorem 10.5 (Weierstrass M-test) implies that the series
$$f(x) = \sum_{n=1}^{\infty} \frac{x^3 \sin nx}{n^2}$$

converges uniformly on $(-N, N)$.

Suppose that
$$f_n(x) = \sum_{k=1}^{n} \frac{x^3 \sin kx}{k^2}$$
on $(-N, N)$. Then the previous paragraph verifies immediately that $f_n \to f$ uniformly on $(-N, N)$. Since each f_n is continuous on $(-N, N)$, Theorem 10.6 (Uniform Convergence and Continuity) implies that f is continuous on $(-N, N)$. Since N is arbitrary, f is then continuous on $\mathbb{R}$. We have completed the proof of the problem. ∎

> **Problem 10.21**
>
> (★)(★) Suppose that $\{a_n\}$ is a sequence of nonzero real numbers. Prove that
> $$\sum_{n=1}^{\infty} \frac{e^{ia_n x}}{n^2}$$
> converges uniformly on $\mathbb{R}$ to a continuous function $f : \mathbb{R} \to \mathbb{C}$. Evaluate the limit
> $$\lim_{T \to \infty} \frac{1}{2T} \int_{-T}^{T} f(x) \, dx.$$

Proof. Clearly, we have
$$\left| \frac{e^{ia_n x}}{n^2} \right| \le \frac{1}{n^2}$$
for all $x \in \mathbb{R}$ so that Theorem 10.5 (Weierstrass M-test) implies that the series converges uniformly on $\mathbb{R}$ to a function $f : \mathbb{R} \to \mathbb{C}$. Since the partial sum of the series is obviously continuous on $\mathbb{R}$, Theorem 10.6 (Uniform Convergence and Continuity) ensures that the function f must be continuous on $\mathbb{R}$. Furthermore, since the convergence is uniform, Theorem 10.7 (Uniform Convergence and Riemann-Stieltjes Integration) gives
$$\frac{1}{2T} \int_{-T}^{T} f(x) \, dx = \frac{1}{2T} \int_{-T}^{T} \sum_{n=1}^{\infty} \frac{e^{ia_n x}}{n^2} \, dx = \frac{1}{2T} \sum_{n=1}^{\infty} \int_{-T}^{T} \frac{e^{ia_n x}}{n^2} \, dx = \sum_{n=1}^{\infty} \frac{\sin a_n T}{n^2 \cdot a_n T}. \quad (10.31)$$

By the Mean Value Theorem for Derivatives, it is true that $|\sin x| \le |x|$ for all $x \in \mathbb{R}$ so that
$$\left| \frac{\sin a_n T}{n^2 \cdot a_n T} \right| \le \frac{1}{n^2}$$
for all $n \in \mathbb{N}$ and all $T \in \mathbb{R}$. Next, we follow from Theorem 10.5 (Weierstrass M-test) that the infinite series
$$\sum_{n=1}^{\infty} \frac{\sin a_n T}{n^2 \cdot a_n T}$$
converges uniformly on $\mathbb{R}$ (with respect to T). Therefore, it deduces from the equation (10.31) and then using [23, Exercise 13, p. 198] to conclude that
$$\lim_{T \to \infty} \frac{1}{2T} \int_{-T}^{T} f(x) \, dx = \lim_{T \to \infty} \sum_{n=1}^{\infty} \frac{\sin a_n T}{n^2 \cdot a_n T}$$

10.3. Uniform Convergence for Series of Functions

$$= \sum_{n=1}^{\infty} \frac{1}{n^2} \left(\lim_{T \to \infty} \frac{\sin a_n T}{a_n T} \right)$$

$$= \sum_{n=1}^{\infty} \frac{1}{n^2}$$

$$= \frac{\pi^2}{6},$$

completing the proof of the problem. ∎

Problem 10.22

(⋆) Suppose that $f : \mathbb{R} \to \mathbb{R}$ is continuous and $-\infty < a < b < \infty$. For each $n \in \mathbb{N}$, we define $f_n : \mathbb{R} \to \mathbb{R}$ by

$$f_n(x) = \frac{1}{n} \sum_{k=0}^{n-1} f\left(x + \frac{k}{n}\right).$$

Prove that f_n converges uniformly on $[a, b]$.

Proof. Given $\epsilon > 0$. Since f is continuous on $\mathbb{R}$, it is uniformly continuous on $[a, b+1]$. Then there exists a $\delta > 0$ such that

$$|f(x) - f(y)| < \epsilon \tag{10.32}$$

for every $x, y \in [a, b+1]$ and $|x - y| < \delta$. Particularly, we may choose an $n \in \mathbb{N}$ such that $\frac{1}{n} < \delta$. Fix $x \in [a, b]$. Then we have

$$\left| \int_x^{x+1} f(t)\,dt - f_n(x) \right| = \left| \sum_{k=0}^{n-1} \int_{x+\frac{k}{n}}^{x+\frac{k+1}{n}} f(t)\,dt - \frac{1}{n} \sum_{k=0}^{n-1} f\left(x + \frac{k}{n}\right) \right|. \tag{10.33}$$

By the First Mean Value Theorem for Integrals, for each $k = 1, 2, \ldots, n-1$, there exists a $y_k \in (x + \frac{k}{n}, x + \frac{k+1}{n}) \subseteq [a, b+1]$ such that

$$\int_{x+\frac{k}{n}}^{x+\frac{k+1}{n}} f(t)\,dt = \frac{f(y_k)}{n}. \tag{10.34}$$

Substituting the value (10.34) into the equation (10.33), we obtain

$$\left| \int_x^{x+1} f(t)\,dt - f_n(x) \right| = \left| \sum_{k=0}^{n-1} \frac{f(y_k)}{n} - \frac{1}{n} \sum_{k=0}^{n-1} f\left(x + \frac{k}{n}\right) \right|$$

$$\leq \frac{1}{n} \sum_{k=0}^{n-1} \left| f(y_k) - f\left(x + \frac{k}{n}\right) \right|. \tag{10.35}$$

Finally, by using the uniform continuity (10.32) to the inequality (10.35), we see that

$$\left| \int_x^{x+1} f(t)\,dt - f_n(x) \right| < \epsilon$$

for all $x \in [a,b]$ and all positive integers $n > \frac{1}{\delta}$. Now if we define $F : [a,b] \to \mathbb{R}$ by

$$F(x) = \int_x^{x+1} f(t)\,dt,$$

then we conclude immediately from Definition 10.2 (Uniform Convergence) that $\{f_n\}$ converges uniformly to F on $[a,b]$. This completes the proof of the problem. ∎

Problem 10.23

(⋆)(⋆) Show that the infinite series
$$\sum_{n=1}^{\infty} \frac{\sin nx}{n}$$
converges uniformly on $[\delta, 2\pi - \delta]$, where $\delta \in (0,\pi)$.

Proof. Given $\epsilon > 0$. Fix $\delta \in (0,\pi)$. We claim that the series

$$\sum_{k=1}^{\infty} \frac{e^{ikx}}{k} \tag{10.36}$$

converges uniformly on $[\delta, 2\pi - \delta]$ first. To this end, we note that if $x \in [\delta, 2\pi - \delta]$, then

$$|1 - e^{ix}| = \sqrt{2(1-\cos x)} \geq \sqrt{2(1-\cos \delta)} > 0.$$

Therefore, we have

$$|e^{inx} + e^{i(n+1)x} + \cdots + e^{imx}| = \left|\frac{e^{inx} - e^{i(m+1)x}}{1 - e^{ix}}\right| \leq \sqrt{\frac{2}{1-\cos\delta}}, \tag{10.37}$$

where $m \geq n \geq 0$. Recall the summation by parts [23, Theorem 3.41, p. 70] that

$$\sum_{k=n}^{m} a_k b_k = A_m b_m - A_{n-1} b_n + \sum_{k=n}^{m-1} A_k (b_k - b_{k+1}), \tag{10.38}$$

where $A_k = a_0 + a_1 + a_2 + \cdots + a_k$. Hence we deduce from the inequality (10.37) and the formula (10.38) with $a_k = e^{ikx}$ that

$$\left|\sum_{k=n}^{m} \frac{e^{ikx}}{k}\right| \leq \left|\frac{A_m}{m} - \frac{A_{n-1}}{n} + \sum_{k=n}^{m-1} A_k \left(\frac{1}{k} - \frac{1}{k+1}\right)\right|$$

$$\leq \sqrt{\frac{2}{1-\cos\delta}} \cdot \left(\frac{1}{m} + \frac{1}{n}\right) + \sqrt{\frac{2}{1-\cos\delta}} \cdot \sum_{k=n}^{m-1} \left|\frac{1}{k} - \frac{1}{k+1}\right|$$

$$\leq \frac{2}{n} \cdot \sqrt{\frac{2}{1-\cos\delta}} + \sqrt{\frac{2}{1-\cos\delta}} \cdot \left(\frac{1}{n} - \frac{1}{m}\right)$$

$$\leq \frac{4}{n} \cdot \sqrt{\frac{2}{1-\cos\delta}}. \tag{10.39}$$

10.3. Uniform Convergence for Series of Functions

Consequently, if n is large enough, then we get immediately from the inequality (10.39) that

$$\Big|\sum_{k=n}^{m} \frac{e^{ikx}}{k}\Big| < \epsilon.$$

Thus we conclude from Theorem 10.3 (Cauchy Criterion for Uniform Convergence) that the series (10.36) converges uniformly on $[\delta, 2\pi - \delta]$ as desired. Finally, we get from the definition of the modulus easily that

$$\Big|\sum_{k=n}^{m} \frac{e^{ikx}}{k}\Big| \geq \Big|\sum_{k=n}^{m} \frac{\sin kx}{k}\Big|,$$

so we may obtain the same conclusion and thus we complete the analysis of the problem. ∎

Problem 10.24

(⋆)(⋆) *Suppose that $f : \mathbb{R} \to \mathbb{R}$ is given by*

$$f(x) = \sum_{n=1}^{\infty} \frac{1}{n^2 + x^2}.$$

Prove that f is differentiable in $\mathbb{R}$ and find $f'(x)$.

Proof. Let $f_n(x) = \frac{1}{n^2+x^2}$ and M be a positive constant. Recall that

$$\sum_{n=1}^{\infty} f_n(0) = \sum_{n=1}^{\infty} \frac{1}{n^2} = \frac{\pi^2}{6}.$$

In addition, if $|x| \leq M$, then we have

$$|f_n'(x)| \leq \Big|\frac{-2x}{(n^2+x^2)^2}\Big| \leq \frac{2M}{n^4}$$

so that

$$\Big|\sum_{n=1}^{\infty} f_n'(x)\Big| \leq \sum_{n=1}^{\infty} |f_n'(x)| \leq 2M \cdot \sum_{n=1}^{\infty} \frac{1}{n^4} < \infty.$$

By Theorem 10.5 (Weierstrass M-test), the series

$$\sum_{n=1}^{\infty} f_n'(x)$$

converges uniformly on $[-M, M]$. Now it deduces from Theorem 10.8 (Uniform Convergence and Differentiability) that f is differentiable in $[-M, M]$ and

$$f'(x) = \sum_{n=1}^{\infty} f_n'(x) = \sum_{n=1}^{\infty} \frac{-2x}{(n^2+x^2)^2} \qquad (10.40)$$

Since M is arbitrary, the formula (10.40) holds for all $x \in \mathbb{R}$ and we complete the proof of the problem. ∎

Problem 10.25

(★)(★) Let $f(x) = \int_0^x \frac{\sin t}{t}\,dt$, where $x \in \mathbb{R}$. Prove that

$$f(x) = \sum_{n=0}^{\infty} \frac{(-1)^n x^{2n+1}}{(2n+1)!(2n+1)}$$

for $x \in \mathbb{R}$.

Proof. Since the power series representation of $\sin t$ is given by

$$\sin t = \sum_{n=0}^{\infty} \frac{(-1)^n}{(2n+1)!} t^{2n+1},$$

we have

$$\frac{\sin t}{t} = \sum_{n=0}^{\infty} \frac{(-1)^n}{(2n+1)!} t^{2n}. \tag{10.41}$$

Let $M > 0$. Since

$$\left|\frac{(-1)^n t^{2n}}{(2n+1)!}\right| \leq \frac{M^{2n}}{(2n+1)!}$$

for $t \in [-M, M]$ and the series

$$\sum_{n=0}^{\infty} \frac{M^{2n}}{(2n+1)!}$$

converges, it yields from Theorem 10.5 (Weierstrass M-test) that the series on the right-hand side of the representation (10.41) converges uniformly on $[-M, M]$. It is definitely that each $\frac{(-1)^n}{(2n+1)!} t^{2n} \in \mathscr{R}$ on $[-M, M]$, it follows from Theorem 10.7 (Uniform Convergence and Riemann-Stieltjes Integration) that

$$\begin{aligned}
f(x) &= \int_0^x \frac{\sin t}{t}\,dt \\
&= \int_0^x \sum_{n=0}^{\infty} \frac{(-1)^n}{(2n+1)!} t^{2n}\,dt \\
&= \sum_{n=0}^{\infty} \int_0^x \frac{(-1)^n}{(2n+1)!} t^{2n}\,dt \\
&= \sum_{n=0}^{\infty} \frac{(-1)^n}{(2n+1)!(2n+1)} x^{2n+1}
\end{aligned}$$

for all $x \in [-M, M]$. Since M is arbitrary, our desired result follows. This completes the proof of the problem. ∎

10.4 Equicontinuous Families of Functions

> **Problem 10.26**
>
> (★)(★) Let $-\infty < a < b < \infty$. Suppose that $\{f_n\}$ is pointwise convergent and equicontinuous on $[a,b]$. Prove that $\{f_n\}$ converges uniformly on $[a,b]$.

Proof. Given $\epsilon > 0$. Let f be the limit function of $\{f_n\}$ on $[a,b]$. By the hypotheses, there is a $\delta > 0$ such that
$$|f_n(x) - f_n(y)| < \frac{\epsilon}{3} \tag{10.42}$$
for all $x, y \in [a,b]$ with $|x-y| < \delta$ and all $n \in \mathbb{N}$. Taking $n \to \infty$ in the inequality (10.42), we get
$$|f(x) - f(y)| \leq \frac{\epsilon}{3} \tag{10.43}$$
for all $x, y \in [a,b]$ with $|x-y| < \delta$. By the definition, the inequality (10.43) shows that f is uniformly continuous on $[a,b]$. Since $[a,b]$ is compact, there exists a set of finite points $\{x_1, x_2, \ldots, x_k\} \subseteq [a,b]$ such that
$$[a,b] \subseteq \bigcup_{i=1}^{k}(x_i - \delta, x_i + \delta). \tag{10.44}$$
Since $f_n \to f$ pointwise on $[a,b]$, there exists an $N_i \in \mathbb{N}$ such that $n \geq N_i$ implies
$$|f_n(x_i) - f(x_i)| < \frac{\epsilon}{3}, \tag{10.45}$$
where $i = 1, 2, \ldots, k$. Put $N = \max(N_1, N_2, \ldots, N_k)$. If $x \in [a,b]$, then the finite open covering property (10.44) ensures that there exists an $i \in \{1, 2, \ldots, k\}$ such that $|x - x_i| < \delta$. Therefore, for all $x \in [a,b]$, it follows from the inequalities (10.42), (10.43) and (10.45) that $n \geq N$ implies
$$|f_n(x) - f(x)| \leq |f_n(x) - f_n(x_i)| + |f_n(x_i) - f(x_i)| + |f(x_i) - f(x)| < \epsilon.$$
By the definition, $\{f_n\}$ converges uniformly on $[a,b]$. This completes the proof of the problem. ∎

> **Problem 10.27**
>
> (★) Let $-\infty < a < b < \infty$. Suppose that $\{f_n\}$ is a sequence of continuous functions on $[a,b]$ and each f_n is differentiable in (a,b). If $\{f_n\}$ is pointwise convergent on $[a,b]$ and $\{f_n'\}$ is uniformly bounded on (a,b), prove that $\{f_n\}$ is uniformly convergent on $[a,b]$.

Proof. For all $x, y \in [a,b]$ and all $n \in \mathbb{N}$, we get from the Mean Value Theorem for Derivatives that
$$|f_n(x) - f_n(y)| = |f_n'(\xi)| \cdot |x - y| \tag{10.46}$$
for some $\xi \in (x, y)$. Since there is a positive constant M such that $|f_n'(x)| \leq M$ for all $x \in [a,b]$ and $n \in \mathbb{N}$, the equation (10.46) gives
$$|f_n(x) - f_n(y)| \leq M|x - y|.$$

By Definition 10.10 (Equicontinuity), we conclude that $\{f_n\}$ is equicontinuous and our desired result follows immediately from Problem 10.26. This completes the proof of the problem. ∎

Problem 10.28

(⋆) Let $-\infty < a < b < \infty$. Suppose that $\{f_n\}$ is a sequence of twice differentiable functions on $[a,b]$. Furthermore, we have $f_n(a) = f_n'(a) = 0$ and $|f_n''(x)| \leq 1$ for all $x \in [a,b]$ and $n \in \mathbb{N}$. Prove that $\{f_n\}$ has a uniformly convergent subsequence on $[a,b]$.

Proof. Applying the Mean Value Theorem for Derivatives to the sequence $\{f_n'\}$, we see that there exists a $\xi \in (x,y)$ such that

$$|f_n'(x) - f_n'(y)| = |f_n''(\xi)| \cdot |x - y| \leq |x - y| \tag{10.47}$$

for all $x, y \in [a,b]$ and all $n \in \mathbb{N}$. Put $y = a$ in the inequality (10.47), we get

$$|f_n'(x)| \leq |x - a| \leq b - a$$

for all $x \in [a,b]$ and all $n \in \mathbb{N}$. Similarly, we apply the Mean Value Theorem for Derivatives to the sequence $\{f_n\}$, so there is a $\theta \in (x,y)$ such that

$$|f_n(x) - f_n(y)| = |f_n'(\theta)| \cdot |x - y| \leq (b - a) \cdot |x - y| \tag{10.48}$$

for all $x, y \in [a,b]$ and all $n \in \mathbb{N}$. Given $\epsilon > 0$, if we take $\delta = \frac{\epsilon}{2(b-a)}$, then it follows from the estimate (10.48) that

$$|f_n(x) - f_n(y)| \leq \frac{\epsilon}{2} < \epsilon$$

holds for all $x, y \in [a,b]$ satisfying $|x-y| < \delta$ and all $n \in \mathbb{N}$. By Definition 10.10 (Equicontinuity), the family $\{f_n\}$ is equicontinuous. By the estimate (10.48) again, if $y = a$, then

$$|f_n(x)| = |f_n(x) - f_n(a)| \leq (b - a) \cdot |x - a| \leq (b - a)^2$$

for all $x \in [a,b]$ and all $n \in \mathbb{N}$. Thus $\{f_n\}$ is uniformly bounded. Hence we follow from the Arzelà-Ascoli Theorem that $\{f_n\}$ has a uniformly convergent subsequence on $[a,b]$, completing the proof of the problem. ∎

Problem 10.29

(⋆) Suppose that $\{f_n\}$ is a family of continuously differentiable functions on $[0,1]$ and it satisfies

$$|f_n'(x)| \leq x^{-\frac{1}{p}}$$

for all $x \in (0,1]$, where $p > 1$. Furthermore, we have

$$\int_0^1 f_n(x)\,dx = 0. \tag{10.49}$$

Prove that $\{f_n\}$ has a uniformly convergent subsequence.

10.4. Equicontinuous Families of Functions

Proof. For $x, y \in [0,1]$ with $0 \leq x < y \leq 1$ and $n \in \mathbb{N}$, we have

$$|f_n(x) - f_n(y)| = \left| \int_x^y f_n'(t) \, dt \right| \leq \int_x^y |f_n'(t)| \, dt \leq \int_x^y t^{-\frac{1}{p}} \, dt = \frac{p}{p-1}(y^{1-\frac{1}{p}} - x^{1-\frac{1}{p}}). \quad (10.50)$$

Define the function $g : [0,1] \to \mathbb{R}$ by $g(x) = x^{1-\frac{1}{p}}$ which is clearly uniformly continuous on $[0,1]$. In other words, given $\epsilon > 0$, there is a $\delta > 0$ such that

$$|x^{1-\frac{1}{p}} - y^{1-\frac{1}{p}}| = |g(x) - g(y)| < \frac{(p-1)\epsilon}{p} \quad (10.51)$$

for all $x, y \in [0,1]$ with $|x - y| < \delta$. Combining the inequalities (10.50) and (10.51), we establish that

$$|f_n(x) - f_n(y)| < \epsilon$$

for all $x, y \in [0,1]$ with $|x - y| < \delta$ and all $n \in \mathbb{N}$. By Definition 10.10 (Equicontinuity), the sequence $\{f_n\}$ is equicontinuous.

By the integral (10.49), it is impossible that $f_n(x) > 0$ or $f_n(x) < 0$ for all $x \in [0,1]$. In other words, for each positive integer n, this observation and the continuity of f_n guarantee that there exists a $x_n \in [0,1]$ such that $f_n(x_n) = 0$. Using the estimate (10.50) with $y = x_n$, we have

$$|f_n(x)| \leq \frac{p}{p-1} |x^{1-\frac{1}{p}} - x_n^{1-\frac{1}{p}}| \leq \frac{2p}{p-1}$$

for all $x \in [0,1]$, i.e., $\{f_n\}$ is uniformly bounded. Now the application of the Arzelà-Ascoli Theorem implies that it contains a uniformly convergent subsequence on $[0,1]$. We have completed the proof of the problem. ∎

Problem 10.30

(⋆) Suppose that $\{f_n\}$ is a collection of continuous functions on $[0,1]$. In addition, suppose that $\{f_n\}$ pointwise converges to 0 on $[0,1]$ and there exists a positive constant M such that

$$\left| \int_0^1 f_n(x) \, dx \right| \leq M \quad (10.52)$$

for all $n \in \mathbb{N}$. Is it true that

$$\lim_{n \to \infty} \int_0^1 f_n(x) \, dx = 0? \quad (10.53)$$

Proof. The answer is negative. In fact, for every $n \in \mathbb{N}$, we consider $f_n : [0,1] \to \mathbb{R}$ by

$$f_n(x) = \begin{cases} 2n^2 x, & \text{if } x \in [0, \frac{1}{2n}); \\ -2n^2 x + 2n, & \text{if } x \in [\frac{1}{2n}, \frac{1}{n}); \\ 0, & \text{if } x \in [\frac{1}{n}, 1]. \end{cases}$$

It is easy to check that each f_n is continuous on $[0,1]$. Now we have

$$f_n(0) = f_n(1) = 0$$

for every $n \in \mathbb{N}$. If $x \in (0, 1)$, then one can find an $N \in \mathbb{N}$ such that $x \geq \frac{1}{N}$ which implies that

$$f_n(x) = 0$$

for all $n \geq N$. Therefore, it means that $\{f_n\}$ pointwise converges to 0 on $[0, 1]$. Since the graph of the f_n is an isosceles triangle with vertices $(0, 0)$, $(\frac{1}{2n}, n)$ and $(\frac{1}{n}, 0)$, we have

$$\int_0^1 f_n(x) \, dx = \frac{1}{2}$$

which implies the truth of the bound (10.52), but the failure of the limit (10.53). Hence we have completed the proof of the problem. ∎

Problem 10.31 (Abel's Test for Uniform Convergence)

(⋆)(⋆) Suppose that $\{g_n\}$ is a sequence of real-valued functions defined on $E \subseteq \mathbb{R}$ such that $g_{n+1}(x) \leq g_n(x)$ for every $x \in E$ and every $n \in \mathbb{N}$. If the family $\{g_n\}$ is uniformly bounded on E and if $\sum f_n$ converges uniformly on E, prove that the series

$$\sum_{n=1}^{\infty} f_n(x) g_n(x)$$

also converges uniformly on E.

Proof. Since $\{g_n\}$ is uniformly bounded on E, there is a positive constant M such that

$$|g_n(x)| \leq M \tag{10.54}$$

for all $x \in E$ and all $n \in \mathbb{N}$. Since $\sum f_n$ converges uniformly on E, given $\epsilon > 0$, there is an $N \in \mathbb{N}$ such that $n > m \geq N$ implies

$$\Bigl| \sum_{k=m}^{n} f_k(x) \Bigr| < \frac{\epsilon}{3M} \tag{10.55}$$

for all $x \in E$. Let $F_{m,n}(x) = \sum_{k=m}^{n} f_k(x)$. Then we have

$$\sum_{k=m}^{n} f_k g_k = f_m g_m + f_{m+1} g_{m+1} + \cdots + f_n g_n$$
$$= F_{m,m} g_m + (F_{m,m+1} - F_{m,m}) g_{m+1} + \cdots + (F_{m,n} - F_{m,n-1}) g_n$$
$$= F_{m,m}(g_m - g_{m+1}) + F_{m,m+1}(g_{m+1} - g_{m+2}) + \cdots + F_{m,n-1}(g_{n-1} - g_n)$$
$$+ F_{m,n} g_n. \tag{10.56}$$

Since $g_{n+1}(x) \leq g_n(x)$ for every $x \in E$ and every $n \in \mathbb{N}$, we have

$$g_k(x) - g_{k+1}(x) \geq 0$$

10.4. Equicontinuous Families of Functions

for all $x \in E$ and $k = m, m+1, \ldots, n-1$. Combining the estimate (10.55) and the formula (10.56), if $n > m \geq N$, then

$$\begin{aligned}
\Big|\sum_{k=m}^{n} f_k(x)g_k(x)\Big| &\leq |F_{m,m}(x)| \cdot [g_m(x) - g_{m+1}(x)] + |F_{m,m+1}(x)| \cdot [g_{m+1}(x) - g_{m+2}(x)] \\
&\quad + \cdots + |F_{m,n-1}(x)| \cdot [g_{n-1}(x) - g_n(x)] + |F_{m,n}(x)| \cdot |g_n(x)| \\
&< \frac{\epsilon}{3M} \cdot [g_m(x) - g_{m+1}(x)] + \cdots + \frac{\epsilon}{3M} \cdot [g_{n-1}(x) - g_n(x)] + \frac{\epsilon}{3M} \cdot |g_n(x)| \\
&= \frac{\epsilon}{3M} \cdot [g_m(x) - g_n(x)] + \frac{\epsilon}{3M} \cdot |g_n(x)|. \tag{10.57}
\end{aligned}$$

Obviously, we have $0 \leq g_m(x) - g_n(x) \leq 2M$. Therefore, we follow from the estimates (10.54) and (10.57) that

$$\Big|\sum_{k=m}^{n} f_k(x)g_k(x)\Big| < \epsilon$$

for all $x \in E$. By Theorem 10.3 (Cauchy Criterion for Uniform Convergence), we conclude that $\sum f_n(x)g_n(x)$ converges uniformly on E. This completes the analysis of the proof. ∎

Problem 10.32

(⋆) (⋆) Let K be a compact metric space and $\mathscr{C}(K)$ be equicontinuous on K. Prove that $\overline{\mathscr{C}(K)}$ is also equicontinuous on K.

Proof. Since K is compact, every element $f \in \mathscr{C}(K)$ must be bounded by the Extreme Value Theorem. Given $\epsilon > 0$. Fix $\theta \in (0, \epsilon)$. Since $\mathscr{C}(K)$ is equicontinuous on K, there exists a $\delta > 0$ such that

$$|f(x) - f(y)| < \frac{\epsilon}{3} \tag{10.58}$$

for all $f \in \mathscr{C}(K)$ and all $x, y \in K$ with $|x - y| < \delta$. Let $F \in \overline{\mathscr{C}(K)}$. Then there is a sequence $\{f_n\} \subseteq \mathscr{C}(K)$ such that

$$\sup_{x \in K} |f_n(x) - F(x)| = \|f_n - F\| \to 0 \tag{10.59}$$

as $n \to \infty$. In other words, there is an $N \in \mathbb{N}$ such that $n \geq N$ implies

$$|f_n(x) - F(x)| < \frac{\epsilon}{3}$$

for all $x \in K$.

Now if $x, y \in K$ and $|x - y| < \delta$, then we follow from the inequality (10.58) and the limit (10.59) that

$$|F(x) - F(y)| \leq |F(x) - f_N(x)| + |f_N(x) - f_N(y)| + |f_N(y) - F(y)| < \frac{\epsilon}{3} + \frac{\epsilon}{3} + \frac{\epsilon}{3} = \epsilon.$$

By Definition 10.10 (Equicontinuity), $\overline{\mathscr{C}(K)}$ is equicontinuous on K, completing the proof of the problem. ∎

10.5 Approximation by Polynomials

> **Problem 10.33**
>
> (★) Suppose that every f_i ($i = 1, 2, \ldots, m$) is a real-valued function on $[a,b]$ such that, for each i, there is a sequence of polynomials converges uniformly to f_i on $[a,b]$. Prove that there also exists a sequence of polynomials converges uniformly to their product $f_1 f_2 \cdots f_m$ on $[a,b]$.

Proof. Since a polynomial is continuous on $[a,b]$, Theorem 10.6 (Uniform Convergence and Continuity) guarantees that each f_i is also continuous on $[a,b]$. Consequently, the product

$$f = f_1 f_2 \cdots f_m$$

is also continuous on $[a,b]$. Now an application of the Weierstrass Approximation Theorem implies that there exists a sequence of polynomials $\{P_n\}$ defined on $[a,b]$ such that $P_n \to f$ uniformly on $[a,b]$ which completes the proof of the problem. ∎

> **Problem 10.34**
>
> (★) Suppose that $f, g : [a,b] \to \mathbb{R}$ are continuous. Let $F = \max(f, g)$. Prove that there exists a sequence of $\{P_n\}$ defined on $[a,b]$ such that $P_n \to F$ uniformly on $[a,b]$.

Proof. Since f, g are continuous on $[a,b]$, $|f - g|$ is also continuous on $[a,b]$. Since we have

$$F = \frac{1}{2}(f + g + |f - g|),$$

F is continuous on $[a,b]$ and our desired result follows directly from the Weierstrass Approximation Theorem. This has completed the proof of the problem. ∎

> **Problem 10.35**
>
> (★)(★) Suppose that f is a continuously differentiable function on $[a,b]$. Prove that there exists a sequence of polynomials $\{P_n\}$ such that $P_n \to f$ and $P'_n \to f$ uniformly on $[a,b]$.

Proof. Since f is continuously differentiable on $[a,b]$, the Weierstrass Approximation Theorem implies the existence of a sequence of polynomials $\{Q_n\}$ such that given $\epsilon > 0$, one can find an $N \in \mathbb{N}$ such that $n \geq N$ implies

$$|Q_n(t) - f'(t)| < \frac{\epsilon}{b - a} \tag{10.60}$$

for all $t \in [a,b]$. Now for each $n \in \mathbb{N}$, we define $P_n : [a,b] \to \mathbb{R}$ by

$$P_n(x) = f(a) + \int_a^x Q_n(t)\, dt.$$

10.5. Approximation by Polynomials

It is clear that every P_n is a polynomial on $[a,b]$. Furthermore, we deduce from the First and the Second Fundamental Theorem of Calculus that

$$\begin{aligned} f(x) - P_n(x) &= [f(x) - f(a)] - [P_n(x) - P_n(a)] \\ &= \int_a^x f'(t)\,dt - \int_a^x P_n'(t)\,dt \\ &= \int_a^x [f'(t) - Q_n(t)]\,dt. \end{aligned}$$

It follows from the estimate (10.60) that for every $x \in [a,b]$ and $n \geq N$, we have

$$|f(x) - P_n(x)| \leq \int_a^x |f'(t) - Q_n(t)|\,dt < \frac{\epsilon}{b-a}(x-a) < \epsilon.$$

By Definition 10.2 (Uniform Convergence), $P_n \to f$ uniformly on $[a,b]$. We have completed the proof of the problem. ∎

Problem 10.36

(★)(★) Let $f \in \mathscr{C}([0,1])$ and $\epsilon > 0$. Prove that there exists a polynomial P with rational coefficients on $[0,1]$ such that
$$\|f - P\| < \epsilon.$$

Proof. By the Weierstrass Approximation Theorem, we know that there exists a polynomial p on $[0,1]$ such that

$$\|f - p\| < \frac{\epsilon}{2}. \tag{10.61}$$

Let
$$p(x) = a_0 x^n + a_1 x^{n-1} + \cdots + a_{n-1} x + a_n,$$

where $a_0, a_1, \ldots, a_n$ are constants. For each a_k, there exists a $q_k \in \mathbb{Q}$ such that

$$|a_k - q_k| < \frac{\epsilon}{2(n+1)}.$$

Now we define
$$P(x) = q_0 x^n + q_1 x^{n-1} + \cdots + q_{n-1} x + q_n.$$

Then it is easy to check that for $x \in [0,1]$, we have

$$|p(x) - P(x)| \leq \sum_{k=0}^n |a_k - q_k| \cdot |x^{n-k}| < \sum_{k=0}^n \frac{\epsilon}{2(n+1)} = \frac{\epsilon}{2}$$

which means that

$$\|p - P\| \leq \frac{\epsilon}{2}. \tag{10.62}$$

Combining the inequalities (10.61) and (10.62), we see that

$$\|f - P\| \leq \|f - p\| + \|p - P\| < \epsilon.$$

This completes the proof of the problem. ∎

Problem 10.37

(★)(★)(★) Let $g_{n,m}(x) = C_m^n x^m (1-x)^{n-m}$, where $0 \le m \le n$. Define the nth **Bernstein polynomial** of a continuous function $f : [0,1] \to \mathbb{R}$ by

$$B_n(x) = \sum_{m=0}^{n} f\left(\frac{m}{n}\right) g_{n,m}(x).$$

Prove that $B_n \to f$ uniformly on $[0,1]$.

Proof. Recall from Theorem 7.10 that f is uniformly continuous on $[0,1]$. In other words, given $\epsilon > 0$, there is a $\delta > 0$ such that

$$|f(x) - f(y)| < \frac{\epsilon}{2} \tag{10.63}$$

for all $x, y \in [0,1]$ with $|x - y| < \delta$. It is clear that

$$1 = [x + (1-x)]^n = \sum_{m=0}^{n} C_m^n x^m (1-x)^{n-m}, \tag{10.64}$$

so we have

$$\begin{aligned}
f(x) - B_n(x) &= \sum_{m=0}^{n} \left[f(x) - f\left(\frac{m}{n}\right) \right] g_{n,m}(x) \\
&= \sum_{1} \left[f(x) - f\left(\frac{m}{n}\right) \right] g_{n,m}(x) + \sum_{2} \left[f(x) - f\left(\frac{m}{n}\right) \right] g_{n,m}(x),
\end{aligned} \tag{10.65}$$

where the first summation on the right-hand side of (10.65) consists of those values of m for which $|x - \frac{m}{n}| < \delta$ and the second summation takes the rest of the values of m.

Now we are going to find bounds of the two summations in the equation (10.65). To this end, we notice from the inequality (10.63) that

$$\left| \sum_{1} \left[f(x) - f\left(\frac{m}{n}\right) \right] g_{n,m}(x) \right| \le \sum_{1} \left| f(x) - f\left(\frac{m}{n}\right) \right| \cdot g_{n,m}(x) < \frac{\epsilon}{2} \sum_{1} g_{n,m}(x) \le \frac{\epsilon}{2}. \tag{10.66}$$

By the Extreme Value Theorem, there exists a $p \in [0,1]$ such that $f(p) = \sup\limits_{x \in [0,1]} |f(x)|$ and this implies that

$$\begin{aligned}
\left| \sum_{2} \left[f(x) - f\left(\frac{m}{n}\right) \right] g_{n,m}(x) \right| &\le \sum_{2} \left| f(x) - f\left(\frac{m}{n}\right) \right| g_{n,m}(x) \\
&\le 2M \sum_{2} g_{n,m}(x) \\
&\le 2M \sum_{2} \frac{(nx - m)^2}{n^2 \delta^2} g_{n,m}(x) \\
&\le \frac{2M}{n^2 \delta^2} \sum_{m=0}^{n} (nx - m)^2 g_{n,m}(x). \tag{10.67}
\end{aligned}$$

10.5. Approximation by Polynomials

Note that
$$(e^y + 1 - x)^n = \sum_{m=0}^{n} C_m^n e^{my}(1-x)^{n-m}. \tag{10.68}$$

Differentiating the equation (10.68) with respect to y twice, we obtain
$$n(e^y + 1 - x)^{n-1} e^y = \sum_{m=0}^{n} C_m^n m e^{my}(1-x)^{n-m} \tag{10.69}$$
$$n(e^y + 1 - x)^{n-1} e^y + n(n-1)(e^y + 1 - x)^{n-2} e^{2y} = \sum_{m=0}^{n} C_m^n m^2 e^{my}(1-x)^{n-m}. \tag{10.70}$$

By putting $e^y = x$ into the formulas (10.69) and (10.70), they yield that
$$nx = \sum_{m=0}^{n} C_m^n m x^m (1-x)^{n-m} \quad \text{and} \quad nx + n(n-1)x^2 = \sum_{m=0}^{n} C_m^n m^2 x^m (1-x)^{n-m}. \tag{10.71}$$

If we rewrite the identities (10.64) and (10.71) as
$$1 = \sum_{m=0}^{n} g_{n,m}(x), \quad nx = \sum_{m=0}^{n} m g_{n,m}(x) \quad \text{and} \quad nx + n(n-1)x^2 = \sum_{m=0}^{n} m^2 g_{n,m}(x),$$

then they imply that
$$\sum_{m=0}^{n}(nx-m)^2 g_{n,m}(x) = \sum_{m=0}^{n}(n^2 x^2 - 2nmx + m^2) g_{n,m}(x)$$
$$= n^2 x^2 \sum_{m=0}^{n} g_{n,m}(x) - 2nx \sum_{m=0}^{n} m g_{n,m}(x) + \sum_{m=0}^{n} m^2 g_{n,m}(x)$$
$$= n^2 x^2 - 2n^2 x^2 + nx + n(n-1)x^2$$
$$= nx(1-x). \tag{10.72}$$

By substituting the identity (10.72) into the inequality (10.67) and using the fact that the function $h(x) = x(1-x)$ defined on $[0,1]$ attains its absolute maximum $\frac{1}{4}$ at $x = \frac{1}{2}$, we derive
$$\left|\sum_2 \left[f(x) - f\left(\frac{m}{n}\right)\right] g_{n,m}(x)\right| \leq \frac{2M}{n^2 \delta^2} \cdot nx(1-x) \leq \frac{M}{2n\delta^2}. \tag{10.73}$$

Recall that δ is *fixed* so that we can choose $n \in \mathbb{N}$ such that $n > \frac{M}{\epsilon \delta^2}$. Then the inequality (10.73) gives
$$\left|\sum_2 \left[f(x) - f\left(\frac{m}{n}\right)\right] g_{n,m}(x)\right| < \frac{\epsilon}{2}. \tag{10.74}$$

Hence we put the bounds (10.66) and (10.74) back into the expression (10.65) to conclude that
$$|f(x) - B_n(x)| < \epsilon$$

for all $x \in [0,1]$ and $n > \frac{M}{\epsilon \delta^2}$. By Definition 10.2 (Uniform Convergence), $B_n \to f$ uniformly on $[0,1]$ and we complete the analysis of the proof. ∎

CHAPTER 11

Improper Integrals

11.1 Fundamental Concepts

In Chapter 9, we study an integral whose integrand is a function f **defined** and **bounded** on a **finite** interval $[a,b]$. However, we can extend the concepts of integration to include functions tending to infinity at some certain points or unbounded intervals. This is the content of the so-called **improper integrals**. The main references we have used here are [2, §10.23], [3, §10.13], [21, Chap. 1] and [33, §6.5].

11.1.1 Improper Integrals of the First Kind

Definition 11.1. Let $a \in \mathbb{R}$. Suppose that $f : [a, +\infty) \to \mathbb{R}$ is a function integrable on every closed and bounded interval $[a, b]$, where $a \leq b < +\infty$, i.e., $f \in \mathscr{R}$ on $[a, b]$. We set

$$\int_a^{+\infty} f(x)\,\mathrm{d}x = \lim_{b \to +\infty} \int_a^b f(x)\,\mathrm{d}x, \tag{11.1}$$

where the left-hand side is called the **improper integral of the first kind**. If the limit (11.1) exists and is finite, then we call the improper integral **convergent**. Otherwise, the improper integral is **divergent**.

Similarly, we may define the following improper integral if the limit exists and is finite:

$$\int_{-\infty}^b f(x)\,\mathrm{d}x = \lim_{a \to -\infty} \int_a^b f(x)\,\mathrm{d}x.$$

Finally, if both

$$\int_{-\infty}^a f(x)\,\mathrm{d}x \quad \text{and} \quad \int_a^{+\infty} f(x)\,\mathrm{d}x$$

are convergent *for some* $a \in \mathbb{R}$, then we have

$$\int_{-\infty}^{+\infty} f(x)\,\mathrm{d}x = \int_{-\infty}^a f(x)\,\mathrm{d}x + \int_a^{+\infty} f(x)\,\mathrm{d}x \tag{11.2}$$

so that the improper integral on the left-hand side of the expression (11.2) is convergent. If one of the integrals on the right-hand side of the expression (11.2) is divergent, then the corresponding improper integral is **divergent**.

> **Remark 11.1**
>
> We must notice that it can be shown that the choice of a in the expression (11.2) is *not* important.

11.1.2 Improper Integrals of the Second Kind

Definition 11.2. Let $a, b \in \mathbb{R}$. Suppose that $f : [a, b) \to \mathbb{R}$ is a function integrable on every closed and bounded interval $[a, c]$, where $a \leq c < b$, i.e., $f \in \mathscr{R}$ on $[a, c]$. We set

$$\int_a^b f(x)\,\mathrm{d}x = \lim_{c \to b-} \int_a^c f(x)\,\mathrm{d}x, \tag{11.3}$$

where the left-hand side is called the **improper integral of the second kind**. If the limit (11.3) exists and is finite, then we call the improper integral **convergent**. Otherwise, the improper integral is **divergent**.

The key of the definition is that f may be **unbounded** in a neighborhood of b. Similarly, if we have $f : (a, b] \to \mathbb{R}$, then we define

$$\int_a^b f(x)\,\mathrm{d}x = \lim_{c \to a+} \int_c^b f(x)\,\mathrm{d}x.$$

Definition 11.3. Let $a, b \in \mathbb{R}$ and $a < c < b$. Suppose that $f : [a, c) \cup (c, b] \to \mathbb{R}$ is a function integrable on every closed and bounded interval of $[a, c) \cup (c, b]$ and $f(x) \to \pm\infty$ as $x \to c$. Then we have

$$\int_a^b f(x)\,\mathrm{d}x = \int_a^c f(x)\,\mathrm{d}x + \int_c^b f(x)\,\mathrm{d}x \tag{11.4}$$

and the improper integral on the left-hand side of (11.4) exists if and only if the two improper integrals on its right-hand side exist.

Examples of improper integrals of the first and the second kinds can be found in Problems 11.1 and 11.2 respectively.

11.1.3 Cauchy Principal Value

We can define the so-called **Cauchy Principal Value**, or simply **principal value**, of an improper integral in some cases.

Definition 11.4 (Cauchy Principal Value)**.** Let $a, b \in \mathbb{R}$ and $a < c < b$. We suppose that $f : [a, c) \cup (c, b] \to \mathbb{R}$ is a function integrable on every closed and bounded interval of $[a, c) \cup (c, b]$ and $f(x) \to \pm\infty$ as $x \to c$. Then the principal value of the improper integral of f on $[a, b]$ is given by

$$\text{P.V.} \int_a^b f(x)\,\mathrm{d}x = \lim_{\epsilon \to 0+} \left(\int_a^{c-\epsilon} f(x)\,\mathrm{d}x + \int_{c+\epsilon}^b f(x)\,\mathrm{d}x \right). \tag{11.5}$$

11.1. Fundamental Concepts

Now if the improper integral (11.4) exists, then it is easy to see that it equals to its principal value (11.5). However, the converse is not true, see Problem 11.4 for an example.

▌ 11.1.4 Properties of Improper Integrals

Theorem 11.5. *Suppose that $a \in \mathbb{R}$ and $f, g : [a, b) \to \mathbb{R}$ are integrable on every closed and bounded interval of $[a, b)$, where b is either finite or $+\infty$.*

(a) *For any $A, B \in \mathbb{R}$, the function $Af + Bg$ is also integrable on every closed and bounded interval of $[a, b)$ and furthermore,*

$$\int_a^b (Af + Bg)(x)\, dx = A \int_a^b f(x)\, dx + B \int_a^b g(x)\, dx.$$

(b) *Let $\varphi : [\alpha, \beta) \to [a, b)$ be a strictly increasing differentiable function such that $\varphi(x) \to b$ as $x \to \beta$. Then the function $F : [\alpha, \beta) \to \mathbb{R}$ defined by $F(t) = f(\varphi(t))\varphi'(t)$ satisfies*

$$\int_a^b f(x)\, dx = \int_\alpha^\beta F(t)\, dt = \int_\alpha^\beta f(\varphi(t))\varphi'(t)\, dt.$$

Theorem 11.6. *Suppose that $a \in \mathbb{R}$ and $f : (a, b) \to \mathbb{R}$ is integrable on every closed and bounded interval of (a, b), where b is either finite or $+\infty$. For $a < c < b$, we have*

$$\int_a^b f(x)\, dx = \int_a^c f(x)\, dx + \int_c^b f(x)\, dx. \tag{11.6}$$

Hence the improper integral

$$\int_a^b f(x)\, dx \tag{11.7}$$

converges if and only if both improper integrals

$$\int_a^c f(x)\, dx \quad \text{and} \quad \int_c^b f(x)\, dx$$

converge.

In particular, if $b = +\infty$ in the formula (11.6), then the first and the second improper integrals on its right-hand side are of the first kind and the second kind respectively. In this case, the integral (11.7) is sometimes called the **improper integral of the third kind**. An example of this is the classical **Gamma function** Γ defined by

$$\Gamma(x) = \int_0^{+\infty} t^{x-1} e^{-t}\, dt,$$

where $x > 0$. See Problem 11.15 for the proof of its convergence.

11.1.5 Criteria for Convergence of Improper Integrals

Theorem 11.7 (Comparison Test). Suppose that $f, g : [a, b) \to [0, +\infty)$, $0 \leq f(x) \leq g(x)$ on $[a, b)$, where b is either finite or $+\infty$, and

$$\int_a^c f(x)\,\mathrm{d}x$$

exists for every $a \leq c < b$. Then we have

$$0 \leq \int_a^b f(x)\,\mathrm{d}x \leq \int_a^b g(x)\,\mathrm{d}x.$$

Furthermore, the convergence of the improper integral of g implies that of f and the divergence of the improper integral of f implies that of g.

Theorem 11.8 (Limit Comparison Test). Suppose that $f, g : [a, b) \to [0, +\infty)$ and both

$$\int_a^c f(x)\,\mathrm{d}x \quad \text{and} \quad \int_a^c g(x)\,\mathrm{d}x$$

exist for every $a \leq c < b$. Furthermore, if

$$\lim_{x \to b} \frac{f(x)}{g(x)} = A \neq 0, \tag{11.8}$$

where b is either finite or $+\infty$, then the improper integrals

$$\int_a^b f(x)\,\mathrm{d}x \quad \text{and} \quad \int_a^b g(x)\,\mathrm{d}x$$

both converge or both diverge. If the value of A in the limit (11.8) is 0, then the convergence of $\int_a^b g(x)\,\mathrm{d}x$ implies the convergence of $\int_a^b f(x)\,\mathrm{d}x$ only.

Theorem 11.9 (Absolute Convergence Test). Suppose that $f : [a, b) \to \mathbb{R}$ is a function such that

$$\int_a^c f(x)\,\mathrm{d}x$$

exists for every $a \leq c < b$ and the improper integral

$$\int_a^b |f(x)|\,\mathrm{d}x$$

converges, where b is either finite or $+\infty$. Then we have

$$\left| \int_a^b f(x)\,\mathrm{d}x \right| \leq \int_a^b |f(x)|\,\mathrm{d}x$$

and the improper integral

$$\int_a^b f(x)\,\mathrm{d}x$$

converges.

11.2. Evaluations of Improper Integrals

Theorem 11.10 (Integral Test for Convergence of Series). *Let N be a positive integer. Suppose that $f : [N, +\infty) \to [0, +\infty)$ decreases on $[N, +\infty)$ and*
$$\int_N^b f(x)\, dx$$
converges for every $N \leq b < +\infty$. Then we have
$$\sum_{k=N+1}^{\infty} f(k) \leq \int_N^{+\infty} f(x)\, dx \leq \sum_{k=N}^{+\infty} f(k).$$

Particularly,
$$\int_N^{+\infty} f(x)\, dx$$
converges if and only if
$$\sum_{k=N}^{+\infty} f(k)$$
converges.

> **Remark 11.2**
>
> Of course, there are analogies of convergence theorems for functions $f : (a, b] \to [0, +\infty)$, where a is either finite or $-\infty$, but we won't repeat the statements here.

11.2 Evaluations of Improper Integrals

> **Problem 11.1**
>
> (★) *Evaluate the following improper integrals:*
>
> (a) $\int_1^{+\infty} \dfrac{dx}{1+x^2}$.
>
> (b) $\int_1^{+\infty} \dfrac{dx}{\sqrt{x}}$.

Proof.

(a) By Definition 11.1, we obtain
$$\int_1^{+\infty} \frac{dx}{1+x^2} = \lim_{b \to +\infty} \int_1^b \frac{dx}{1+x^2} = \lim_{b \to +\infty} [\arctan x]_1^b = \lim_{b \to +\infty} \left(\arctan b - \frac{\pi}{4} \right) = \frac{\pi}{4}.$$

(b) By Definition 11.1, we have
$$\int_1^{+\infty} \frac{dx}{\sqrt{x}} = \lim_{b \to +\infty} \int_1^b \frac{dx}{\sqrt{x}} = \lim_{b \to +\infty} [2\sqrt{x}]_1^b = \lim_{b \to +\infty} (2\sqrt{b} - 2) = \infty.$$

This completes the proof of the problem. ∎

> **Problem 11.2**
>
> (★) Evaluate the following improper integrals:
>
> (a) $\displaystyle\int_0^1 \frac{\mathrm{d}x}{\sqrt{x}}$.
>
> (b) $\displaystyle\int_1^3 \frac{1}{\sqrt{9-x^2}}\,\mathrm{d}x$.

Proof.

(a) By Definition 11.2, it is easy to check that

$$\int_0^1 \frac{\mathrm{d}x}{\sqrt{x}} = \lim_{\epsilon \to 0+} \int_\epsilon^1 \frac{\mathrm{d}x}{\sqrt{x}} = \lim_{\epsilon \to 0+} [2\sqrt{x}]_\epsilon^1 = \lim_{\epsilon \to 0+} (2 - 2\sqrt{\epsilon}) = 2.$$

(b) By Definition 11.2 and then the substitution $x = 3\sin\theta$, we know that

$$\int_1^3 \frac{1}{\sqrt{9-x^2}}\,\mathrm{d}x = \lim_{c \to 3-} \int_1^c \frac{1}{\sqrt{9-x^2}}\,\mathrm{d}x = \lim_{c \to 3-} \int_0^{\arcsin \frac{c}{3}} \mathrm{d}\theta = \lim_{c \to 3-} \arcsin \frac{c}{3} = \frac{\pi}{2}.$$

Hence we complete the proof of the problem. ∎

> **Problem 11.3**
>
> (★) Evaluate the improper integral
>
> $$\int_1^{+\infty} \frac{1}{x^\alpha}\,\mathrm{d}x,$$
>
> where α is real.

Proof. Clearly, if $b \geq 1$, then we have

$$\int_1^b \frac{1}{x^\alpha}\,\mathrm{d}x = \begin{cases} \left[\dfrac{x^{1-\alpha}}{1-\alpha}\right]_1^b, & \text{if } \alpha \neq 1; \\ [\ln x]_1^b, & \text{otherwise,} \end{cases}$$

$$= \begin{cases} \dfrac{b^{1-\alpha} - 1}{1-\alpha}, & \text{if } \alpha \neq 1; \\ \ln b, & \text{otherwise.} \end{cases}$$

On the one hand, if $\alpha \neq 1$, then we get from Definition 11.1 that

$$\int_1^{+\infty} \frac{1}{x^\alpha}\,\mathrm{d}x = \lim_{b \to +\infty} \frac{b^{1-\alpha} - 1}{1-\alpha} = \begin{cases} \dfrac{1}{\alpha - 1}, & \text{if } \alpha > 1; \\ +\infty, & \text{otherwise.} \end{cases} \qquad (11.9)$$

11.2. Evaluations of Improper Integrals

On the other hand, if $\alpha = 1$, then we ahve

$$\int_1^{+\infty} \frac{1}{x}\,dx = \lim_{b\to+\infty} \ln b = +\infty. \tag{11.10}$$

Combining the two results (11.9) and (11.10), we see that

$$\int_1^{+\infty} \frac{1}{x^\alpha}\,dx = \begin{cases} \dfrac{1}{\alpha-1}, & \text{if } \alpha > 1; \\ +\infty, & \text{otherwise.} \end{cases}$$

We have completed the proof of the problem. ∎

Problem 11.4

(⋆) *Evaluate the improper integral of*

$$\int_{-1}^{1} \frac{dx}{x} \tag{11.11}$$

and its principal value.

Proof. We write

$$\int_{-1}^{1} \frac{dx}{x} = \int_{-1}^{0} \frac{dx}{x} + \int_{0}^{1} \frac{dx}{x}. \tag{11.12}$$

Since the two improper integrals on the right-hand side of (11.12) are divergent, the improper integral (11.11) is divergent too. However, we have

$$\text{P.V.} \int_{-1}^{1} \frac{dx}{x} = \lim_{\epsilon \to 0+} \left(\int_{-1}^{-\epsilon} \frac{dx}{x} + \int_{\epsilon}^{1} \frac{dx}{x} \right) = \lim_{\epsilon \to 0+} \left(\ln|x| \Big|_{-1}^{-\epsilon} + \ln|x| \Big|_{\epsilon}^{1} \right) = 0.$$

This completes the proof of the problem. ∎

Problem 11.5

(⋆) *Let $-\infty < a < b < +\infty$. Evaluate the following improper integral*

$$\int_a^b \frac{dx}{(b-x)^\alpha}$$

if $\alpha < 1$.

Proof. For $a \leq c < b$, we have

$$\int_a^c \frac{dx}{(b-x)^\alpha} = \left[\frac{(b-x)^{1-\alpha}}{\alpha-1} \right]_a^c = \frac{(b-c)^{1-\alpha} - (b-a)^{1-\alpha}}{\alpha-1}.$$

Since $\alpha < 1$, we have

$$\lim_{c\to b-} (b-c)^{1-\alpha} = 0$$

and then we get from Definition 11.2 that

$$\int_a^b \frac{dx}{(b-x)^\alpha} = \lim_{c \to b-} \frac{(b-c)^{1-\alpha} - (b-a)^{1-\alpha}}{\alpha - 1} = \frac{(b-a)^{1-\alpha}}{1-\alpha}.$$

We complete the proof of the problem. ∎

Problem 11.6

(★) Prove that

$$\int_0^{+\infty} f\left(ax + \frac{b}{x}\right) dx = \frac{1}{a} \int_0^{+\infty} f(\sqrt{y^2 + 4ab})\, dx,$$

where a and b are positive. (Assume the improper integrals in the question are well-defined.)

Proof. Suppose that

$$y = ax - \frac{b}{x}. \tag{11.13}$$

Then we have

$$ax + \frac{b}{x} = \sqrt{y^2 + 4ab}. \tag{11.14}$$

Adding the two expressions (11.13) and (11.14) to get

$$x = \frac{1}{2a}(y + \sqrt{y^2 + 4ab}) = \varphi(y)$$

so that

$$\frac{dx}{dy} = \varphi'(y) = \frac{1}{2a} \cdot \frac{y + \sqrt{y^2 + 4ab}}{\sqrt{y^2 + 4ab}}.$$

When $x \to 0+$, $y \to -\infty$; when $x \to +\infty$, $y \to +\infty$. By Theorem 11.5(b), we have

$$\int_0^{+\infty} f\left(ax + \frac{b}{x}\right) dx = \frac{1}{2a} \int_{-\infty}^{+\infty} f(\sqrt{y^2 + 4ab}) \cdot \frac{y + \sqrt{y^2 + 4ab}}{\sqrt{y^2 + 4ab}}\, dy$$

$$= \frac{1}{2a} \int_{-\infty}^{0} f(\sqrt{y^2 + 4ab}) \cdot \frac{y + \sqrt{y^2 + 4ab}}{\sqrt{y^2 + 4ab}}\, dy$$

$$+ \frac{1}{2a} \int_0^{+\infty} f(\sqrt{y^2 + 4ab}) \cdot \frac{y + \sqrt{y^2 + 4ab}}{\sqrt{y^2 + 4ab}}\, dy$$

$$= \frac{1}{2a} \int_0^{+\infty} f(\sqrt{y^2 + 4ab}) \cdot \frac{\sqrt{y^2 + 4ab} - y}{\sqrt{y^2 + 4ab}}\, dy$$

$$+ \frac{1}{2a} \int_0^{+\infty} f(\sqrt{y^2 + 4ab}) \cdot \frac{y + \sqrt{y^2 + 4ab}}{\sqrt{y^2 + 4ab}}\, dy$$

$$= \frac{1}{a} \int_0^{+\infty} f(\sqrt{y^2 + 4ab})\, dy.$$

This completes the proof of the problem. ∎

11.2. Evaluations of Improper Integrals

Problem 11.7

(⋆) Suppose that $f : (0, 1) \to \mathbb{R}$ is an increasing function and $\int_0^1 f(x)\,dx$ exists. Prove that

$$\int_0^1 f(x)\,dx = \lim_{n \to +\infty} \frac{1}{n} \sum_{k=1}^n f\left(\frac{k}{n}\right).$$

Proof. Let $n \in \mathbb{N}$. Since f is increasing, we see that

$$\int_0^{1-\frac{1}{n}} f(x)\,dx = \sum_{k=1}^{n-1} \int_{\frac{k-1}{n}}^{\frac{k}{n}} f(x)\,dx \le \sum_{k=1}^{n-1} \int_{\frac{k-1}{n}}^{\frac{k}{n}} f\left(\frac{k}{n}\right)\,dx = \frac{1}{n}\sum_{k=1}^{n-1} f\left(\frac{k}{n}\right)$$

and

$$\int_{\frac{1}{n}}^{1} f(x)\,dx = \sum_{k=1}^{n-1} \int_{\frac{k}{n}}^{\frac{k+1}{n}} f(x)\,dx \ge \sum_{k=1}^{n-1} \int_{\frac{k}{n}}^{\frac{k+1}{n}} f\left(\frac{k}{n}\right)\,dx = \frac{1}{n}\sum_{k=1}^{n-1} f\left(\frac{k}{n}\right).$$

Hence we conclude that

$$\int_0^{1-\frac{1}{n}} f(x)\,dx \le \frac{1}{n}\sum_{k=1}^{n-1} f\left(\frac{k}{n}\right) \le \int_{\frac{1}{n}}^1 f(x)\,dx. \tag{11.15}$$

Since $\int_0^1 f(x)\,dx$ exists, Definition 11.2 ensures that

$$\int_0^1 f(x)\,dx = \lim_{n \to +\infty} \int_0^{1-\frac{1}{n}} f(x)\,dx = \lim_{n \to +\infty} \int_{\frac{1}{n}}^1 f(x)\,dx. \tag{11.16}$$

Using this fact (11.16) and applying Theorem 5.6 (Squeeze Theorem for Convergent Sequence) to the inequalities (11.15), we have the desired result which completes the proof of the problem. ∎

Problem 11.8

(⋆)(⋆) Suppose that $0 < \theta < 1$. We define

$$f_\theta(x) = \left[\frac{\theta}{x}\right] - \theta\left[\frac{1}{x}\right]$$

for all $x \in (0, 1)$. Evaluate the improper integral

$$\int_0^1 f_\theta(x)\,dx.$$

Proof. We notice that

$$f_\theta(x) = -\left(\frac{\theta}{x} - \left[\frac{\theta}{x}\right]\right) + \theta\left(\frac{1}{x} - \left[\frac{1}{x}\right]\right)$$

which implies that

$$\int_0^1 f_\theta(x)\,dx = -\int_0^1 \left(\frac{\theta}{x} - \left[\frac{\theta}{x}\right]\right)dx + \int_0^1 \theta\left(\frac{1}{x} - \left[\frac{1}{x}\right]\right)dx$$

$$= -\int_0^\theta \left(\frac{\theta}{x} - \left[\frac{\theta}{x}\right]\right)dx - \int_\theta^1 \left(\frac{\theta}{x} - \left[\frac{\theta}{x}\right]\right)dx + \int_0^1 \theta\left(\frac{1}{x} - \left[\frac{1}{x}\right]\right)dx. \quad (11.17)$$

If $\theta < x \leq 1$, then $\left[\frac{\theta}{x}\right] = 0$ so that

$$\int_\theta^1 \left(\frac{\theta}{x} - \left[\frac{\theta}{x}\right]\right)dx = -\int_\theta^1 \frac{\theta}{x}\,dx = -\theta \ln x \Big|_\theta^1 = \theta \ln \theta. \quad (11.18)$$

Furthermore, the first integral on the right-hand side of (11.17) can be written as

$$\int_0^\theta \left(\frac{\theta}{x} - \left[\frac{\theta}{x}\right]\right)dx = \lim_{\epsilon \to 0+} \int_\epsilon^\theta \left(\frac{\theta}{x} - \left[\frac{\theta}{x}\right]\right)dx. \quad (11.19)$$

Applying the substitution $x = \theta y$ to the integral on the right-hand side of (11.19), we obtain

$$\int_\epsilon^\theta \left(\frac{\theta}{x} - \left[\frac{\theta}{x}\right]\right)dx = \theta \int_{\epsilon\theta^{-1}}^1 \left(\frac{1}{y} - \left[\frac{1}{y}\right]\right)dy.$$

Therefore, we have

$$\int_0^\theta \left(\frac{\theta}{x} - \left[\frac{\theta}{x}\right]\right)dx = \lim_{\epsilon \to 0+} \theta \int_{\epsilon\theta^{-1}}^1 \left(\frac{1}{y} - \left[\frac{1}{y}\right]\right)dy = \theta \int_0^1 \left(\frac{1}{y} - \left[\frac{1}{y}\right]\right)dy. \quad (11.20)$$

By putting the expression (11.18) and (11.20) back into the expression (11.17), we establish immediately that

$$\int_0^1 f_\theta(x)\,dx = \theta \ln \theta,$$

completing the analysis of the problem. ∎

11.3 Convergence of Improper Integrals

Problem 11.9

(★) Prove that

$$\int_0^{+\infty} e^{-x^2}\,dx$$

is convergent.

Proof. Let $f(x) = e^{-x^2}$ on $[0, +\infty)$ and

$$g(x) = \begin{cases} f(x), & \text{if } x \in [0,1]; \\ e^{-x}, & \text{otherwise.} \end{cases}$$

11.3. Convergence of Improper Integrals

Since $e^{-x^2} \leq e^{-x}$ for all $x \geq 1$, we have
$$0 \leq f(x) \leq g(x)$$
on $[0, +\infty)$. Since f is bounded on $[0, 1]$, the proper integral
$$\int_0^1 e^{-x^2}\,dx$$
is finite. Furthermore, we have
$$\int_0^{+\infty} g(x)\,dx = \int_0^1 e^{-x^2}\,dx + \int_1^{+\infty} e^{-x}\,dx = \int_0^1 e^{-x^2}\,dx - [e^{-x}]_1^{+\infty} = \int_0^1 e^{-x^2}\,dx + e^{-1}$$
which means that the improper integral
$$\int_0^{+\infty} g(x)\,dx$$
converges. By Theorem 11.7 (Comparison Test), the improper integral
$$\int_0^{+\infty} e^{-x^2}\,dx$$
is convergent. We end the analysis of the proof of the problem. ∎

Problem 11.10

(⋆) Let $p \in \mathbb{R}$. Prove that
$$\int_1^{+\infty} x^p e^{-x}\,dx$$
is convergent.

Proof. Consider $f(x) = x^p e^{-x}$ and $g(x) = x^{-2}$. Obviously, by repeated use of L'Hôspital's Rule, we know that
$$\lim_{x \to +\infty} \frac{f(x)}{g(x)} = \lim_{x \to +\infty} \frac{x^{p+2}}{e^x} = 0.$$
Since we have
$$\int_1^{+\infty} x^{-2}\,dx = -[x^{-1}]_1^{+\infty} = 1,$$
Theorem 11.8 (Limit Comparison Test) implies the required result. Hence we have completed the proof of the problem. ∎

Problem 11.11

(⋆) Prove Theorem 11.9 (Absolute Convergence Test).

Proof. Suppose that $\int_a^b |f(x)|\,dx$ converges and $\int_a^x f(t)\,dt$ exists for all $x \in [a,b)$. We know that
$$0 \le f(x) + |f(x)| \le 2|f(x)|,$$
so Theorem 11.7 (Comparison Test) shows that
$$\int_a^b [f(x) + |f(x)|]\,dx$$
converges. Hence we immediately deduce from Theorem 11.5(a) that the improper integral
$$\int_a^b f(x)\,dx$$
converges. This completes the proof of the problem. ∎

Problem 11.12

(★) *Prove that the improper integral*
$$\int_1^{+\infty} \frac{\sin x}{x^2}\,dx$$
converges.

Proof. We notice that
$$\left|\frac{\sin x}{x^2}\right| \le \frac{1}{x^2}$$
for all $x \in [1, +\infty)$. By Problem 11.3, the improper integral
$$\int_1^{+\infty} \frac{1}{x^2}\,dx$$
is convergent. By Theorem 11.7 (Comparison Test),
$$\int_1^{+\infty} \left|\frac{\sin x}{x^2}\right|\,dx$$
is also convergent. Finally, it follows from Theorem 11.9 (Absolute Convergence Test) that the improper integral
$$\int_1^{+\infty} \frac{\sin x}{x^2}\,dx$$
converges. We have completed the proof of the problem. ∎

Problem 11.13

(★) *Prove that the improper integral*
$$\int_0^1 (-\ln x)^\alpha\,dx$$
converges if and only if $\alpha > -1$.

11.3. Convergence of Improper Integrals

Proof. Let $t = -\ln x$. Then we have $dt = -\frac{1}{x} dx$. When $x \to 0+$, $t \to +\infty$; when $x = 1$, $t = 0$. Thus Theorem 11.5(b) implies that

$$\int_0^1 (-\ln x)^\alpha \, dx = -\int_{+\infty}^0 t^\alpha e^{-t} \, dt$$
$$= \int_0^{+\infty} t^\alpha e^{-t} \, dt$$
$$= \int_0^1 t^\alpha e^{-t} \, dt + \int_1^{+\infty} t^\alpha e^{-t} \, dt. \tag{11.21}$$

Now we are going to investigate the convergence of the two improper integrals of the expression (11.21).

Obviously, we have

$$\lim_{t \to +\infty} \frac{t^\alpha e^{-t}}{t^{-2}} = \lim_{t \to +\infty} \frac{t^{\alpha+2}}{e^t} = 0$$

for every $\alpha \in \mathbb{R}$. Therefore, the last integral of the expression (11.21) converges by Theorem 11.8 (Limit Comparison Test). Next, since

$$\lim_{t \to 0+} \frac{t^\alpha e^{-t}}{t^\alpha} = \lim_{t \to 0+} e^{-t} = 1,$$

Theorem 11.8 (Limit Comparison Test) again shows that the first integral of the expression (11.21) converges if and only if the improper integral

$$\int_0^1 t^\alpha \, dt \tag{11.22}$$

converges.

If $\alpha = -1$, then the integral (11.22) becomes

$$\int_0^1 \frac{dt}{t} = \lim_{\epsilon \to 0+} \int_\epsilon^1 \frac{dt}{t} = \lim_{\epsilon \to 0+} \ln t \Big|_\epsilon^1 = -\lim_{\epsilon \to 0+} \ln \epsilon$$

which is definitely divergent. For $\alpha \neq -1$, we have

$$\int_0^1 t^\alpha \, dt = \lim_{\epsilon \to 0+} \frac{1 - \epsilon^{1+\alpha}}{1 + \alpha}$$

which is convergent if and only if $\alpha > -1$. Hence the improper integral

$$\int_0^1 (-\ln x)^\alpha \, dx$$

converges if and only if $\alpha > -1$, completing the proof of the problem. ∎

> **Problem 11.14**
>
> (⋆)(⋆) Suppose that $f \in \mathscr{R}([0,1])$, f is periodic with period 1 and
> $$\int_0^1 f(x)\, dx = 0. \tag{11.23}$$
> If $\alpha > 0$, prove that the improper integral
> $$\int_1^{+\infty} x^{-\alpha} f(x)\, dx$$
> exists. Hint: It is true that
> $$\int_1^b x^{-\alpha} f(x)\, dx = \int_1^b x^{-\alpha}\, dg,$$
> where $b \geq 1$ and $g(x) = \int_1^x f(t)\, dt$.

Proof. Let $1 \leq x \leq b < +\infty$ and set
$$g(x) = \int_1^x f(t)\, dt.$$
By the First Fundamental Theorem of Calculus, g is continuous on $[1, b]$. By the hypothesis (11.23) and the periodicity of f, we get
$$g(x+1) = \int_1^{x+1} f(t)\, dt = \int_0^x f(t+1)\, dt = \int_0^1 f(t)\, dt + \int_0^x f(t)\, dt = g(x)$$
so that g is also periodic with period 1. Consequently, g is bounded by a positive constant M on $[1, +\infty)$. Next, it follows from the hint that, for every $b \geq 1$, we have
$$\int_1^b x^{-\alpha} f(x)\, dx = \int_1^b x^{-\alpha}\, dg$$
$$= [x^{-\alpha} g(x)]_1^b + \alpha \int_1^b x^{-\alpha-1} g(x)\, dx$$
$$= b^{-\alpha} g(b) - g(1) + \alpha \int_1^b x^{-\alpha-1} g(x)\, dx. \tag{11.24}$$
Since g is bounded by M on $[1, +\infty)$ and $\alpha > 0$, we have
$$\lim_{b \to +\infty} b^{-\alpha} g(b) = 0. \tag{11.25}$$
In addition, we have
$$|x^{-\alpha-1} g(x)| \leq M x^{-\alpha-1}$$
for all $x \in [1, +\infty)$, so Problem 11.3, Theorems 11.7 (Comparison Test) and 11.9 (Absolute Convergence Test) imply that
$$\int_1^{+\infty} x^{-\alpha-1} g(x)\, dx \tag{11.26}$$

11.3. Convergence of Improper Integrals

is convergent for every $\alpha > 0$. Applying the results (11.24) and (11.25) to the expression (11.26) to conclude that
$$\int_1^{+\infty} x^{-\alpha} f(x) \, dx$$
exists, completing the proof of the problem. ∎

Problem 11.15

(★)(★) Prove that
$$\Gamma(x) = \int_0^{+\infty} t^{x-1} e^{-t} \, dt \qquad (11.27)$$
converges for all $x > 0$.

Proof. We write
$$\Gamma(x) = \int_0^1 t^{x-1} e^{-t} \, dt + \int_1^{+\infty} t^{x-1} e^{-t} \, dt. \qquad (11.28)$$
Now Problem 11.10 ensures that the second integral of the expression (11.28) converges for every $x \in \mathbb{R}$. For the first integral, we notice that if $a > 0$ and $t = \frac{1}{s}$, then we obtain
$$\int_a^1 t^{x-1} e^{-t} \, dt = \int_1^{\frac{1}{a}} s^{-x-1} e^{-\frac{1}{s}} \, ds.$$
For $s > 1$, since
$$0 \leq s^{-x-1} e^{-\frac{1}{s}} \leq s^{-x-1}$$
and we know from Problem 11.3 that
$$\int_1^{+\infty} s^{-x-1} \, ds$$
is convergent for all $x > 0$, Theorem 11.7 (Comparison Test) now implies that
$$\int_1^{+\infty} s^{-x-1} e^{-\frac{1}{s}} \, ds$$
converges for all $x > 0$. Hence the first integral of the expression (11.28) and then the improper integral (11.27) both converge for all $x > 0$. This completes the proof of the problem. ∎

Problem 11.16 (Cauchy Criterion for Improper Integrals)

(★)(★) Suppose that $a \in \mathbb{R}$ and $f : [a,b) \to \mathbb{R}$ is integrable on every closed and bounded interval of $[a,b)$, where b is finite or $+\infty$. Then the improper integral
$$\int_a^b f(x) \, dx \qquad (11.29)$$
converges if and only if for every $\epsilon > 0$, there exists a $M \geq a$ such that for all $M \leq A, B < b$, we have
$$\left| \int_A^B f(x) \, dx \right| < \epsilon. \qquad (11.30)$$

Proof. Suppose that the improper integral (11.29) converges to L. Given $\epsilon > 0$. By Definition 11.1 or 11.2, there exists a $M \geq a$ such that $M \leq c < b$ implies

$$\left| \int_a^c f(x)\,\mathrm{d}x - L \right| < \frac{\epsilon}{2}. \tag{11.31}$$

If $M \leq A, B < b$, then the estimate (11.31) gives

$$\left| \int_A^B f(x)\,\mathrm{d}x \right| = \left| \int_a^B f(x)\,\mathrm{d}x - L + L - \int_a^A f(x)\,\mathrm{d}x \right|$$
$$\leq \left| \int_a^B f(x)\,\mathrm{d}x - L \right| + \left| L - \int_a^A f(x)\,\mathrm{d}x \right|$$
$$< \epsilon.$$

Conversely, we define $F : [a, b) \to \mathbb{R}$ by

$$F(r) = \int_a^r f(x)\,\mathrm{d}x.$$

Then our hypothesis implies that F exists on $[a, b)$ and with this notation, the condition (11.30) is equivalent to

$$|F(B) - F(A)| < \epsilon. \tag{11.32}$$

It is easy to see that there exists a $N \in \mathbb{N}$ such that $\frac{1}{N} < b - M$. Therefore, if we write $A_n = b - \frac{1}{n}, B_m = b - \frac{1}{m}, a_n = F(A_n)$ and $a_m = F(B_m)$, then we have $M \leq A_n, B_m < b$ for all $n, m \geq N$ and so the inequality (11.32) means that

$$|a_m - a_n| < \epsilon$$

for all $n, m \geq N$. In other words, $\{a_n\}$ is Cauchy and it converges to the real number L. Hence this fact implies that

$$\int_a^b f(x)\,\mathrm{d}x = \lim_{n \to +\infty} \int_a^{b - \frac{1}{n}} f(x)\,\mathrm{d}x = \lim_{n \to +\infty} F\left(b - \frac{1}{n}\right) = \lim_{n \to +\infty} a_n = L.$$

This completes the proof of the problem. ∎

Problem 11.17 (Abel's Test for Improper Integrals)

(★)(★) Suppose that $a \in \mathbb{R}$ and $f : [a, +\infty) \to \mathbb{R}$ is a function which is integrable on every closed and bounded interval of $[a, +\infty)$ and the improper integral $\int_a^{+\infty} f(x)\,\mathrm{d}x$ converges. Let $g : [a, +\infty) \to \mathbb{R}$ be a monotonic bounded function. Then the improper integral

$$\int_a^{+\infty} f(x)g(x)\,\mathrm{d}x$$

converges.

11.3. Convergence of Improper Integrals

Proof. Let $a \leq A \leq B < +\infty$. Without loss of generality, we may assume that g is increasing on $[A, B]$. Thus the Second Mean Value Theorem for Integrals [10] implies the existence of a $p \in [A, B]$ such that

$$\int_A^B f(x)g(x)\,dx = g(A)\int_A^p f(x)\,dx + g(B)\int_p^B f(x)\,dx. \tag{11.33}$$

By the hypothesis, there exists a positive constant M_1 such that

$$|g(x)| \leq M_1$$

on $[a, +\infty)$. Given $\epsilon > 0$. By Problem 11.16 (Cauchy Criterion for Improper Integrals), there is a positive constant $M_2 \geq a$ such that

$$\left|\int_A^p f(x)\,dx\right| \leq \frac{\epsilon}{2M_1} \quad \text{and} \quad \left|\int_p^B f(x)\,dx\right| \leq \frac{\epsilon}{2M_1} \tag{11.34}$$

for all $A \geq M_2 \geq a$. Combining the formula (11.33) and the inequalities (11.34), we see that

$$\left|\int_A^B f(x)g(x)\,dx\right| \leq |g(A)| \cdot \left|\int_A^p f(x)\,dx\right| + |g(B)| \cdot \left|\int_p^B f(x)\,dx\right|$$
$$\leq M_1 \cdot \frac{\epsilon}{2M_1} + M_1 \cdot \frac{\epsilon}{2M_1}$$
$$= \epsilon$$

for all $A, B \geq M_2 \geq a$. By Problem 11.16 (Cauchy Criterion for Improper Integrals) again, we conclude that the improper integral

$$\int_a^{+\infty} f(x)g(x)\,dx$$

converges, completing the proof of the problem. ∎

Problem 11.18 (Dirichlet's Test for Improper Integrals)

(⋆)(⋆) Suppose that $a \in \mathbb{R}$ and $f : [a, +\infty) \to \mathbb{R}$ is a function which is integrable on every closed and bounded interval of $[a, +\infty)$ and there exists a positive constant M such that

$$\left|\int_a^b f(x)\,dx\right| \leq M \tag{11.35}$$

for all $b > a$. Let $g : [a, +\infty) \to \mathbb{R}$ be a monotonic function such that $g(x) \to 0$ as $x \to +\infty$. Then the improper integral

$$\int_a^{+\infty} f(x)g(x)\,dx$$

converges.

Proof. Given that $\epsilon > 0$. Since $g(x) \to 0$ as $x \to +\infty$, there exists a $M_1 \geq a$ such that

$$|g(x)| < \frac{\epsilon}{4M}$$

for all $x \geq M_1$. Let $A, B \geq M_1 \geq a$. By the Second Mean Value Theorem for Integrals [10] and the hypothesis (11.35), there exists a p between A and B such that

$$\left|\int_A^B f(x)g(x)\,dx\right| \leq |g(A)| \cdot \left|\int_A^p f(x)\,dx\right| + |g(B)| \cdot \left|\int_p^B f(x)\,dx\right|$$

$$\leq |g(A)| \cdot \left(\left|\int_a^p f(x)\,dx\right| + \left|\int_a^A f(x)\,dx\right|\right)$$

$$+ |g(B)| \cdot \left(\left|\int_a^B f(x)\,dx\right| + \left|\int_a^p f(x)\,dx\right|\right)$$

$$< \frac{\epsilon}{4M} \cdot 2M + \frac{\epsilon}{4M} \cdot 2M$$

$$= \epsilon.$$

Hence Problem 11.16 (Cauchy Criterion for Improper Integrals) shows that the improper integral

$$\int_a^{+\infty} f(x)g(x)\,dx$$

converges. We have completed the proof of the problem. ∎

11.4 Miscellaneous Problems on Improper Integrals

Problem 11.19

(★) Suppose that $\int_a^{+\infty} f(x)\,dx$ converges. Is it true that $f(x) \to 0$ as $x \to +\infty$?

Proof. The answer is negative. We consider the convergence of the improper integral

$$\int_0^{+\infty} \sin x^2 \,dx.$$

By the substitution $t = x^2$, we have

$$\int_0^{+\infty} \sin x^2 \,dx = \int_0^{+\infty} \frac{\sin t}{2\sqrt{t}}\,dt. \tag{11.36}$$

Since $g(t) = \frac{1}{2\sqrt{t}} \to 0$ as $t \to +\infty$, it follows from Problem 11.18 (Dirichlet's Test for Improper Integrals) that the integral (11.36) converges. However, it is easy to see that the function $f(x) = \sin x^2$ *does not* have limit as $x \to +\infty$ so that $f(x) \to 0$. This completes the proof of the problem. ∎

11.4. Miscellaneous Problems on Improper Integrals

Problem 11.20

$(\star)(\star)$ Prove that the improper integral

$$\int_2^{+\infty} \frac{1}{(\ln x)^{\ln x}}\,dx \qquad (11.37)$$

converges.

Proof. We consider the function $f : [2,+\infty) \to (0,+\infty)$ defined by

$$f(x) = \frac{1}{(\ln x)^{\ln x}}.$$

By differentiation, we know that

$$f'(x) = -\frac{1}{x(\ln x)^{\ln x}}$$

which is negative on $[2,+\infty)$, i.e, f is decreasing on $[2,+\infty)$.

Next, we consider the convergence of the series

$$\sum_{n=2}^{+\infty} \frac{1}{(\ln n)^{\ln n}}. \qquad (11.38)$$

Let $a_n = \frac{1}{(\ln n)^{\ln n}}$. Now the decreasing property of f ensures that the sequence $\{a_n\}$ is decreasing. Since

$$\alpha = \limsup_{n\to+\infty} \sqrt[k]{|2^k a_{2^k}|} = \limsup_{n\to+\infty} \sqrt[k]{\frac{2^k}{(\ln 2^k)^{\ln 2^k}}} = \limsup_{n\to+\infty} \frac{2}{(k\ln 2)^{\ln 2}} = 0,$$

we obtain from Theorem 6.7 (Root Test) that the series

$$\sum_{k=1}^{+\infty} 2^k \cdot \frac{1}{(\ln 2^k)^{\ln 2^k}}$$

converges. Thus it follows from [23, Theorem 3.27, p. 61] that the series (11.38) is also convergent. Hence Theorem 11.10 (Integral Test for Convergence of Series) ensures that the improper integral (11.37) converges, completing the proof of the problem. ∎

Problem 11.21

$(\star)$ Suppose that $\int_a^{+\infty} f(x)\,dx$ converges absolutely and the function g is bounded on $[a,+\infty)$. Prove that the improper integral

$$\int_a^{+\infty} f(x)g(x)\,dx \qquad (11.39)$$

converges.

Proof. We have $|g(x)| \leq M$ on $[a, +\infty)$ for some positive constant M. Since

$$0 \leq |f(x)g(x)| \leq M|f(x)|$$

on $[a, +\infty)$ and $\int_a^{+\infty} |f(x)|\,\mathrm{d}x$ converges, Theorem 11.7 (Comparison Test) shows that

$$\int_a^{+\infty} |f(x)g(x)|\,\mathrm{d}x$$

converges. Finally, the convergence of the improper integral (11.39) follows immediately from Theorem 11.9 (Absolute Convergence Test). Hence we have completed the proof of the problem. ∎

Problem 11.22

(★)(★) Suppose that $\int_a^{+\infty} f(x)\,\mathrm{d}x$ is convergent and f is monotonic on $[a, +\infty)$. Verify that

$$\lim_{x \to +\infty} xf(x) = 0.$$

Proof. Without loss of generality, we may assume that f is monotonically decreasing on $[a, +\infty)$. We claim that $f(x) \geq 0$ for all $x \geq a$. Otherwise, there exists a $p \geq a$ such that $f(p) < 0$. By our assumption, for all $x \geq p$, we have $f(x) \leq f(p) < 0$ and so

$$\int_p^{+\infty} f(x) \leq \int_p^{+\infty} f(p)\,\mathrm{d}x = -\infty,$$

a contradiction. Thus this proves the claim.

Given that $\epsilon > 0$. Since $\int_a^{+\infty} f(x)\,\mathrm{d}x$ converges, Problem 11.16 (Cauchy Criterion for Improper Integrals) guarantees that there exists a $M \geq a$ such that if $\frac{x}{2} \geq M$, then

$$\left| \int_{\frac{x}{2}}^{x} f(t)\,\mathrm{d}t \right| < \frac{\epsilon}{2}. \tag{11.40}$$

Furthermore, we have

$$\left| \int_{\frac{x}{2}}^{x} f(t)\,\mathrm{d}t \right| = \int_{\frac{x}{2}}^{x} f(t)\,\mathrm{d}t \geq f(x) \cdot \left(x - \frac{x}{2}\right) = \frac{xf(x)}{2}. \tag{11.41}$$

Combining the inequalities (11.40) and (11.41), we conclude that

$$|xf(x)| < \epsilon$$

which is equivalent to saying that $xf(x) \to 0$ as $x \to +\infty$. We complete the proof of the problem. ∎

11.4. Miscellaneous Problems on Improper Integrals

Problem 11.23 (Frullani's Integral)

(★)(★) Suppose that f is continuous on $[0, +\infty)$ and the limit

$$\lim_{x \to +\infty} f(x)$$

exists. Let $a, b > 0$. Compute the improper integral

$$\int_0^{+\infty} \frac{f(ax) - f(bx)}{x}\, dx.$$

Proof. For $0 < r < R < +\infty$, using the method of substitution, we have

$$\begin{aligned}
\int_r^R \frac{f(ax) - f(bx)}{x}\, dx &= \int_r^R \frac{f(ax)}{x}\, dx - \int_r^R \frac{f(bx)}{x}\, dx \\
&= \int_{ar}^{aR} \frac{f(x)}{x}\, dx - \int_{br}^{bR} \frac{f(x)}{x}\, dx \\
&= \int_{ar}^{br} \frac{f(x)}{x}\, dx - \int_{aR}^{bR} \frac{f(x)}{x}\, dx \\
&= \int_{ar}^{br} f(x)\, d\alpha - \int_{aR}^{bR} f(x)\, d\alpha,
\end{aligned} \tag{11.42}$$

where $\alpha(x) = \ln x$. It is clear that α is monotonically increasing on $[ar, br]$ and $[aR, bR]$. Since f is continuous on $[ar, br]$ and $[aR, bR]$, it is bounded on these closed intervals and we follow from Theorem 9.3(a) that $f \in \mathscr{R}(\alpha)$ on $[ar, br]$ and $[aR, bR]$. Using the form of the First Mean Value Theorem for Integrals [3, Theorem 7.30, p. 160], we see that there exist $p \in [ar, br]$ and $q \in [aR, bR]$ such that

$$\int_{ar}^{br} f(x)\, d\alpha = f(p) \ln \frac{br}{ar} = f(p) \ln \frac{b}{a} \quad \text{and} \quad \int_{aR}^{bR} f(x)\, d\alpha = f(q) \ln \frac{bR}{aR} = f(q) \ln \frac{b}{a}. \tag{11.43}$$

If we put the two formulas (11.43) back into the expression (11.42), then we get

$$\int_r^R \frac{f(ax) - f(bx)}{x}\, dx = [f(p) - f(q)] \ln \frac{b}{a}. \tag{11.44}$$

As $r \to 0+$ and $R \to +\infty$, we obtain $p \to 0+$ and $q \to +\infty$. Recall that $f(0+) = f(0)$ and $f(+\infty)$ is a real number, so we may take $r \to 0+$ and $R \to +\infty$ in the expression (11.44), we achieve

$$\int_0^{+\infty} \frac{f(ax) - f(bx)}{x}\, dx = [f(0) - f(+\infty)] \ln \frac{b}{a},$$

completing the proof of the problem. ∎

> **Problem 11.24**
>
> (★) Suppose that $f : [0, +\infty) \to [0, +\infty)$ is continuous and
> $$\int_0^{+\infty} f(x)\,dx \tag{11.45}$$
> converges. Prove that
> $$\lim_{n \to +\infty} \frac{1}{n} \int_0^n x f(x)\,dx = 0.$$

Proof. Suppose that
$$M = \frac{1}{1 + \int_0^{+\infty} f(x)\,dx}.$$
Given $\epsilon > 0$. By the hypothesis (11.45), there is an $N \in \mathbb{N}$ such that $n \geq N$ implies that
$$0 \leq \int_n^{+\infty} f(x)\,dx < \frac{\epsilon}{M}. \tag{11.46}$$
Now for all sufficiently large enough n such that $n > n\epsilon \geq N$, we know from the inequality (11.46) that
$$0 \leq \frac{1}{n} \int_0^n x f(x)\,dx = \frac{1}{n} \int_0^{n\epsilon} x f(x)\,dx + \frac{1}{n} \int_{n\epsilon}^n x f(x)\,dx$$
$$< \frac{\epsilon}{M} \int_0^{n\epsilon} f(x)\,dx + \int_{n\epsilon}^n f(x)\,dx$$
$$< \frac{\epsilon}{M} \int_0^{n\epsilon} f(x)\,dx + \frac{\epsilon}{M}$$
$$< \frac{\epsilon}{M} \cdot \Big(\int_0^{+\infty} f(x)\,dx + 1 \Big)$$
$$= \epsilon.$$
This means that
$$\lim_{n \to +\infty} \frac{1}{n} \int_0^n x f(x)\,dx = 0,$$
completing the proof of the problem. ∎

> **Problem 11.25**
>
> (★)(★) For every $x > 1$, prove that
> $$\text{P.V.} \int_0^x \frac{dt}{\ln t}$$
> exists.

Proof. Given $\epsilon > 0$ and $x \in [0, 1)$. Now we have
$$\lim_{t \to 0^+} \frac{1}{\ln t} = 0,$$

11.4. Miscellaneous Problems on Improper Integrals

so if we define
$$f(t) = \begin{cases} \dfrac{1}{\ln t}, & \text{if } t > 0; \\ 0, & \text{if } t = 0, \end{cases}$$

then f is well-defined and continuous on $[0, x]$ so that
$$\int_0^x \frac{dt}{\ln t} = \int_0^x f(t)\, dt$$

exists.

When $x > 1$, we notice that
$$\text{P.V.} \int_0^x \frac{dt}{\ln t} = \lim_{\epsilon \to 0+} \left(\int_0^{1-\epsilon} \frac{dt}{\ln t} + \int_{1+\epsilon}^x \frac{dt}{\ln t} \right). \tag{11.47}$$

By Taylor's Theorem, we can show that
$$\ln t = (t-1) - \frac{(t-1)^2}{2} + \frac{(t-1)^3}{3\xi^3} = (t-1) + \left[\frac{2(t-1)}{3\xi^3} - 1 \right] \cdot \frac{(t-1)^2}{2},$$

where $\xi \in (1, t)$. Let $\alpha(t) = \frac{2(t-1)}{3\xi^3}$. Then it is clear that
$$\lim_{t \to 1} \alpha(t) = 0.$$

Besides, we have the identity
$$\frac{1}{\ln t} = \frac{1}{t-1} - \frac{\alpha(t) - 1}{2 + [\alpha(t) - 1](t-1)} \tag{11.48}$$

so that
$$\lim_{t \to 1} \frac{\alpha(t) - 1}{2 + [\alpha(t) - 1](t-1)} = 0.$$

Therefore, the second term on the right-hand side of the equation (11.48) is continuous and thus integrable around $t = 1$. This fact and the expression (11.48) indicate that the principal value (11.47) depends on the evaluation of the principal value of $\frac{1}{t-1}$ on $[0, x]$ which is
$$\text{P.V.} \int_0^x \frac{dt}{t-1} = \lim_{\epsilon \to 0+} \left(\int_0^{1-\epsilon} \frac{dt}{t-1} + \int_{1+\epsilon}^x \frac{dt}{t-1} \right)$$
$$= \lim_{\epsilon \to 0+} \left(\ln|t-1|\Big|_0^{1-\epsilon} + \ln|t-1|\Big|_{1+\epsilon}^x \right)$$
$$= \ln(x-1).$$

Hence we conclude that
$$\text{P.V.} \int_0^x \frac{dt}{\ln t}$$

exists, completing the proof of the problem. ∎

> **Problem 11.26**
>
> (★)(★) Suppose that the improper integral $\int_0^{+\infty} f(x)\,dx$ converges. Prove that the improper integral
> $$\int_0^{+\infty} e^{-\epsilon x} f(x)\,dx$$
> converges for every $\epsilon > 0$.

Proof. Let $1 < A < C$ and fix $\epsilon > 0$. Then $e^{-\epsilon x} \geq 0$ on $[A, C]$. By the Second Mean Value Theorem for Integrals [10], we see that

$$\int_A^C e^{-\epsilon x} f(x)\,dx = e^{-A\epsilon} \int_A^B f(x)\,dx$$

for some $B \in [A, C]$. Thus we have

$$\left| \int_A^C e^{-\epsilon x} f(x)\,dx \right| \leq \left| \int_A^B f(x)\,dx \right|. \tag{11.49}$$

Since $\int_0^{+\infty} f(x)\,dx$ converges, Problem 11.16 (Cauchy Criterion for Improper Integrals) ensures that

$$\left| \int_A^B f(x)\,dx \right| < \epsilon \tag{11.50}$$

for large enough A. Combining the inequalities (11.49) and (11.50), we obtain

$$\left| \int_A^C e^{-\epsilon x} f(x)\,dx \right| < \epsilon$$

for large enough A and our conclusion follows immediately from Problem 11.16 (Cauchy Criterion for Improper Integrals) again. This ends the analysis of the problem. ∎

CHAPTER 12

Lebesgue Measure

12.1 Fundamental Concepts

In Chapter 9, we recall the basic theory and main results of the Riemann integral of a bounded function f over a bounded and closed interval $[a, b]$ with approximations associated with f and partitions of $[a, b]$. However, the techniques developed so far for Riemann integration is unsatisfactory for general sets. Consequently, only a small class of functions can be "integrable" in the sense of Riemann. This leads to the theories and the notions of the **Lebesgue measure**, **Lebesgue measurable functions** and **Lebesgue integration**. The discussion of these components will be presented in the successive three chapters and for simplicity, we only discuss the Lebesgue theory on the real line. The main references for this chapter are [7, Chap. 1], [12, Chap. 2 - 5], [22, Chap. 2], [26, Chap. 1] and [27, Chap. 7].

12.1.1 Lebesgue Outer Measure

Definition 12.1 (Length of Intervals). *Let I be a nonempty interval of $\mathbb{R}$ and a and b are the end-points of I.*[a] *We define its length, denoted by $\ell(I)$, to be infinite if I is unbounded. Otherwise, we define*
$$\ell(I) = b - a.$$

Definition 12.2 (Lebesgue Outer Measure). *Let $E \subseteq \mathbb{R}$ and $\{I_k\}$ be a countable collection of nonempty open and bounded intervals covering E, i.e.,*
$$E \subseteq \bigcup_{k=1}^{\infty} I_k.$$

Then the **Lebesgue outer measure** (or simply **outer measure**) of E, denoted by $m^*(E)$, is given by
$$m^*(E) = \inf \Big\{ \sum_{k=1}^{\infty} \ell(I_k) \,\Big|\, E \subseteq \bigcup_{k=1}^{\infty} I_k \Big\},$$

where the infimum takes over all countable collections of nonempty open and bounded intervals covering E.

---
[a] Note that $|a|$ or $|b|$ can be infinite.

Theorem 12.3 (Properties of Outer Measure). *The outer measure m^* satisfies the following properties:*

(a) $m^*(\varnothing) = 0$.

(b) $m^*(I) = \ell(I)$ for every interval I.

(c) *(Monotonicity)* If $E \subseteq F$, then $m^*(E) \leq m^*(F)$.

(d) *(Countably Subadditivity)* If $\{E_k\}$ is a countable collection of sets of $\mathbb{R}$, then we have
$$m^*\Big(\bigcup_{n=1}^{\infty} E_k\Big) \leq \sum_{k=1}^{\infty} m^*(E_k).$$

(e) *(Translation invariance)* For every $E \subseteq \mathbb{R}$ and $p \in \mathbb{R}$, we have $m^*(p+E) = m^*(E)$, where $p + E = \{p + x \mid x \in E\}$.

> **Remark 12.1**
>
> The outer measure m^*, however, is *not* **countable additive**. In other words, there exists a countable collection $\{E_k\}$ of disjoint subsets of $\mathbb{R}$ such that
> $$m^*\Big(\bigcup_{k=1}^{\infty} E_k\Big) \neq \sum_{k=1}^{\infty} m^*(E_k).$$

12.1.2 Lebesgue Measurable Sets

Let S be any subset of $\mathbb{R}$. If $E \subseteq S$, then we denote $E^c = S \setminus E$.

Definition 12.4 (Lebesgue Measurability). *A set $E \subseteq \mathbb{R}$ is **Lebesgue measurable**, or simply **measurable**, if for every set $A \subseteq \mathbb{R}$, we have*[b]
$$m^*(A) = m^*(A \cap E) + m^*(A \cap E^c).$$

If E is measurable, then we define the **Lebesgue measure** of E to be
$$m(E) = m^*(E).$$

If all the sets considered in Theorem 12.3(c) and (d) (Properties of Outer Measure) are measurable, then the corresponding results also hold with m^* replaced by m. That is, if $E \subseteq F$, then
$$m(E) \leq m(F)$$
and if all E_k are measurable, then
$$m\Big(\bigcup_{n=1}^{\infty} E_k\Big) \leq \sum_{n=1}^{\infty} m(E_k). \tag{12.1}$$

Further properties of measurable sets are listed in the following theorem.

---

[b]This definition is due to Carathèodory.

12.1. Fundamental Concepts

Theorem 12.5 (Properties of Measurable Sets). *Let $E \subseteq \mathbb{R}$. Then we have the following results:*

(a) *If $m^*(E) = 0$, then it is measurable.*

(b) *Every interval I is measurable.*

(c) *If E is measurable, then E^c is also measurable.*

(d) *If E is measurable and $p \in \mathbb{R}$, then $p + E$ is also measurable and $m(p+E) = m(E)$.*

(e) *The finite union or intersection of measurable sets is also measurable.*

> **Remark 12.2**
>
> We notice that any set $E \subseteq \mathbb{R}$ with positive outer measure contains a nonmeasurable subset. See, for examples, [12, Chap. 4, pp. 81 - 83] and [22, Theorem 17, p. 48].

Definition 12.6 (Almost Everywhere). *Let $E \subseteq \mathbb{R}$ be a measurable set. We way that a property P holds **almost everywhere** on E if there exists a subset $A \subseteq E$ such that $m(A) = 0$ and P holds for all $x \in E \setminus A$.*

12.1.3 σ-algebras and Borel Sets

In the following discussion, we suppose that $S \subseteq \mathbb{R}$.

Definition 12.7 (σ-algebra). *Let $\mathfrak{M}$ be a collection of subsets of S. Then $\mathfrak{M}$ is called an σ-algebra in S if $\mathfrak{M}$ satisfies the following conditions:*

(a) $S \in \mathfrak{M}$.

(b) *If $E \in \mathfrak{M}$, then $E^c \in \mathfrak{M}$.*

(c) *If $E = \bigcup_{k=1}^{\infty} E_k$ and if $E_k \in \mathfrak{M}$ for all $k = 1, 2, \ldots$, then $E \in \mathfrak{M}$.*

It follows easily from Definition 12.7 (σ-algebra) that if $E, F, E_k \in \mathfrak{M}$ for all $k = 1, 2, \ldots$, then we have

$$\bigcap_{k=1}^{\infty} E_k^c \in \mathfrak{M} \quad \text{and} \quad E \setminus F \in \mathfrak{M}. \tag{12.2}$$

Theorem 12.8. *The union of a countable collection of measurable sets is also measurable, i.e., the set*

$$E = \bigcup_{k=1}^{\infty} E_k$$

is measurable if all E_k are measurable. Particularly, the collection $\mathfrak{M}$ of all measurable sets of S is an σ-algebra in S.

> **Remark 12.3**
>
> Hence we deduce immediately from the first relation (12.2) and Theorem 12.8 that the intersection of a countable collection of measurable sets is also measurable.

Theorem 12.9. *Every open or closed set in S is measurable.*

Theorem 12.10. *Let $\mathscr{T}$ be a topology of $\mathbb{R}$.*

(a) *The intersection of all σ-algebras in $\mathbb{R}$ containing $\mathscr{T}$ is a smallest σ-algebra $\mathscr{B}$ in $\mathbb{R}$ such that $\mathscr{T} \subseteq \mathscr{B}$.*[c]

(b) *Members of $\mathscr{B}$ are called **Borel sets** of $\mathbb{R}$ and they are **Borel measurable**.*[d]

(c) *The σ-algebra $\mathscr{B}$ is generated by the collection of all open intervals or half-open intervals or closed intervals or open rays or closed rays of $\mathbb{R}$.*[e]

It is well-known that every Borel measurable set is a Lebesgue measurable set, but there exist Lebesgue measurable sets which are *not* Borel measurable, see [7, Exercise 9, p. 48] and [24, Remarks 2.21, p. 53].

12.1.4 More Properties of Lebesgue Measure

Theorem 12.11 (Countably Additivity)**.** *If $\{E_k\}$ is a countable collection of **disjoint** measurable sets, then $\bigcup_{k=1}^{\infty} E_k$ is measurable and*

$$m\Big(\bigcup_{k=1}^{\infty} E_k\Big) = \sum_{k=1}^{\infty} m(E_k).$$

Theorem 12.12 (Continuity of Lebesgue Measure)**.** *Suppose that $\{E_k\}$ is a sequence of measurable sets.*

(a) *If $E_k \subseteq E_{k+1}$ for all $k \in \mathbb{N}$, then we have*

$$m\Big(\bigcup_{k=1}^{\infty} E_k\Big) = \lim_{k \to \infty} m(E_k).$$

(b) *If $E_{k+1} \subseteq E_k$ for all $k \in \mathbb{N}$ and $m(E_1) < \infty$, then we have*

$$m\Big(\bigcap_{k=1}^{\infty} E_k\Big) = \lim_{k \to \infty} m(E_k).$$

---

[c] The Borel σ-algebra $\mathscr{B}$ is sometimes called the σ-algebra generated by $\mathscr{T}$.
[d] See also Problem 12.35.
[e] In fact, it is [7, Propositin 1.2, p. 22].

12.2 Lebesgue Outer Measure

> **Problem 12.1**
>
> (★) Prove Theorem 12.5(a).

Proof. Let $m^*(E) = 0$ and A be any set of $\mathbb{R}$. Since $A \cap E \subseteq E$ and $A \cap E^c \subseteq A$, we know from Theorem 12.3(c) (Properties of Outer Measure) that

$$m^*(A \cap E) \leq m^*(E) = 0 \quad \text{and} \quad m^*(A \cap E^c) \leq m^*(A).$$

Therefore, we get
$$m^*(A) \geq m^*(A \cap E) + m^*(A \cap E^c). \tag{12.3}$$

Since $A = (A \cap E) \cup (A \cap E^c)$, we always have

$$m^*(A) \leq m^*(A \cap E) + m^*(A \cap E^c) \tag{12.4}$$

by Theorem 12.3(d) (Properties of Outer Measure). Now the inequalities (12.3) and (12.3) give

$$m^*(A) = m^*(A \cap E) + m^*(A \cap E^c).$$

By Definition 12.4 (Lebesgue Measurability), E is measurable and this completes the proof of the problem. ∎

> **Problem 12.2**
>
> (★) Let E be open in $\mathbb{R}$. If F and G are subsets of $\mathbb{R}$ such that $F \subseteq E$ and $G \cap E = \emptyset$, prove that
> $$m^*(F \cup G) = m^*(F) + m^*(G).$$

Proof. By Theorem 12.9, E is measurable. Since $F \subseteq E$, we have $F \cap E^c = \emptyset$. Since $G \cap E = \emptyset$, we have $G \subseteq E^c$. Hence we see from Definition 12.4 (Lebesgue Measurability) that

$$m^*(F \cup G) = m^*((F \cup G) \cap E) + m^*((F \cup G) \cap E^c) = m^*(F) + m^*(G),$$

as desired. We have completed the proof of the problem. ∎

> **Problem 12.3**
>
> (★) Construct an unbounded subset of $\mathbb{R}$ with Lebesgue outer measure 0.

Proof. We claim that the set $\mathbb{Q}$ satisfies our requirements. In fact, $\mathbb{Q}$ is clearly unbounded. Furthermore, if $r \in \mathbb{Q}$, then $m^*(\{r\}) = 0$ by Definition 12.2 (Lebesgue Outer Measure). Since we have

$$\mathbb{Q} = \bigcup_{r \in \mathbb{Q}} \{r\},$$

Theorem 12.3(d) (Properties of Outer Measure) implies that

$$m^*(\mathbb{Q}) \le \sum_{r \in \mathbb{Q}} m^*(\{r\}) = 0,$$

completing the proof of the problem. ∎

Problem 12.4

(⋆) *Suppose that E is the set of irrationals in the interval $(0, 1)$. Prove that $m^*(E) = 1$.*

Proof. By Theorem 12.3(b) (Properties of Outer Measure), we see that

$$m^*((0,1)) = \ell((0,1)) = 1.$$

Let $F = (0,1) \cap \mathbb{Q}$. By Problem 12.3, we have

$$m^*(F) = 0. \tag{12.5}$$

Since $(0,1) = E \cup F$, it deduces from Theorem 12.3(c) (Properties of Outer Measure) and the result (12.5) that

$$1 = m^*((0,1)) \le m^*(E) + m^*(F) = m^*(E) \le 1$$

so that $m^*(E) = 1$. Hence we have the required result and it completes the proof of the problem. ∎

Problem 12.5

(⋆) *Suppose that $E \subseteq \mathbb{R}$ and $m^*(E) < \infty$. Prove that for every $\epsilon > 0$, there exists an open set $V \subseteq \mathbb{R}$ containing E such that*

$$m^*(V) - m^*(E) < \epsilon. \tag{12.6}$$

Proof. Given $\epsilon > 0$. Then it follows from Definition 12.2 (Lebesgue Outer Measure) that there is a countable collection of intervals $\{I_k\}$ covering E such that

$$\sum_{k=1}^{\infty} \ell(I_k) < m^*(E) + \epsilon. \tag{12.7}$$

Let $V = \bigcup_{k=1}^{\infty} I_k$. We know that V is open in $\mathbb{R}$ and $E \subseteq V$. Using Theorems 12.9, 12.3(b) and (d) (Properties of Outer Measure) and the inequality (12.7), we have

$$m(V) = m^*(V) = m^*\Big(\bigcup_{k=1}^{\infty} I_k\Big) \le \sum_{k=1}^{\infty} m^*(I_k) = \sum_{k=1}^{\infty} \ell(I_k) < m^*(E) + \epsilon$$

which implies the inequality (12.6). We complete the proof of the problem. ∎

12.2. Lebesgue Outer Measure

Problem 12.6

(★) A set $V \subseteq \mathbb{R}$ is called a G_δ set if it is the intersection of a countable collection of open sets of $\mathbb{R}$. Suppose that $E \subseteq \mathbb{R}$ and $m^*(E) < \infty$. Prove that there exists a G_δ set V containing E such that
$$m^*(V) = m^*(E).$$

Proof. By Problem 12.5, for each $k \in \mathbb{N}$, there exists an open set V_k in $\mathbb{R}$ containing E such that
$$m^*(V_k) < m^*(E) + \frac{1}{k}.$$

Let $V = \bigcap_{k=1}^{\infty} V_k$. Then it is clearly a G_δ set by the definition. Furthermore, since $E \subseteq V$, for each $k \in \mathbb{N}$, we must have
$$E \subseteq V \subseteq V_k,$$
so Theorems 12.3(c) (Properties of Outer Measure) and 12.9 imply that
$$m^*(E) \leq m^*(V) \leq m^*(V_k) < m^*(E) + \frac{1}{k}. \tag{12.8}$$

Taking $k \to \infty$ in the inequalities (12.8), we obtain immediately that
$$m^*(V) = m^*(E).$$

This completes the proof of the problem. ∎

Problem 12.7

(★)(★) Let $n \in \mathbb{N}$. Suppose that $\{E_1, E_2, \ldots, E_n\}$ is a finite sequence of disjoint measurable subsets of $\mathbb{R}$. For an arbitrary set $A \subseteq \mathbb{R}$, prove that
$$m^*\Big(A \cap \bigcup_{i=1}^{n} E_i\Big) = \sum_{i=1}^{n} m^*(A \cap E_i). \tag{12.9}$$

Proof. If $n = 1$, then the formula (12.9) is obviously true. Assume that the formula (12.9) is true for $n = k$ for some positive integer k, i.e.,
$$m^*\Big(A \cap \bigcup_{i=1}^{k} E_i\Big) = \sum_{i=1}^{k} m^*(A \cap E_i).$$

Suppose that $\{E_1, E_2, \ldots, E_k, E_{k+1}\}$ is a finite sequence of disjoint measurable sets. Now we follow from the assumption that
$$m^*\Big(A \cap \bigcup_{i=1}^{k} E_i\Big) + m^*(A \cap E_{k+1}) = \sum_{i=1}^{k} m^*(A \cap E_i) + m^*(A \cap E_{k+1}) = \sum_{i=1}^{k+1} m^*(A \cap E_i). \tag{12.10}$$

Since $\bigcup_{i=1}^{k} E_i$ and E_{k+1} are disjoint, we have

$$\Big(\bigcup_{i=1}^{k} E_i\Big) \cap E_{k+1} = \varnothing \quad \text{and} \quad \Big(\bigcup_{i=1}^{k} E_i\Big) \cap E_{k+1}^c = \bigcup_{i=1}^{k} E_i$$

which imply that

$$\Big(\bigcup_{i=1}^{k+1} E_i\Big) \cap E_{k+1} = E_{k+1} \quad \text{and} \quad \Big(\bigcup_{i=1}^{k+1} E_i\Big) \cap E_{k+1}^c = \bigcup_{i=1}^{k} E_i. \tag{12.11}$$

Substituting the identities (12.11) into the left-hand side of the equation (12.10) and then using the fact that E_{k+1} is measurable, we obtain from Definition 12.4 (Lebesgue Measurability) that

$$m^*\Big(A \cap \bigcup_{i=1}^{k} E_i\Big) + m^*(A \cap E_{k+1}) = m^*\Big(A \cap \Big(\Big(\bigcup_{i=1}^{k+1} E_i\Big) \cap E_{k+1}^c\Big)\Big)$$

$$+ m^*\Big(A \cap \Big(\Big(\bigcup_{i=1}^{k+1} E_i\Big) \cap E_{k+1}\Big)\Big)$$

$$= m^*\Big(\underbrace{\Big(A \cap \bigcup_{i=1}^{k+1} E_i\Big)}_{\text{The new } A.} \cap E_{k+1}^c\Big) + m^*\Big(\underbrace{\Big(A \cap \bigcup_{i=1}^{k+1} E_i\Big)}_{\text{The new } A.} \cap E_{k+1}\Big)$$

$$= m^*\Big(A \cap \bigcup_{i=1}^{k+1} E_i\Big). \tag{12.12}$$

Compare the equations (12.10) and (12.12) to get

$$m^*\Big(A \cap \bigcup_{i=1}^{k+1} E_i\Big) = \sum_{i=1}^{k+1} m^*(A \cap E_i).$$

By induction, our formula (12.9) is true for every $n \in \mathbb{N}$ and we end the analysis of the problem. ∎

Problem 12.8

(⋆) Suppose that $E \subseteq \mathbb{R}$ has positive outer measure. Prove that there exists a bounded subset $F \subseteq E$ such that $m^*(F) > 0$.

Proof. For each $n \in \mathbb{Z}$, we denote $I_n = [n, n+1]$. Then we have

$$E = \bigcup_{n \in \mathbb{Z}} (E \cap I_n).$$

By Theorem 12.3(d) (Properties of Outer Measure), we see that

$$0 < m^*(E) \le \sum_{n \in \mathbb{Z}} m^*(E \cap I_n)$$

12.3 Lebesgue Measurable Sets

Problem 12.9

(★) Prove that if E and F are measurable, then
$$m(E \cup F) + m(E \cap F) = m(E) + m(F).$$

Proof. Notice that
$$E \cup F = (E \setminus F) \cup (E \cap F) \cup (F \setminus E).$$
Obviously, $E \setminus F$, $E \cap F$ and $F \setminus E$ are pairwise disjoint. By Theorem 12.11 (Countably Additive), it is true that
$$m(E \cup F) + m(E \cap F) = [m(E \setminus F) + m(E \cap F)] + [m(F \setminus E) + m(E \cap F)]. \tag{12.13}$$
Since $E = (E \setminus F) \cup (E \cap F)$ and $F = (F \setminus E) \cup (E \cap F)$, Theorem 12.11 (Countably Additivity) again shows that the equation (12.13) reduces to
$$m(E \cup F) + m(E \cap F) = m(E) + m(F).$$
This completes the proof of the problem. ∎

Problem 12.10

(★) Let $E \subseteq [0,1]$ and $m(E) > 0$. Prove that there exist $x, y \in E$ such that $|x - y|$ is irrational.

Proof. Assume that $|x - y| \in \mathbb{Q}$ for all $x, y \in E$. Then it means that $E \subseteq x + \mathbb{Q}$ for any $x \in E$. By Problem 12.3 and Theorem 12.3(e) (Properties of Outer Measure), we have
$$m^*(x + \mathbb{Q}) = m^*(\mathbb{Q}) = 0,$$
where $x \in E$. Thus Theorem 12.5(a) (Properties of Measurable Sets) implies that E is measurable and then
$$0 < m(E) = m^*(E) \leq m^*(x + \mathbb{Q}) = m(x + \mathbb{Q}) = 0,$$
a contradiction. Hence there exist $x, y \in E$ such that $|x - y|$ is irrational. We end the proof of the problem. ∎

Problem 12.11

(★) Let $E \subseteq [0,1]$ and $m(E) > 0$. Prove that there exist $x, y \in E$ with $x \neq y$ such that $x - y$ is rational.

Proof. Denote the rationals in $[-1, 1]$ by $r_1, r_2, \ldots$. Define $E_n = r_n + E$ for every $n \in \mathbb{N}$. By Theorem 12.5(d) (Properties of Measurable Sets), we see that

$$m(E_n) = m(r_n + E) = m(E) > 0.$$

Thus we must have

$$\sum_{n=1}^{\infty} m(E_n) = \infty. \qquad (12.14)$$

Clearly, we have $E_n \subseteq [-1, 2]$ for every $n \in \mathbb{N}$ so that

$$\bigcup_{n=1}^{\infty} E_n \subseteq [-1, 2]. \qquad (12.15)$$

Assume that $E_i \cap E_j = \emptyset$ if $i \neq j$. Then Theorem 12.11 (Countably Additivity) and the set relation (12.15) imply that

$$\sum_{n=1}^{\infty} m(E_n) = m\Big(\bigcup_{n=1}^{\infty} E_n \Big) \leq m([-1, 2]) = 3$$

which contradicts the result (12.14). In other words, there exist $n, m \in \mathbb{N}$ such that

$$E_n \cap E_m \neq \emptyset.$$

Let $z \in E_n \cap E_m$. By the definition, there are $x, y \in E$ such that $z = x + r_n = y + r_m$ which gives

$$x - y = r_m - r_n \in \mathbb{Q}.$$

We have completed the proof of the problem. ∎

Problem 12.12

(⋆) *Suppose that $\{r_n\}$ is an enumeration of rationals $\mathbb{Q}$. Prove that*

$$\mathbb{R} \setminus \bigcup_{n=1}^{\infty} \Big(r_n - \frac{1}{n^2}, r_n + \frac{1}{n^2} \Big) \neq \emptyset. \qquad (12.16)$$

Proof. Since each $(r_n - \frac{1}{n^2}, r_n + \frac{1}{n^2})$ is measurable, Theorem 12.8 and then Theorem 12.5(c) (Properties of Measurable Sets) imply that the set

$$\mathbb{R} \setminus \bigcup_{n=1}^{\infty} \Big(r_n - \frac{1}{n^2}, r_n + \frac{1}{n^2} \Big)$$

is measurable. By the inequality (12.1), we obtain

$$m\Big(\bigcup_{n=1}^{\infty} \Big(r_n - \frac{1}{n^2}, r_n + \frac{1}{n^2} \Big) \Big) \leq \sum_{n=1}^{\infty} m\Big(\Big(r_n - \frac{1}{n^2}, r_n + \frac{1}{n^2} \Big) \Big) = \sum_{n=1}^{\infty} \frac{2}{n^2} < \infty$$

so that the set relation (12.16) holds. This completes the proof of the problem. ∎

12.3. Lebesgue Measurable Sets

Problem 12.13

(⋆)(⋆) Suppose that $E \subseteq \mathbb{R}$ and $F \subseteq \mathbb{R}$ are measurable and $E \subseteq F$. Show that $F \setminus E$ is also measurable and
$$m(F \setminus E) = m(F) - m(E). \tag{12.17}$$

Proof. We note that
$$F = F \cap \mathbb{R} = F \cap (E \cup E^c) = (F \cap E) \cup (F \cap E^c) = E \cup (F \cap E^c),$$
so we have
$$F \setminus E = F \cap E^c, \tag{12.18}$$
Next, it is clear that
$$E \cap (F \cap E^c) = (E \cap F) \cap (E \cap E^c) = (E \cap F) \cap \varnothing = \varnothing,$$
i.e., E and $F \cap E^c$ are disjoint.

We observe that Definition 12.4 (Lebesgue Measurability) is symmetric in E and E^c, so E^c is measurable and thus the second relation (12.2) guarantees that $F \setminus E$ is also measurable. Since E and $F \cap E^c$ are disjoint, we apply Theorem 12.11 (Countably Additivity) and the set relation (12.18) to obtain
$$m(F) = m(E \cup (F \cap E^c)) = m(E) + m(F \cap E^c) = m(E) + m(F \setminus E)$$
which certainly gives the desired formula (12.17). We complete the proof of the problem. ∎

Problem 12.14

(⋆) Let N be a positive integer. Suppose that $E_1, E_2, \ldots, E_N \subseteq [0,1]$ are measurable sets and $\sum_{n=1}^{N} m(E_n) > N - 1$. Prove that
$$m\Big(\bigcap_{n=1}^{N} E_n\Big) > 0.$$

Proof. We have the identity
$$\bigcap_{n=1}^{N} E_n = [0,1] \setminus \bigcup_{n=1}^{N} E'_n,$$
where $E'_n = [0,1] \setminus E_n$. Since every E'_n ($n = 1, \ldots, N$) is measurable, the definition and the inequality (12.1) give
$$m\Big(\bigcup_{n=1}^{N} E'_n\Big) = m\Big(\bigcup_{n=1}^{N} ([0,1] \setminus E_n)\Big) \leq \sum_{n=1}^{N} m([0,1] \setminus E_n). \tag{12.19}$$

Furthermore, we apply the identity (12.17) to the right-hand side of the inequality (12.19) to get

$$m\Big(\bigcup_{n=1}^{N} E_n'\Big) \leq \sum_{n=1}^{N} m([0,1]) - \sum_{n=1}^{N} m(E_n) < N - (N-1) = 1. \tag{12.20}$$

Therefore, using the identity (12.17) and the result (12.20), we establish

$$m\Big(\bigcap_{n=1}^{N} E_n\Big) = m\Big([0,1] \setminus \bigcup_{n=1}^{N} E_n'\Big) = m([0,1]) - m\Big(\bigcup_{n=1}^{N} E_n'\Big) > 1 - 1 = 0.$$

We complete the proof of the problem. ∎

> **Problem 12.15** (The Borel-Cantelli Lemma)
>
> (⋆) Suppose that $\{E_k\}$ is a countable collection of measurable sets and
>
> $$\sum_{k=1}^{\infty} m(E_k) < \infty.$$
>
> Prove that almost all $x \in \mathbb{R}$ lie in at most finitely many of the sets E_k.

Proof. For each $n \in \mathbb{N}$, let $A_n = \bigcup_{k=n}^{\infty} E_k$. Then we have $A_1 \supseteq A_2 \supseteq \cdots$ and Theorem 12.3(d) (Properties of Outer Measure) implies that

$$m(A_1) = m\Big(\bigcup_{k=1}^{\infty} E_k\Big) \leq \sum_{k=1}^{\infty} m(E_k) < \infty. \tag{12.21}$$

Apply Theorem 12.12 (Continuity of Lebesgue Measure), we acquire that

$$m\Big(\bigcap_{n=1}^{\infty} \Big(\bigcup_{k=n}^{\infty} E_k\Big)\Big) = m\Big(\bigcap_{n=1}^{\infty} A_n\Big) = \lim_{n \to \infty} m(A_n) \leq \lim_{n \to \infty} \sum_{k=n}^{\infty} m(E_k) = 0.$$

Hence almost all $x \in \mathbb{R}$ fail to lie in the set

$$\bigcap_{n=1}^{\infty} \Big(\bigcup_{k=n}^{\infty} E_k\Big).$$

In other words, it means that for almost all $x \in \mathbb{R}$ lie in at most finitely many E_k, completing the proof of the problem. ∎

> **Problem 12.16**
>
> (⋆) Let $\{E_k\}$ be a sequence of measurable sets such that
>
> $$m\Big(\bigcup_{k=1}^{\infty} E_k\Big) < \infty \quad \text{and} \quad \inf_{k \in \mathbb{N}}\{m(E_k)\} = \alpha > 0.$$
>
> Suppose that E is the set of points that lie in an infinity of sets E_k. Prove that E is measurable and $m(E) \geq \alpha$.

12.3. Lebesgue Measurable Sets

Proof. With the same notations as in the proof of Problem 12.15, we see that

$$E = \bigcap_{n=1}^{\infty} A_n$$

and so

$$m(E) = m\Big(\bigcap_{n=1}^{\infty} A_n\Big) = \lim_{n \to \infty} m(A_n). \tag{12.22}$$

Obviously, we have

$$m(A_n) = m\Big(\bigcup_{k=n}^{\infty} E_k\Big) \geq m(E_n) \geq \alpha > 0 \tag{12.23}$$

for every $n \in \mathbb{N}$. Now if we combine the results (12.22) and (12.23), then we get immediately that

$$m(E) \geq \alpha.$$

We complete the proof of the problem. ∎

Problem 12.17

(⋆) Prove Theorem 12.10(a) and (b).

Proof. Let Ω be the family of all σ-algebras $\mathfrak{M}$ in $\mathbb{R}$ containing $\mathscr{T}$. It is clear that the collection of all subsets of $\mathbb{R}$ satisfies Definition 12.7 (σ-algebra), so $\Omega \neq \varnothing$. Let

$$\mathscr{B} = \bigcap_{\mathfrak{M} \in \Omega} \mathfrak{M}. \tag{12.24}$$

Since $\mathscr{T} \subseteq \mathfrak{M}$ for all $\mathfrak{M} \in \Omega$, $\mathscr{T} \subseteq \mathscr{B}$. If $E_n \in \mathscr{B}$ for every $n = 1, 2, \ldots$, then the definition (12.24) implies that $E_n \in \mathfrak{M}$ for every $\mathfrak{M} \in \Omega$. Since every $\mathfrak{M}$ is an σ-algebra, we have $\bigcup_{n=1}^{\infty} E_n \in \mathfrak{M}$ for all $\mathfrak{M} \in \Omega$. By the definition (12.24) again, we achieve

$$\bigcup_{n=1}^{\infty} E_n \in \mathscr{B}$$

so that $\mathscr{B}$ satisfies Definition 12.7(c) (σ-algebra). Now we can show the other two parts in a similar way. Hence we conclude that $\mathscr{B}$ is an σ-algebra and it follows from the definition (12.24) that $\mathscr{B}$ is actually a *smallest* σ-algebra containing $\mathscr{T}$.

For the second assertion, by Theorem 12.8, we know that the collection $\mathfrak{M}'$ of all Lebesgue measurable sets of $\mathbb{R}$ is an σ-algebra in $\mathbb{R}$ and by Theorem 12.9, $\mathfrak{M}'$ contains $\mathscr{T}$. In other words, it means that

$$\mathfrak{M}' \in \Omega$$

and thus $\mathscr{B} \subseteq \mathfrak{M}'$ by the definition (12.24). Consequently, a Borel set must be Lebesgue measurable. Hence we have completed the proof of the problem. ∎

Problem 12.18

(★) Given $\epsilon > 0$. Construct a dense open set V in $\mathbb{R}$ such that $m(V) < \epsilon$.

Proof. Let $\{r_n\}$ be a sequence of rational numbers of $\mathbb{R}$. For every n, we consider the open interval I_n given by
$$I_n = (r_n - \epsilon \cdot 2^{-(n+1)}, r_n + \epsilon \cdot 2^{-(n+1)}).$$
Define $V = \bigcup_{n=1}^{\infty} I_n$ which is evidently open in $\mathbb{R}$. Since $\mathbb{Q} \subseteq V$ and $\mathbb{Q}$ is dense in $\mathbb{R}$, V is also dense in $\mathbb{R}$. Finally, we note from Theorem 12.3(d) (Properties of Outer Measure) that
$$m(V) = m\Big(\bigcup_{n=1}^{\infty} I_n\Big) \le \sum_{n=1}^{\infty} m(I_n) = \epsilon \cdot \sum_{n=1}^{\infty} \frac{1}{2^n} = \epsilon,$$
completing the proof of the problem. ∎

Problem 12.19

(★) Let C be the Cantor set. Prove that $m(C) = 0$.

Proof. In Problem 4.28, we know that
$$C = \bigcap_{n=1}^{\infty} E_n, \tag{12.25}$$
where
$$E_n = \bigcup_{k=0}^{2^{n-1}-1} \Big(\Big[\frac{3k+0}{3^n}, \frac{3k+1}{3^n}\Big] \cup \Big[\frac{3k+2}{3^n}, \frac{3k+3}{3^n}\Big]\Big).$$
Since each E_n is the union of closed intervals, E_n is measurable by Theorem 12.8. Thus it follows from Theorem 12.3(d) (Properties of Outer Measure) that
$$m(E_n) \le \sum_{k=1}^{2^{n-1}-1} \Big\{ m\Big(\Big[\frac{3k+0}{3^n}, \frac{3k+1}{3^n}\Big]\Big) + m\Big(\Big[\frac{3k+2}{3^n}, \frac{3k+3}{3^n}\Big]\Big)\Big\}$$
$$= \sum_{k=1}^{2^{n-1}-1} \frac{2}{3^n}$$
$$= \frac{2(2^{n-1}-1)}{3^n}$$
$$< \Big(\frac{2}{3}\Big)^n.$$
Hence we know from the representation (12.25) that
$$m(C) < \Big(\frac{2}{3}\Big)^n \tag{12.26}$$
for every $n \in \mathbb{N}$. Taking $n \to \infty$ in the inequality (12.26), we conclude that $m(C) = 0$ which completes the analysis of the problem. ∎

12.3. Lebesgue Measurable Sets

Problem 12.20

(⋆) Construct an open subset V of $\mathbb{R}$ such that $m(V) \neq m(\overline{V})$.

Proof. Let V be the dense open set constructed in Problem 12.18. Then $m(V) < \epsilon$, but $\overline{V} = \mathbb{R}$ which shows that $m(\overline{V}) = \infty$, completing the proof of the problem. ∎

Problem 12.21

(⋆)(⋆) Suppose that E and F are subsets of $\mathbb{R}$ with finite outer measure. Prove that
$$m^*(E \cup F) = m^*(E) + m^*(F) \tag{12.27}$$
if and only if there are measurable sets $E \subseteq E'$ and $F \subseteq F'$ such that
$$m(E' \cap F') = 0. \tag{12.28}$$

Proof. Suppose that there are measurable sets E' and F' with $E \subseteq E'$ and $F \subseteq F'$ such that the equation (12.28) holds. Recall that we always have
$$m^*(E \cup F) \leq m^*(E) + m^*(F).$$

Given $\epsilon > 0$. By Problem 12.5 and Theorem 12.9, there exists an open set V containing $E \cup F$ such that
$$m(V) = m^*(V) < m^*(E \cup F) + \epsilon. \tag{12.29}$$
Since $E \subseteq V \cap E'$, $F \subseteq V \cap F'$ and both $V \cap E'$ and $V \cap F'$ are measurable, we have
$$m^*(E) + m^*(F) \leq m^*(V \cap E') + m^*(V \cap F') = m(V \cap E') + m(V \cap F'). \tag{12.30}$$
By Problem 12.9 and the inequality (12.29), we further reduce the inequality (12.30) to
$$\begin{aligned} m^*(E) + m^*(F) &\leq m((V \cap E') \cup (V \cap F')) + m((V \cap E') \cap (V \cap F')) \\ &\leq m(V \cap (E' \cup F')) + m(E' \cap F') \\ &\leq m(V) + 0 \\ &< m^*(E \cup F) + \epsilon. \end{aligned}$$
Since ϵ is arbitrary, we obtain $m^*(E) + m^*(F) \leq m^*(E \cup F)$ which implies the equation (12.27).

Conversely, we suppose that the equation (12.27) holds. By Problem 12.6, there are G_δ sets E' and F' containing E and F respectively such that
$$m(E') = m^*(E') = m^*(E) \quad \text{and} \quad m(F') = m^*(F') = m^*(F).$$
Assume that $m(E' \cap F') > 0$. Then Problem 12.9 implies that
$$\begin{aligned} m^*(E \cup F) &= m^*(E) + m^*(F) \\ &= m(E') + m(F') \\ &= m(E' \cup F') + m(E' \cap F') \end{aligned}$$

$$> m(E' \cup F')$$
$$\geq m^*(E' \cup F')$$

which is a contradiction. Consequently, we obtain $m(E' \cap F') = 0$. This completes the proof of the problem. ∎

Problem 12.22

(⋆) Prove that $\mathbb{Q}$ and $\mathbb{R} \setminus \mathbb{Q}$ are Borel sets in $\mathbb{R}$.

Proof. Let $q \in \mathbb{Q}$. Since $\mathscr{T} \subseteq \mathscr{B}$ and $\mathscr{B}$ is an σ-algebra, Definition 12.7 (σ-algebra) shows that $\mathscr{B}$ also contains all closed subsets of $\mathbb{R}$. Since $\{q\}$ is closed in $\mathbb{R}$, we have $\{q\} \in \mathscr{B}$ and the countability of $\mathbb{Q}$ implies that

$$\mathbb{Q} \in \mathscr{B}.$$

By Definition 12.7 (σ-algebra) again, we have

$$\mathbb{R} \setminus \mathbb{Q} \in \mathscr{B}$$

and hence we complete the proof of the problem. ∎

Problem 12.23

(⋆)(⋆) Define
$$\mathfrak{M} = \{E \subseteq \mathbb{R} \mid E \text{ or } E^c \text{ is at most countable}\}.$$
Prove that $\mathfrak{M}$ is an σ-algebra in $\mathbb{R}$.

Proof. We check Definition 12.7 (σ-algebra). Since $\mathbb{R}^c = \varnothing$, we have $\mathbb{R} \in \mathfrak{M}$. Let $E \in \mathfrak{M}$. If E^c is at most countable, then $E^c \in \mathfrak{M}$. Otherwise, E is at most countable. In this case, since $(E^c)^c = E$ is at most countable, we still have $E^c \in \mathfrak{M}$. Suppose that $E_k \in \mathfrak{M}$ for all $k = 1, 2, \ldots$. If *all* E_k are at most countable, then the set

$$E = \bigcup_{k=1}^{\infty} E_k$$

must be at most countable which means that $E \in \mathfrak{M}$. Next, suppose that *there exists* an uncountable E_k. Without loss of generality, we may assume that it is E_1. Since $E_1 \in \mathfrak{M}$, E_1^c is at most countable. Since

$$E^c = \bigcap_{k=1}^{\infty} E_k^c \subseteq E_1^c,$$

E^c must be at most countable. In other words, $E \in \mathfrak{M}$ and we follow from Definition 12.7 (σ-algebra) that $\mathfrak{M}$ is an σ-algebra. This ends the proof of the problem. ∎

12.3. Lebesgue Measurable Sets

Problem 12.24

(★)(★) Suppose that $E \subset \mathbb{R}$ is compact and $K_n = \{x \in \mathbb{R} \mid d(x, E) < \frac{1}{n}\}$ for all $n \in \mathbb{N}$, where $d(x, E) = \inf\{d(x, y) \mid y \in E\}$. Prove that

$$m(E) = \lim_{n \to \infty} m(K_n).$$

Proof. It is clear that

$$E \subseteq \bigcap_{n=1}^{\infty} K_n.$$

Let $x \in \bigcap_{n=1}^{\infty} K_n \setminus E \subseteq E^c$. Since E is compact, it is closed in $\mathbb{R}$ and then E^c is open in $\mathbb{R}$. By the definition, there is a $\epsilon > 0$ such that $N_\epsilon(x) \subseteq E^c$ and this means that

$$d(x, E) \geq \epsilon. \tag{12.31}$$

Pick an $N \in \mathbb{N}$ such that $N > \frac{1}{\epsilon}$. Recall that $x \in \bigcap_{n=1}^{\infty} K_n$, so $x \in K_N$. However, the definition of K_N shows that

$$d(x, E) < \frac{1}{N} < \epsilon$$

which contradicts the inequality (12.31). Therefore, no such x exist and it is equivalent to

$$E = \bigcap_{n=1}^{\infty} K_n.$$

Next, since E is bounded, every K_n is also bounded. In particular, we have $m(K_1) < \infty$. It is obvious that

$$K_{n+1} \subseteq K_n$$

for every $n \in \mathbb{N}$. Hence we deduce from Theorem 12.12 (Continuity of Lebesgue Measure), we conclude that

$$m(E) = m\Big(\bigcap_{n=1}^{\infty} K_n\Big) = \lim_{n \to \infty} m(K_n),$$

completing the proof of the problem. ∎

Problem 12.25

(★)(★) Suppose that $E_1, E_2, \ldots$ are measurable subsets of $\mathbb{R}$ and $m(E_i \cap E_j) = 0$ for all $i \neq j$. Let $E = \bigcup_{n=1}^{\infty} E_n$. Prove that

$$m(E) = \sum_{n=1}^{\infty} m(E_n).$$

Proof. Define $F_1 = E_1, F_2 = E_2 \setminus E_1, F_3 = E_3 \setminus (E_1 \cup E_2), \ldots$ and in general

$$F_n = E_n \setminus \bigcup_{k=1}^{n-1} E_k \tag{12.32}$$

for every $n = 2, 3, \ldots$. By Theorem 12.8, the collection $\mathfrak{M}$ of all measurable sets of $\mathbb{R}$ is an σ-algebra in $\mathbb{R}$. Recall from the proof of Problem 12.13 that

$$A \setminus B = A \cap B^c,$$

so the representation (12.32) can be rewritten as

$$F_n = E_n \cap \Big(\bigcup_{k=1}^{n-1} E_k\Big)^c = E_n \cap \bigcap_{k=1}^{n-1} E_k^c$$

for each $n = 2, 3, \ldots$. Since $E_n, E_1^c, E_2^c, \ldots, E_{n-1}^c \in \mathfrak{M}$, we see that $F_n \in \mathfrak{M}$ for each $n = 2, 3, \ldots$.

Clearly, we have $F_i \cap F_j = \varnothing$ for $i \neq j$ and

$$F_1 \cup F_2 \cup \cdots \cup F_n = E_1 \cup E_2 \cup \cdots \cup E_n$$

for all $n \in \mathbb{N}$. As a result, we have

$$E = \bigcup_{n=1}^{\infty} E_n = \bigcup_{n=1}^{\infty} F_n.$$

Since each $E_n \in \mathfrak{M}$, the set

$$E_n \cap \Big(\bigcup_{k=1}^{n-1} E_k\Big)$$

is also an element of $\mathfrak{M}$. Thus it follows from Theorem 12.11 (Countably Additivity) that

$$m(E) = m\Big(\bigcup_{n=1}^{\infty} F_n\Big) = \sum_{n=1}^{\infty} m(F_n) \tag{12.33}$$

and

$$m(E_n) = m\Big(E_n \cap \Big(\bigcup_{k=1}^{n-1} E_k\Big)\Big) + m\Big(E_n \cap \Big(\bigcup_{k=1}^{n-1} E_k\Big)^c\Big)$$

$$= m\Big(E_n \cap \Big(\bigcup_{k=1}^{n-1} E_k\Big)\Big) + m(F_n). \tag{12.34}$$

Furthermore, since

$$E_n \cap \Big(\bigcup_{k=1}^{n-1} E_k\Big) = \bigcup_{k=1}^{n-1} (E_n \cap E_k),$$

Theorem 12.3(d) (Properties of Outer Measure) and our hypothesis give

$$m\Big(E_n \cap \Big(\bigcup_{k=1}^{n-1} E_k\Big)\Big) \leq \sum_{k=1}^{n-1} m(E_n \cap E_k) = 0. \tag{12.35}$$

12.3. Lebesgue Measurable Sets

Combining the equation (12.34) and the result (12.35), we conclude immediately that
$$m(F_n) = m(E_n)$$
for every $n = 2, 3, \ldots$. It is trivial that $m(F_1) = m(E_1)$. Hence our desired result derives from these and the formula (12.33). This completes the proof of the problem. ∎

Problem 12.26

(⋆)(⋆)(⋆) Suppose that E is Lebesgue measurable and $m(E) < \infty$. Prove that there exists a $p \in \mathbb{R}$ such that
$$m(E \cap (-\infty, p)) = m(E \cap (p, \infty)).$$

Proof. Define $f : \mathbb{R} \to [0, \infty)$ by
$$f(x) = m(E \cap (-\infty, x)).$$

Now f is an increasing function. Let $x < y$. Since E and (x, y) are measurable, their intersection $E \cap (x, y)$ is also measurable. Since $(x, y) = (-\infty, y) \setminus (-\infty, x]$, we have
$$E \cap (x, y) = E \cap [(-\infty, y) \setminus (-\infty, x]] = [E \cap (-\infty, y)] \setminus [E \cap (-\infty, x]].$$

Obviously, $E \cap (-\infty, x] \subseteq E \cap (-\infty, y)$, so Problem 12.13 implies that
$$\begin{aligned} m(E \cap (x, y)) &= m(E \cap (-\infty, y)) - m(E \cap (-\infty, x]) \\ &= m(E \cap (-\infty, y)) - m(E \cap (-\infty, x)) \\ &= f(y) - f(x). \end{aligned}$$

By Theorem 12.3(c) (Properties of Outer Measure), we see that
$$0 \le f(y) - f(x) \le m((x,y)) = y - x$$
which implies that
$$|f(x) - f(y)| \le |x - y|$$
for every $x, y \in \mathbb{R}$. In other words, f is a continuous function in $\mathbb{R}$. Since
$$\lim_{x \to -\infty} f(x) = 0 \quad \text{and} \quad \lim_{x \to \infty} f(x) = m(E),$$
the Intermediate Value Theorem ensures that there exists a $p \in \mathbb{R}$ such that
$$\frac{1}{2} m(E) = f(p) = m(E \cap (-\infty, p)). \tag{12.36}$$

We note that $E = [E \cap (-\infty, p)] \cup [E \cap [p, \infty)]$ and $[E \cap (-\infty, p)] \cap [E \cap [p, \infty)] = \varnothing$, so we have
$$m(E) = m(E \cap (-\infty, p)) + m(E \cap [p, \infty))$$
and then
$$m(E \cap (p, \infty)) = m(E \cap [p, \infty)) = \frac{1}{2} m(E). \tag{12.37}$$

Therefore, the two expressions (12.36) and (12.37) give the desired result which completes the analysis of the problem. ∎

> **Problem 12.27**
>
> (★)(★) Suppose that E is a bounded measurable set with $m(E) > 0$. Let $0 < \alpha < m(E)$. Prove that there exists a measurable set $F \subset E$ such that $m(F) = \alpha$.

Proof. Since E is bounded, there exist real numbers a and b with $a < b$ such that $E \subseteq [a,b]$. Define $f : [a,b] \to [0,\infty)$ by
$$f(x) = m([a,x] \cap E).$$

Using similar idea as in the proof of Problem 12.26, it can be shown easily that f is an increasing continuous function. Furthermore, we know that
$$f(a) = 0 \quad \text{and} \quad f(b) = m(E),$$
so the Intermediate Value Theorem implies that there is a $p \in [a,b]$ such that
$$f(p) = m([a,p] \cap E) = \alpha.$$

Since $[a,p]$ is measurable, the set $F = [a,p] \cap E \subset E$ is also measurable. This completes the proof of the problem. ∎

> **Problem 12.28**
>
> (★)(★) Prove that for every $\epsilon > 0$, there exists an open set $W \subseteq [0,1]$ such that
> $$\mathbb{Q} \cap [0,1] \subset W \quad \text{and} \quad m(W) < \epsilon.$$
> Then construct a closed subset K of $[0,1]$ such that
> $$m(K) > 0 \quad \text{and} \quad K^\circ = \varnothing.$$

Proof. By Problem 12.18, there exists an open set V containing $\mathbb{Q}$ in $\mathbb{R}$ such that $m(V) < \epsilon$. If we consider $W = V \cap [0,1]$, then it is open in $[0,1]$ and it contains $\mathbb{Q} \cap [0,1]$. It is clear that
$$m(W) \leq m(V) < \epsilon. \tag{12.38}$$
This prove the first assertion.

For the second assertion, we suppose that $K = [0,1] \setminus W$. Then K is certainly closed in $[0,1]$. By the estimate (12.38) and using Problem 12.13, we see that
$$m(K) = m([0,1] \setminus W) = m([0,1]) - m(W) \geq 1 - \epsilon > 0.$$

Assume that $K^\circ \neq \varnothing$. Then we pick $p \in K^\circ$. Since K° is open in $[0,1]$ by Problem 4.11, there exists a $\delta > 0$ such that
$$N_\delta(p) \subseteq K^\circ \subseteq K \subseteq [0,1].$$
However, the neighborhood $N_\delta(p)$ must contain a rational r of $[0,1]$ which contradicts the fact that K contains no rationals in $[0,1]$. Hence we complete the proof of the problem. ∎

12.3. Lebesgue Measurable Sets

> **Remark 12.4**
>
> By the second assertion of Problem 12.28, we see that the concept of Lebesgue measure zero is different from empty interior.

> **Problem 12.29**
>
> (⋆)(⋆) Suppose that $\{a_n\}$ is a sequence of real numbers and $\{\alpha_n\}$ is a sequence of positive numbers such that $\sum_{n=1}^{\infty} \sqrt{\alpha_n} < \infty$. Prove that there corresponds a measurable set E such that
>
> $$m(E^c) = 0 \quad \text{and} \quad \sum_{n=1}^{\infty} \frac{\alpha_n}{|x - a_n|} < \infty \qquad (12.39)$$
>
> for all $x \in E$.

Proof. Consider the following set

$$E = \{x \in \mathbb{R} \,|\, \text{there is an } N \in \mathbb{N} \text{ such that } n \geq N \text{ implies } |x - a_n| \geq \sqrt{\alpha_n}\}. \qquad (12.40)$$

If $x \in E$, then we have

$$\frac{\alpha_n}{|x - a_n|} \leq \sqrt{\alpha_n}$$

for all $n \geq N$. By the hypothesis, we know that

$$\sum_{n=1}^{\infty} \frac{\alpha_n}{|x - a_n|} \leq \sum_{n=1}^{\infty} \sqrt{\alpha_n} < \infty$$

so that the convergence of the infinite series (12.39) is satisfied. Thus it remains to prove that

$$m(E^c) = 0.$$

To see this, we consider the intervals

$$I_n = (a_n - \sqrt{\alpha_n}, a_n + \sqrt{\alpha_n}),$$

where $n = 1, 2, \ldots$. On the one hand, we notice that

$$\sum_{n=1}^{\infty} m(I_n) = \sum_{n=1}^{\infty} 2\sqrt{\alpha_n} < \infty,$$

so $\{I_n\}$ satisfies the hypothesis of Problem 12.15 (The Borel-Cantelli Lemma). Thus almost all $x \in \mathbb{R}$ lie in *at most* finitely many of the sets I_n. On the other hand, $x \in E^c$ if and only if

$$|x - a_n| < \sqrt{\alpha_n}$$

for infinitely many n by the definition (12.40) if and only if $x \in I_n$ for infinitely many n. Consequently, we conclude that $m(E^c) = 0$ which completes the proof of the problem. ∎

12.4 Necessary and Sufficient Conditions for Measurable Sets

> **Problem 12.30**
>
> (★) Suppose that E and F differ by a set of measure 0. Prove that E is measurable if and only if F is measurable.

Proof. By the hypothesis, we have
$$m((E \setminus F) \cup (F \setminus E)) = 0$$
so that $m(E \setminus F) = 0$. Thus it yields from this and Theorem 12.3(d) (Properties of Outer Measure) that
$$\begin{aligned} m^*(E \cup F) &= m^*((E \setminus F) \cup F) \\ &\leq m^*(E \setminus F) + m^*(F) \\ &= m(E \setminus F) + m^*(F) \\ &= m^*(F) \\ &\leq m^*(E \cup F). \end{aligned}$$
In other words, it means that $m^*(E \cup F) = m^*(F)$. Similarly, we have $m^*(E \cup F) = m^*(E)$. These two relations imply that
$$m^*(E) = m^*(F).$$
Hence it follows from Definition 12.4 (Lebesgue Measurability) that E is measurable if and only if F is measurable. This ends the proof of the problem. ∎

> **Problem 12.31**
>
> (★)(★) The set $E \subseteq \mathbb{R}$ is measurable if and only if for every $\epsilon > 0$, there exists an open set $V \subseteq \mathbb{R}$ containing E such that
> $$m^*(V \setminus E) < \epsilon. \qquad (12.41)$$

Proof. Suppose that E is measurable. Given $\epsilon > 0$. We consider the case that $m(E) < \infty$. Since $m^*(E) = m(E) < \infty$, it follows from Problem 12.5 that there exists an open set $V \subseteq \mathbb{R}$ such that
$$m(V) - m(E) < \epsilon. \qquad (12.42)$$
By Problem 12.13, we have
$$m(V \setminus E) = m(V) - m(E). \qquad (12.43)$$
Combining the inequality (12.42) and the result (12.43), we establish the expected result (12.41).

Next, we consider the case that $m^*(E) = \infty$. Now, for each $k \in \mathbb{N}$, we define
$$E_k = E \cap [k, k+1).$$

12.4. Necessary and Sufficient Conditions for Measurable Sets

By Theorem 12.3(c) (Properties of Outer Measure), we observe that

$$m(E_k) \leq m^*([k, k+1)) = 1$$

so that the result in the previous paragraph can be applied to each E_k. In other words, for each positive integer k, there is an open set V_k containing E_k such that

$$m(V_k \setminus E_k) < \frac{\epsilon}{2^k}.$$

It is clear that the set $V = \bigcup_{k=1}^{\infty} V_k$ is open in $\mathbb{R}$ containing E. Furthermore, we have

$$V \setminus E \subseteq \bigcup_{k=1}^{\infty} (V_k \setminus E_k)$$

and this implies that

$$m(V \setminus E) \leq \sum_{k=1}^{\infty} m(V_k \setminus E_k) < \sum_{k=1}^{\infty} \frac{\epsilon}{2^k} = \epsilon$$

which is the case (12.41) again.

Conversely, we suppose that the hypothesis (12.41) holds. Thus for each $k \in \mathbb{N}$, we can choose an open set V_k containing E such that

$$m^*(V_k \setminus E) < \frac{\epsilon}{2^k}. \tag{12.44}$$

Now if we define $V = \bigcup_{k=1}^{\infty} V_k$, then it is an open set in $\mathbb{R}$ and we have

$$V \setminus E = \bigcup_{k=1}^{\infty} (V_k \setminus E)$$

so that we may apply Theorem 12.3(d) (Properties of Outer Measure) to the inequality (12.44) to get

$$m^*(V \setminus E) \leq \sum_{k=1}^{\infty} m^*(V_k \setminus E) < \epsilon.$$

Since ϵ is arbitrary, we have $m^*(V \setminus E) = 0$ and Theorem 12.5(a) (Properties of Measurable Sets) ensures that $V \setminus E$ is measurable. Finally, we use the fact $E = V \setminus (V \setminus E)$ and the set relation (12.2) to conclude that E is measurable and hence we complete the proof of the problem. ∎

> **Remark 12.5**
>
> Similar to the proof of Problem 12.31, we can show that E is measurable if and only if for every $\epsilon > 0$, there exists a closed set K in $\mathbb{R}$ contained in E such that
>
> $$m^*(E \setminus K) < \epsilon. \tag{12.45}$$

> **Problem 12.32**
>
> (★)(★) Let $E \subseteq \mathbb{R}$. Prove that E is measurable if and only if for every $\epsilon > 0$, there exist an open set V and a closed set K such that
> $$K \subseteq E \subseteq V \quad \text{and} \quad m(V \setminus K) < \epsilon. \tag{12.46}$$

Proof. Suppose that E is measurable. Then we know from the inequalities (12.41) and (12.45) that there exist an open set V and a closed set K such that
$$K \subseteq E \subseteq V, \quad m^*(V \setminus E) < \frac{\epsilon}{2} \quad \text{and} \quad m^*(E \setminus K) < \frac{\epsilon}{2}. \tag{12.47}$$

Since V and K are measurable by Theorem 12.9, both $V \setminus E$ and $E \setminus K$ are measurable and the inequalities (12.47) are also true with m^* replaced by m. Since
$$V \setminus K = (V \setminus E) \cup (E \setminus K),$$
it follows from Theorem 12.3(d) (Properties of Outer Measure) that
$$m(V \setminus K) = m((V \setminus E) \cup (E \setminus K)) \leq m(V \setminus E) + +m(E \setminus K) < \epsilon.$$

Conversely, we suppose that the hypotheses (12.46) hold. Since $V \setminus E \subseteq V \setminus K$, it deduces from Theorem 12.3(c) (Properties of Outer Measure) that
$$m^*(V \setminus E) \leq m^*(V \setminus K) = m(V \setminus K) < \epsilon.$$

Since ϵ is arbitrary, we actually have $m^*(V \setminus E) = 0$. By Theorem 12.5(a) (Properties of Measurable Sets), we see that $V \setminus E$ is measurable. As $E = V \setminus (V \setminus E)$, we conclude that E is measurable. We complete the proof of the problem. ∎

> **Problem 12.33**
>
> (★)(★) The set $E \subseteq \mathbb{R}$ is measurable if and only if there exists a G_δ set $U \subseteq \mathbb{R}$ containing E such that
> $$m^*(U \setminus E) = 0. \tag{12.48}$$

Proof. Suppose that E is measurable. By Problem 12.31, for each positive integer k, we can select an open set U_k containing E and
$$m^*(U_k \setminus E) < \frac{1}{k}.$$
We define
$$U = \bigcap_{k=1}^{\infty} U_k.$$
By the definition, U is a G_δ set containing E. In addition, for each $k \in \mathbb{N}$, we have $U \setminus E \subseteq U_k \setminus E$, so Theorem 12.3(c) (Properties of Outer Measure) shows that
$$m^*(U \setminus E) \leq m^*(U_k \setminus E) < \frac{1}{k}. \tag{12.49}$$

12.4. Necessary and Sufficient Conditions for Measurable Sets

Taking $k \to \infty$ in the inequality (12.49), we conclude that $m^*(U \setminus E) = 0$.

Conversely, we suppose that the hypothesis (12.48) holds for E. By Theorem 12.5(a) (Properties of Measurable Sets), $U \setminus E$ is measurable. Since U is a G_δ set, it is measurable. Since $E = U \setminus (U \setminus E)$, we conclude that E is also measurable and thus we have completed the proof of the problem. ∎

> **Remark 12.6**
>
> Similar to the proof of Problem 12.33, we can show that E is measurable if and only if there exists a F_σ set W of $\mathbb{R}$ contained in E such that
>
> $$m^*(E \setminus W) = 0 \tag{12.50}$$

> **Problem 12.34**
>
> (★)(★) Suppose that $E \subseteq \mathbb{R}$ is measurable with $m^*(E) < \infty$. For every $\epsilon > 0$, there is a finite disjoint collection of open intervals $\{I_1, I_2, \ldots, I_n\}$ such that if $I = \bigcup_{k=1}^{n} I_k$, then
>
> $$m^*(E \setminus I) + m^*(I \setminus E) < \epsilon.$$

Proof. By Problem 12.31, there exists an open set V in $\mathbb{R}$ such that

$$E \subseteq V \quad \text{and} \quad m^*(V \setminus E) < \frac{\epsilon}{2}. \tag{12.51}$$

Since $m^*(E) < \infty$, we get from this, the measurability of E and the inequality (12.51) that

$$m^*(V) = m^*(E) + m^*(V \setminus E) < \infty,$$

i.e., V has finite outer measure. Recall from [23, Exercise 29, p. 45] that there exists a countable disjoint collection of open intervals $\{I_k\}$ whose union is V. Since each I_k is measurable, it follows from Theorems 12.3(b) (Properties of Outer Measure) and 12.11 (Countably Additivity) that

$$\sum_{k=1}^{\infty} \ell(I_k) = \sum_{k=1}^{\infty} m(I_k) = m\Big(\bigcup_{k=1}^{\infty} I_k\Big) = m(V) = m^*(V) < \infty.$$

Now the definition of a series shows that there is a positive integer n such that

$$\sum_{k=n+1}^{\infty} \ell(I_k) < \frac{\epsilon}{2}. \tag{12.52}$$

Therefore, if we define $I = \bigcup_{k=1}^{n} I_k$, then since $I \setminus E \subseteq V \setminus E$, we know from Theorem 12.3(c) (Properties of Outer Measure) and the inequality (12.51) that

$$m^*(I \setminus E) \le m^*(V \setminus E) < \frac{\epsilon}{2}. \tag{12.53}$$

Furthermore, since $E \subseteq V$, it is true that

$$E \setminus I \subseteq V \setminus I = \bigcup_{k=n+1}^{\infty} I_k.$$

Again we apply Theorems 12.3(b), (c), (d) (Properties of Outer Measure) and then using the inequality (12.52) to conclude that

$$m^*(E \setminus I) \leq m^*\Big(\bigcup_{k=n+1}^{\infty} I_k \Big) = \sum_{k=1}^{\infty} m^*(I_k) \leq \sum_{k=n+1}^{\infty} \ell(I_k) < \frac{\epsilon}{2}. \qquad (12.54)$$

Hence our desired result follows directly from the sum of the two estimates (12.53) and (12.54). This completes the proof of the problem.[f] ∎

> **Problem 12.35**
>
> (⋆) (⋆) The set A is said to be **Borel measurable** if A is a Borel set, i.e., $A \in \mathscr{B}$. Let $E \subseteq \mathbb{R}$. Prove that E is Lebesgue measurable if and only if
>
> $$E = A \cup B,$$
>
> where A is Borel measurable and $m^*(B) = 0$.

Proof. Suppose that $E = A \cup B$, where A is Borel measurable and $m^*(B) = 0$. By the paragraph following Theorem 12.10, A is measurable. By Theorem 12.5(a) (Properties of Measurable Sets), B is measurable. Hence E is also measurable because it is an union of two measurable sets.

Conversely, we suppose that E is measurable. By Remark 12.6, there exists a F_σ set W in $\mathbb{R}$ such that $W \subseteq E$ and the formula (12.50) holds. Obviously, we have

$$E = W \cup (E \setminus W).$$

As a F_σ set, Theorem 12.10 guarantees that W belongs to $\mathscr{B}$ and thus W is Borel measurable. Besides, the formula (12.50) ensures that $E \setminus W$ acts as the role of the set B in the problem. Hence we complete the proof of the problem. ∎

---

[f]This result is the first principle of the so-called **Littlewood's Three Principles**. For other two principles, please read Chapter 13.

CHAPTER 13

Lebesgue Measurable Functions

13.1 Fundamental Concepts

In Chapter 12, we review the basic definitions and results of Lebesgue measurable sets. Another building block of the theory is the main focus of this chapter: **Lebesgue measurable functions**. In this chapter, we are going to review the fundamental results of Lebesgue measurable functions, or for short measurable functions. The main references for this chapter are [7, Chap. 2], [12, Chap. 4, 5], [22, Chap. 3], [26, Chap. 1] and [27, Chap. 7].

13.1.1 Lebesgue Measurable Functions

Definition 13.1 (Lebesgue Measurable Functions). *Denote $\mathbb{R}^* = \mathbb{R} \cup \{\pm\infty\}$ to be the extended real number system. The function $f : \mathbb{R} \to \mathbb{R}^*$ is called **Lebesgue measurable**, or simply **measurable**, if the set*
$$f^{-1}((a, \infty]) = \{x \in \mathbb{R} \mid f(x) > a\}$$
is measurable for all $a \in \mathbb{R}$.

Theorem 13.2 (Criteria for Measurable Functions). *The following four conditions are equivalent.*

(a) *For every $a \in \mathbb{R}$, the set $f^{-1}((a, \infty]) = \{x \in \mathbb{R} \mid f(x) > a\}$ is measurable.*

(b) *For every $a \in \mathbb{R}$, the set $f^{-1}([a, \infty]) = \{x \in \mathbb{R} \mid f(x) \geq a\}$ is measurable.*

(c) *For every $a \in \mathbb{R}$, the set $f^{-1}([-\infty, a)) = \{x \in \mathbb{R} \mid f(x) < a\}$ is measurable.*

(d) *For every $a \in \mathbb{R}$, the set $f^{-1}([-\infty, a]) = \{x \in \mathbb{R} \mid f(x) \leq a\}$ is measurable.*

> **Remark 13.1**
>
> If the sets considered in Definition 13.1 (Lebesgue Measurable Functions) and Theorem 13.2 (Criteria for Measurable Functions) are Borel, then the function f is called **Borel measurable** or simply a **Borel function**.

Theorem 13.3. *The function $f : \mathbb{R} \to \mathbb{R}$ is Lebesgue (resp. Borel) measurable if and only if $f^{-1}(V)$ is Lebesgue (resp. Borel) measurable for every open set V in $\mathbb{R}$. Particularly, if f is continuous on $\mathbb{R}$, then f is Borel measurable.*

Similar to the situation of Lebesgue measurable sets and Borel measurable sets, a Borel measurable function must be Lebesgue measurable, but not the converse.

Theorem 13.4 (Properties of Measurable Functions). *Suppose that $f, g : \mathbb{R} \to \mathbb{R}$ are measurable, $\{f_n\}$ is a sequence of measurable functions and $k \in \mathbb{N}$.*

(a) *If $\Phi : E \to \mathbb{R}$ is continuous on E, where $f(\mathbb{R}) \subseteq E$, then the composition $\Phi \circ f : \mathbb{R} \to \mathbb{R}$ is also measurable.*[a]

(b) *f^k, $f + g$ and fg are measurable.*

(c) *The functions*

$$\sup_{n \in \mathbb{N}} f_n(x), \quad \inf_{n \in \mathbb{N}} f_n(x), \quad \limsup_{n \to \infty} f_n(x) \quad \text{and} \quad \liminf_{n \to \infty} f_n(x)$$

are all measurable.

(d) *If $f(x) = h(x)$ a.e. on $\mathbb{R}$, then h is measurable.*

13.1.2 Simple Functions and the Littlewood's Three Principles

Definition 13.5 (Simple Functions). *Let $E \subseteq \mathbb{R}$. The **characteristic function** of E is given by*

$$\chi_E(x) = \begin{cases} 1, & \text{if } x \in E; \\ 0, & \text{otherwise.} \end{cases}$$

If $E_1, E_2, \ldots, E_n$ are pairwise disjoint measurable sets, then the function

$$s(x) = \sum_{k=1}^{n} a_k \chi_{E_k}(x), \tag{13.1}$$

*where $a_1, a_2, \ldots, a_n \in \mathbb{R}$, is called a **simple function**.*

Theorem 13.6 (The Simple Function Approximation Theorem). *Suppose that $f : \mathbb{R} \to \mathbb{R}$ is measurable and $f(x) \geq 0$ on $\mathbb{R}$. Then there exists a sequence $\{s_n\}$ of simple functions such that*

$$0 \leq s_1(x) \leq s_2(x) \leq \cdots \leq f(x) \quad \text{and} \quad f(x) = \lim_{n \to \infty} s_n(x) \tag{13.2}$$

for all $x \in \mathbb{R}$.

> **Remark 13.2**
>
> The condition that f is nonnegative can be omitted in Theorem 13.6 (The Simple Function Approximation Theorem). In this case, the set of inequalities (13.2) are replaced by
>
> $$0 \leq |s_1(x)| \leq |s_2(x)| \leq \cdots \leq |f(x)|.$$

---
[a] Here the result also holds if Φ is a Borel function.

13.2. Lebesgue Measurable Functions

If the measurable sets $E_1, E_2, \ldots, E_n$ considered in the expression (13.1) are closed intervals, then the function s will be called a **step function**. In this case, we have the following approximation theorem in terms of step functions:

Theorem 13.7. *Suppose that $f : \mathbb{R} \to \mathbb{R}$ is measurable on $\mathbb{R}$. Then there exists a sequence $\{\varphi_n\}$ of step functions such that*
$$f(x) = \lim_{n \to \infty} \varphi_n(x)$$
a.e. on $\mathbb{R}$.

Classically, Littlewood says that (i) every measurable set is nearly a finite union of intervals; (ii) every measurable function is nearly continuous and (iii) every pointwise convergent sequence of measurable functions is nearly uniformly convergent. The mathematical formulation of the first principle has been given in Problem 12.34. Now the precise statements of the remaining principles are given in the following two results:

Theorem 13.8 (Egorov's Theorem). *Suppose that $\{f_n\}$ is a sequence of measurable functions defined on a measurable set E with finite measure. Given $\epsilon > 0$. If $f_n \to f$ pointwise a.e. on E, then there exists a closed set $K_\epsilon \subseteq E$ such that*
$$m(E \setminus K_\epsilon) < \epsilon \quad \text{and} \quad f_n \to f \text{ uniformly on } K_\epsilon.$$

Theorem 13.9 (Lusin's Theorem). *Suppose that f is a finite-valued measurable function on the measurable set E with $m(E) < \infty$. Given $\epsilon > 0$. Then there exists a closed set $K_\epsilon \subseteq E$ such that*
$$m(E \setminus K_\epsilon) < \epsilon \quad \text{and} \quad f \text{ is continuous on } K_\epsilon.$$

13.2 Lebesgue Measurable Functions

Problem 13.1

(★) Suppose that $f : \mathbb{R} \to \mathbb{R}$ is Lebesgue measurable. Prove that $|f|$ is also Lebesgue measurable.

Proof. Consider $\Phi : \mathbb{R} \to [0, \infty)$ defined by
$$\Phi(x) = |x|$$
which is clearly continuous. Since $|f| = \Phi \circ f$, Theorem 13.4(a) (Properties of Measurable Functions) ensures that $|f|$ is also Lebesgue measuable and it completes the proof of the problem. ∎

Problem 13.2

(★) Suppose that $f, g : \mathbb{R} \to \mathbb{R}$ are measurable. Prove that $\max(f, g)$ and $\min(f, g)$ are measurable.

Proof. Note that

$$\max(f,g) = \frac{1}{2}(f+g+|f-g|) \quad \text{and} \quad \min(f,g) = \frac{1}{2}(f+g-|f-g|).$$

By Problem 12.1 and Theorem 13.4(b) (Properties of Measurable Functions), we see immediately that both $\max(f,g)$ and $\min(f,g)$ are measurable. This completes the proof of the problem. ∎

> **Remark 13.3**
>
> As a particular case of Problem 13.2, the functions $f^+ = \max(f,0)$ and $f^- = \min(f,0)$ are measurable.

> **Problem 13.3**
>
> (⋆) Does there exist a nonmeasurable nonnegative function f such that $\sqrt{f}$ is measurable?

Proof. The answer is negative. Assume that $\sqrt{f}$ was measurable. We know that $\Phi(x) = x^2$ is continuous in $\mathbb{R}$. Now we have

$$f = (\sqrt{f})^2 = \Phi \circ \sqrt{f},$$

so Theorem 13.4(a) (Properties of Measurable Functions) ensures that f is measurable, a contradiction. Hence this completes the proof of the problem. ∎

> **Problem 13.4**
>
> (⋆) Suppose that $f : [0,\infty) \to \mathbb{R}$ is differentiable. Prove that f' is measurable.

Proof. Let $x \in [0,\infty)$. By the definition, we have

$$f(x) = \lim_{n\to\infty} f_n(x),$$

where $f_n(x) = \frac{f(x+\frac{1}{n})-f(x)}{\frac{1}{n}}$. Since f is differentiable in $[0,\infty)$, it is continuous on $[0,\infty)$. By Theorem 13.3, f is (Borel) measurable.

For each $n \in \mathbb{N}$, we notice that

$$f\left(x+\frac{1}{n}\right) = (f \circ g)(x),$$

where $g(x) = x + \frac{1}{n}$ which is clearly continuous on $[0,\infty)$ and so measurable. Thus we follow from Theorem 13.4(a) (Properties of Measurable Functions)[b] that $f(x+\frac{1}{n})$ is measurable. By Theorem 13.4(b) and then (c) (Properties of Measurable Functions), each f_n and then f, as the limit of $\{f_n\}$, are also measurable. This ends the proof of the problem. ∎

---

[b] With Φ and f replaced by f and g in our question respectively.

13.2. Lebesgue Measurable Functions

> **Problem 13.5**
>
> (⋆) Suppose that $E \subseteq \mathbb{R}$ is measurable and $f : E \to \mathbb{R}$ is continuous on E. Prove that f is measurable.

Proof. Let $a \in \mathbb{R}$ and $A = \{x \in E \mid f(x) > a\}$. If $A = \emptyset$, then there is nothing to prove. Otherwise, pick $x \in A$. Then the Sign-preserving Property implies that there exists a $\delta_x > 0$ such that $f(y) > a$ holds for all $y \in (x - \delta_x, x + \delta_x) \cap E$. Therefore, we have

$$A = \bigcup_{x \in A} [(x - \delta_x, x + \delta_x) \cap E] = \Big[\bigcup_{x \in A} (x - \delta_x, x + \delta_x) \Big] \cap E.$$

It is trivial that the set

$$\bigcup_{x \in A} (x - \delta_x, x + \delta_x)$$

is open in $\mathbb{R}$, so it is measurable by Theorem 12.9. This fact and the measurability of E shows that A is measurable. By Definition 13.1 (Lebesgue Measurable Functions), f is measurable and this completes the proof of the problem. ∎

> **Problem 13.6**
>
> (⋆) If f is measurable on $E \subseteq \mathbb{R}$, prove that $\{x \in E \mid f(x) = a\}$ is measurable for every $a \in \mathbb{R}$.

Proof. We have

$$\{x \in E \mid f(x) = a\} = \{x \in E \mid f(x) \geq a\} \cap \{x \in E \mid f(x) \leq a\}. \tag{13.3}$$

By Theorem 13.4 (Properties of Measurable Functions), the sets on the right-hand side of the expression (13.3) are measurable. Finally, we apply Theorem 12.5(e) (Properties of Measurable Sets) to conclude that

$$\{x \in E \mid f(x) = a\}$$

is also measurable, so it ends the analysis of the problem. ∎

> **Problem 13.7**
>
> (⋆) If f is measurable on $E \subseteq \mathbb{R}$, prove that $\{x \in E \mid f(x) = \infty\}$ is measurable.

Proof. We observe that

$$\{x \in E \mid f(x) = \infty\} = \bigcap_{n=1}^{\infty} \{x \in E \mid f(x) > n\}. \tag{13.4}$$

Since f is measurable on E, each set $\{x \in E \mid f(x) > n\}$ is measurable. Recall from Remark 12.3 that the set on the right-hand side of the expression (13.4) is measurable and our desired result follows. This completes the proof of the problem. ∎

Problem 13.8

(⋆) Suppose that f and g are measurable on $E \subseteq \mathbb{R}$. Prove that the set $\{x \in E \,|\, f(x) > g(x)\}$ is measurable.

Proof. For every $x \in E$, since $f(x) > g(x)$, Theorem 2.2 (Density of Rationals) implies that there is a $r \in \mathbb{Q}$ such that $f(x) > r > g(x)$. Therefore, we have

$$\{x \in E \,|\, f(x) > r > g(x)\} = \{x \in E \,|\, f(x) > r\} \cap \{x \in E \,|\, g(x) < r\}. \tag{13.5}$$

By the hypotheses, both sets $\{x \in E \,|\, f(x) > r\}$ and $\{x \in E \,|\, g(x) < r\}$ are measurable. Thus it follows from this and the expression (13.5) that the set

$$\{x \in E \,|\, f(x) > r > g(x)\}$$

is also measurable. Now we know that

$$\{x \in E \,|\, f(x) > g(x)\} = \bigcup_{r \in \mathbb{Q}} \{x \in E \,|\, f(x) > r > g(x)\}$$

and then Theorem 12.8 guarantees that $\{x \in E \,|\, f(x) > g(x)\}$ is measurable and we complete the analysis of the problem. ∎

Problem 13.9

(⋆) Prove that

(a) a characteristic function χ_E is measurable if and only if E is measurable.

(b) a simple function $s = \sum_{k=1}^{n} a_k \chi_{E_k}$ is measurable.

Proof. In the following discussion, suppose that $E, E_1, \ldots, E_n$ are measurable.

(a) We deduce from Definition 13.5 (Simple Functions) that

$$\{x \in \mathbb{R} \,|\, \chi_E(x) < a\} = \begin{cases} \varnothing, & \text{if } a \leq 0; \\ \mathbb{R}, & \text{if } a > 1; \\ \mathbb{R} \setminus E, & \text{if } 0 < a \leq 1. \end{cases}$$

Consequently, the set $\{x \in \mathbb{R} \,|\, \chi_E(x) < a\}$ is measurable for every $a \in \mathbb{R}$ and Definition 13.1 (Lebesgue Measurable Functions) implies that χ_E is measurable. Conversely, suppose that χ_E is measurable. Since $E = \mathbb{R} \setminus \{x \in \mathbb{R} \,|\, \chi_E(x) < \frac{1}{2}\}$ and $\{x \in \mathbb{R} \,|\, \chi_E(x) < \frac{1}{2}\}$ is measurable, we conclude that E is also measurable.

(b) By part (a), each χ_{E_k} is measurable. Then repeated applications of Theorem 13.4(b) (Properties of Measurable Functions) show that the simple function s is also measurable.

13.2. Lebesgue Measurable Functions

We complete the analysis of the proof. ∎

> **Problem 13.10**
>
> (★) Construct a nonmeasurable function.

Proof. By Remark 12.2, $\mathbb{R}$ contains a nonmeasurable set S. Consider the characteristic function

$$\chi_S(x) = \begin{cases} 1, & \text{if } x \in S; \\ 0, & \text{otherwise.} \end{cases}$$

Then χ_S cannot be measurable by Problem 13.9. This completes the proof of the problem. ∎

> **Problem 13.11**
>
> (★) Suppose that $f : \mathbb{R} \to (0, \infty)$ is measurable. Prove that $\frac{1}{f}$ is measurable.

Proof. Let $g = \frac{1}{f} : \mathbb{R} \to (0, \infty)$. If $a \leq 0$, then $\{x \in \mathbb{R} \,|\, g(x) > a\} = \mathbb{R}$ which is definitely measurable. For $a > 0$, we note that $g(x) > a$ if and only if $0 < f(x) < \frac{1}{a}$ so that

$$\begin{aligned}\{x \in \mathbb{R} \,|\, g(x) > a\} &= \left\{x \in \mathbb{R} \,\Big|\, 0 < f(x) < \frac{1}{a}\right\} \\ &= \{x \in \mathbb{R} \,|\, f(x) > 0\} \cap \left\{x \in \mathbb{R} \,\Big|\, f(x) < \frac{1}{a}\right\}. \end{aligned} \qquad (13.6)$$

Since f is measurable, Theorem 13.2 (Criteria for Measurable Functions) ensures that the two sets on the right-hand side of the expression (13.6) are measurable. Thus the set

$$\{x \in \mathbb{R} \,|\, g(x) > a\}$$

is measurable. By Definition 13.1 (Lebesgue Measurable Functions), the g is a measurable function, completing the proof of the problem. ∎

> **Problem 13.12**
>
> (★)(★) Suppose that $f : \mathbb{R} \to \mathbb{R}$ is measurable and $f(x+1) = f(x)$ a.e. on $\mathbb{R}$. Prove that there exists a function $g : \mathbb{R} \to \mathbb{R}$ such that $f = g$ a.e. and $g(x+1) = g(x)$ on $\mathbb{R}$.

Proof. Suppose that $E = \{x \in \mathbb{R} \,|\, f(x+1) \neq f(x)\}$. By the hypothesis, we have $m(E) = 0$. Next, we define

$$F = \bigcup_{n \in \mathbb{Z}} (E + n) \qquad (13.7)$$

and

$$g(x) = \begin{cases} f(x), & \text{if } x \notin F; \\ 0, & \text{otherwise.} \end{cases} \qquad (13.8)$$

By Theorem 12.5(d) (Properties of Measurable Sets), each $E+n$ is measurable and $m(E+n) = 0$. Applying this fact to the inequality (12.1), we derive

$$m(F) \le \sum_{n=1}^{\infty} m(E+n) = 0.$$

Therefore, we follow from the definition (13.8) that

$$f = g$$

a.e. on $\mathbb{R}$, proving the first assertion.

Next, since $g(x+1) = g(x)$ on F^c, it suffices to prove the second assertion on F. To see this, if $x \in F$, then we note from the definition (13.7) that $x = E + m$ for some $m \in \mathbb{Z}$ which implies that

$$x + 1 = E + m + 1 \in F.$$

Hence it is obvious from the definition (13.8) that for every $x \in F$, we still have

$$g(x+1) = 0 = g(x)$$

which proves the second assertion and this completes the proof of the problem. ∎

Problem 13.13

(⋆)(⋆) *Let $f : \mathbb{R} \to \mathbb{R}$ be continuous a.e. on $\mathbb{R}$. Prove that f is Lebesgue measurable.*

Proof. Let $a \in \mathbb{R}$ and denote $E_a = \{x \in \mathbb{R} \,|\, f(x) > a\}$. Let $x \in E_a$ and A be the set of all discontinuities of f. There are two cases:

- **Case (1): f is discontinuous at x.** In this case, $A \ne \emptyset$. By the hypothesis, we know that $m(A) = 0$ so that A is measurable by Theorem 12.5(a) (Properties of Measurable Sets).

- **Case (2): f is continuous at x.** In this case, the Sign-preserving Property ensures that there is a $\delta_x > 0$ such that $f(y) > a$ for all $y \in (x - \delta_x, x + \delta_x)$. In other words, we have $(x - \delta_x, x + \delta_x) \subseteq E_a$.

Therefore, we are able to write

$$E_a = A \cup \bigcup_{x \in E_a \setminus A} (x - \delta_x, x + \delta_x).$$

Obviously, since every $(x - \delta_x, x + \delta_x)$ is open in $\mathbb{R}$, the union

$$\bigcup_{x \in E_a \setminus A} (x - \delta_x, x + \delta_x) \tag{13.9}$$

is also open in $\mathbb{R}$. By Theorem 12.9, the set (13.9) is measurable. Hence we deduce from **Case (1)** that E_a is also measurable. Recall that a is arbitrary, so Theorem 13.2 (Criteria for Measurable Functions) concludes that f is Lebesgue measurable which completes the proof of the problem. ∎

13.2. Lebesgue Measurable Functions

Problem 13.14

(★)(★) Let $f : \mathbb{R} \to \mathbb{R}$ be monotone. Verify that f is Borel measurable.

Proof. Without loss of generality, we may assume that f is increasing. Let $a \in \mathbb{R}$. Suppose that $p \in \{x \in \mathbb{R} \,|\, f(x) < a\}$. Then $f(p) < a$ and for all $y < p$, since f is increasing, we have

$$f(y) < f(p) < a.$$

Thus we have $y \in \{x \in \mathbb{R} \,|\, f(x) < a\}$. Now we put $\alpha = \sup\{x \in \mathbb{R} \,|\, f(x) < a\}$. On the one hand, if $f(\alpha) < a$, then we have

$$\{x \in \mathbb{R} \,|\, f(x) < a\} = (-\infty, \alpha].$$

On the other hand, if $f(\alpha) \geq a$, then we have

$$\{x \in \mathbb{R} \,|\, f(x) < a\} = (-\infty, \alpha).$$

In any case, we conclude from Theorem 12.10(c) that the set $\{x \in \mathbb{R} \,|\, f(x) < a\}$ is Borel measurable. By Remark 13.1, the function f is a Borel function, completing the proof of the problem. ∎

Problem 13.15

(★)(★) Let S be dense in $\mathbb{R}$ and $f : \mathbb{R} \to \mathbb{R}$. Prove that f is measurable if and only if the set $\{x \in \mathbb{R} \,|\, f(x) > s\}$ is measurable for every $s \in S$.

Proof. Suppose first that f is measurable. Then Theorem 13.2 (Criteria for Measurable Functions) says that $\{x \in \mathbb{R} \,|\, f(x) > a\}$ is Lebesgue measurable for every $a \in \mathbb{R}$. In particular, the set $\{x \in \mathbb{R} \,|\, f(x) > s\}$ is measurable for every $s \in S$.

For the converse direction, it suffices to prove that $\{x \in \mathbb{R} \,|\, f(x) > s\}$ is measurable for every $s \in S^c$. In fact, given $p \in S^c$. Since S is dense in $\mathbb{R}$, there exists a decreasing sequence $\{p_n\} \subseteq S$ such that $p_n \to p$ as $n \to \infty$. Next, we claim that

$$\{x \in \mathbb{R} \,|\, f(x) > p\} = \bigcup_{n=1}^{\infty} \{x \in \mathbb{R} \,|\, f(x) > p_n\}. \tag{13.10}$$

To see this, for each $n \in \mathbb{N}$, if $y \in \{x \in \mathbb{R} \,|\, f(x) > p_n\}$, then since $p_n > p$, we have $f(y) > p$ so that $y \in \{x \in \mathbb{R} \,|\, f(x) > p\}$, i.e.,

$$\bigcup_{n=1}^{\infty} \{x \in \mathbb{R} \,|\, f(x) > p_n\} \subseteq \{x \in \mathbb{R} \,|\, f(x) > p\}. \tag{13.11}$$

On the other hand, if $y \in \{x \in \mathbb{R} \,|\, f(x) > p\}$, then we have $\epsilon = f(y) - p > 0$. Since p_n converges to p decreasingly, there exists an $N \in \mathbb{N}$ such that $n \geq N$ implies $p_n - p < \epsilon$ and this means that

$$f(y) > p_n,$$

i.e., $y \in \{x \in \mathbb{R} \mid f(x) > p_n\}$ for all $n \geq N$. Consequently, we have

$$\{x \in \mathbb{R} \mid f(x) > p\} \subseteq \bigcup_{n=1}^{\infty} \{x \in \mathbb{R} \mid f(x) > p_n\}. \tag{13.12}$$

Now our claim (13.10) follows immediately by combining the set relations (13.11) and (13.12). By the assumption, each $\{x \in \mathbb{R} \mid f(x) > p_n\}$ is measurable, so the expression (13.10) shows that

$$\{x \in \mathbb{R} \mid f(x) > p\}$$

is also measurable, where $p \in S^c$. By Definition 13.1 (Lebesgue Measurable Functions), f is measurable and it completes the analysis of the problem. ■

Problem 13.16

(★)(★) Suppose that $f : \mathbb{R} \to \mathbb{R}$ is bijective and continuous and $\mathscr{B}$ denotes the collection of all Borel sets of $\mathbb{R}$. Prove that $f(\mathscr{B}) \subseteq \mathscr{B}$.

Proof. Suppose that
$$\mathscr{D} = \{E \subseteq \mathbb{R} \mid f(E) \in \mathscr{B}\}.$$

First of all, we want to show that $\mathscr{D}$ is an σ-algebra. To this end, recall a basic fact from [16, Exercise 2(g) and (h), p. 21] that if f is bijective, then

$$f(E \cap F) = f(E) \cap f(F) \quad \text{and} \quad f(E \setminus F) = f(E) \setminus f(F). \tag{13.13}$$

Thus if $E \in \mathscr{D}$, then the second set equation (13.13) implies that

$$f(E^c) = f(\mathbb{R}) \setminus f(E) = \mathbb{R} \setminus f(E).$$

Since $f(E), \mathbb{R} \in \mathscr{B}$, we have $f(E^c) \in \mathscr{B}$ and then $E^c \in \mathscr{D}$. Let $E_k \in \mathscr{D}$ so that

$$f\Big(\bigcup_{k=1}^{\infty} E_k\Big) = \bigcup_{k=1}^{\infty} f(E_k). \tag{13.14}$$

Since $f(E_k) \in \mathscr{B}$ for every $k = 1, 2, \ldots$, we have

$$\bigcup_{k=1}^{\infty} f(E_k) \in \mathscr{B}$$

and then the set relation (13.14) shows that

$$\bigcup_{k=1}^{\infty} E_k \in \mathscr{D}.$$

Therefore, $\mathscr{D}$ is an σ-algebra by Definition 12.7 (σ-algebra) which proves our claim.

Next, for every closed interval $[a, b]$, the continuity of f ensures that $f([a, b])$ is compact and connected by Theorem 7.9 (Continuity and Compactness) and Theorem 7.12 (Continuity and Connectedness). Note that $f([a, b]) \subseteq \mathbb{R}$, so Theorem 4.16 implies that $f([a, b])$ must be a closed

13.2. Lebesgue Measurable Functions

interval. Since f is continuous and one-to-one, we know from Problem 7.33 that f is strictly monotonic and hence either

$$f([a,b]) = [f(a), f(b)] \quad \text{or} \quad f([a,b]) = [f(b), f(a)].$$

Now both $[f(a), f(b)]$ and $[f(b), f(a)]$ are closed in $\mathbb{R}$, so they belong to $\mathscr{B}$. By the definition, we conclude that $[a,b] \in \mathscr{D}$ for every $-\infty < a \leq b < \infty$. In other words, $\mathscr{D}$ contains every closed intervals of $\mathbb{R}$ and Theorem 12.10(c) says that $\mathscr{B} \subseteq \mathscr{D}$. Since $\mathscr{D}$ is an σ-algebra and $\mathscr{B}$ is a **smallest** σ-algebra containing all closed intervals of $\mathbb{R}$, we establish that

$$\mathscr{B} = \mathscr{D}$$

which means $f(\mathscr{B}) \subseteq \mathscr{B}$, as desired. Hence we complete the analysis of the problem. ■

> **Problem 13.17**
>
> ($\star$) Let E be measurable. Prove that $f : E \to \mathbb{R}$ is measurable if and only if $f^{-1}(V)$ is measurable for every open set V in $\mathbb{R}$.

Proof. We suppose that $f^{-1}(V)$ is measurable for every open set V in $\mathbb{R}$. In particular,

$$f^{-1}((a, \infty)) = \{x \in E \mid f(x) > a\}$$

is measurable for every $a \in \mathbb{R}$. By Theorem 13.2 (Criteria for Measurable Functions), f is measurable.

Conversely, suppose that f is measurable. By [23, Exercise 29, p. 45], V is an union of countable collection of disjoint open intervals $\{(a_n, b_n)\}$, i.e.,

$$V = \bigcup_{n=1}^{\infty} (a_n, b_n).$$

Recall the two facts[c]

$$f^{-1}(A \cup B) = f^{-1}(A) \cup f^{-1}(B) \quad \text{and} \quad f^{-1}(A \cap B) = f^{-1}(A) \cap f^{-1}(B),$$

we have

$$\begin{aligned} f^{-1}(V) &= \bigcup_{n=1}^{\infty} f^{-1}((a_n, b_n)) \\ &= \bigcup_{n=1}^{\infty} f^{-1}((-\infty, b_n) \cap (a_n, \infty)) \\ &= \bigcup_{n=1}^{\infty} f^{-1}((-\infty, b_n)) \cap f^{-1}((a_n, \infty)). \end{aligned}$$

By Theorem 13.2 (Criteria for Measurable Functions), both $f^{-1}((-\infty, b_n))$ and $f^{-1}((a_n, \infty))$ are measurable so that every $f^{-1}((a_n, b_n))$ is measurable. Thus Theorem 12.8 implies that $f^{-1}(V)$ is measurable, as desired. Hence we have completed the proof of the problem. ■

---
[c] See [16, Exercise 2(b) and (c), p. 20].

Problem 13.18

$(\star)(\star)$ Suppose that $E \subseteq \mathbb{R}$ is measurable and $m(E) < \infty$. Let f be measurable on E and $|f(x)| < \infty$ a.e. on E. Prove that for every $\epsilon > 0$, there exists a measurable subset $F \subseteq E$ such that
$$m(E \setminus F) < \epsilon \quad \text{and} \quad f|_F \text{ is bounded on } F.$$

Proof. For each $n \in \mathbb{N}$, consider
$$E_n = \{x \in E \mid |f(x)| > n\} \tag{13.15}$$
and $A = \{x \in E \mid f(x) = \pm\infty\}$. Since $|f(x)| < \infty$ a.e. on E, we have $m(A) = 0$. Since f is measurable on E and
$$E_n = \{x \in E \mid f(x) > n\} \cup \{x \in E \mid f(x) < -n\},$$
every E_n is also measurable. Furthermore, it is easy to check from the definition (13.15) that
$$E_1 \supseteq E_2 \supseteq \cdots \quad \text{and} \quad A = \bigcap_{n=1}^{\infty} E_n.$$
Since $m(E_1) \leq m(E) < \infty$, we apply Theorem 12.12 (Continuity of Lebesgue Measure) to conclude that
$$\lim_{n \to \infty} m(E_n) = m(A) = 0. \tag{13.16}$$
Thus given $\epsilon > 0$, there exists an $N \in \mathbb{N}$ such that $m(E_N) < \epsilon$. Put $F = E \setminus E_N$. Then we have
$$E_N = E \setminus F \quad \text{and} \quad F = \{x \in E \mid |f(x)| \leq N\}$$
which mean that $f|_F$ is definitely bounded by N on F. Therefore, this completes the proof of the problem. ∎

Problem 13.19

$(\star)(\star)$ Suppose that $\{f_n\}$ is a sequence of measurable functions on $[0,1]$ with $|f_n(x)| < \infty$ a.e. on $[0,1]$. Prove that there exists a sequence $\{a_n\}$ of positive real numbers such that
$$\frac{f_n(x)}{a_n} \to 0 \tag{13.17}$$
as $n \to \infty$ a.e. on $[0,1]$.

Proof. Given $\epsilon > 0$. Using the same notations and argument as in Problem 13.18, we still have the limit (13.16). In other words, for each $n \in \mathbb{N}$, there corresponds a $k_n \in \mathbb{N}$ such that
$$m(E_k(n)) = m(\{x \in [0,1] \mid |f_n(x)| > k_n\}) < \epsilon, \tag{13.18}$$
where
$$E_k(n) = \{x \in [0,1] \mid |f_n(x)| > k\}.$$

13.2. Lebesgue Measurable Functions

Now we pick $\epsilon = 2^{-n}$ and $a_n = nk_n > 0$ in the estimate (13.18) to get

$$m(E_k(n)) = m\left(\left\{x \in [0,1] \,\bigg|\, \frac{|f_n(x)|}{a_n} > \frac{1}{n}\right\}\right) < 2^{-n}$$

which implies that

$$\sum_{n=1}^{\infty} m(E_k(n)) < \sum_{n=1}^{\infty} 2^{-n} = 1 < \infty.$$

Hence it follows from Problem 12.15 (The Borel-Cantelli Lemma) that the set

$$E = \{x \in [0,1] \,|\, x \in E_k(n) \text{ for infinitely many } n\}$$

has measure 0. In other words, every $x \in E^c$ lies in *finitely many* $E_k(n)$. Thus, if $x \in E^c$, then there exists an $N \in \mathbb{N}$ such that $n \geq N$ implies

$$\frac{|f_n(x)|}{a_n} \leq \frac{1}{n}$$

and this gives the limit (13.17) a.e. on $[0,1]$ because $m(E^c) = 1$. We end the proof of the problem. ∎

Problem 13.20

$(\star)(\star)$ Suppose that $f : \mathbb{R} \to \mathbb{R}$ is measurable and V is an open set in $\mathbb{R}$ containing 0. Prove that there exists a measurable set E such that $m(E) > 0$ and $f(x) - f(y) \in V$ for every pair $x, y \in E$.

Proof. Since V is an open set in $\mathbb{R}$ containing 0, there exists a $\epsilon > 0$ such that $(-\epsilon, \epsilon) \subseteq V$. Let $I = (-\frac{\epsilon}{2}, \frac{\epsilon}{2})$ and

$$E_p = f^{-1}(p + I) \tag{13.19}$$

for every $p \in \mathbb{R}$. Note that every E_p is measurable by Theorem 13.3.

We claim that $\bigcup_{p \in \mathbb{Q}}(p + I) = \mathbb{R}$. Otherwise, there was a $x \in \mathbb{R}$ such that

$$x \notin \bigcup_{p \in \mathbb{Q}}(p + I).$$

However, there exists a rational q such that $q \in (x, x + \frac{\epsilon}{4})$ and this implies that

$$x \in \left(q - \frac{\epsilon}{2}, q + \frac{\epsilon}{2}\right).$$

Therefore, no such x exists and we have the claim. By this, we obtain

$$\bigcup_{p \in \mathbb{Q}} E_p = \bigcup_{p \in \mathbb{Q}} f^{-1}(p + I) = f^{-1}\left(\bigcup_{p \in \mathbb{Q}}(p + I)\right) = f^{-1}(\mathbb{R}) = \mathbb{R}. \tag{13.20}$$

Assume that $m(E_p) = 0$ for all $p \in \mathbb{Q}$. Then it follows from the set relation (13.20) and the inequality (12.1) that

$$\infty = m(\mathbb{R}) \leq \sum_{p \in \mathbb{Q}} m(E_p) = 0,$$

a contradiction. Thus there exists a $p_0 \in \mathbb{Q}$ such that $m(E_{p_0}) > 0$ which implies that $E_{p_0} \neq \emptyset$. Now for every pair $x, y \in E_{p_0}$, the definition (13.19) shows that $f(x), f(y) \in (p_0 - \frac{\epsilon}{2}, p_0 + \frac{\epsilon}{2})$ which certainly gives
$$f(x) - f(y) \in (-\epsilon, \epsilon).$$
It completes the proof of the problem. ∎

13.3 Applications of Littlewood's Three Principles

Problem 13.21

(★)(★)(★) *Prove Theorem 13.8 (Egorov's Theorem).*

Proof. For each pair of $s, t \in \mathbb{N}$, we consider
$$E_t(s) = \left\{ x \in E \,\Big|\, |f_i(x) - f(x)| < \tfrac{1}{s} \text{ for all } i \geq t \right\}.$$
We fix s. Notice that
$$E_1(s) \subseteq E_2(s) \subseteq \cdots \quad \text{and} \quad \lim_{t \to \infty} E_t(s) = E.$$
By Theorem 12.12 (Continuity of Lebesgue Measure), we see that there exists a $t_s \in \mathbb{N}$ such that
$$m(E) - m(E_{t_s}(s)) < \frac{1}{2^s}. \tag{13.21}$$
Using Problem 12.13, we can express the inequality (13.21) as
$$m(E \setminus E_{t_s}(s)) < \frac{1}{2^s}.$$
Now we can choose $N \in \mathbb{N}$ large enough so that
$$\sum_{s=N}^{\infty} \frac{1}{2^s} < \frac{\epsilon}{2}. \tag{13.22}$$
Suppose that
$$K_{\epsilon, N} = \bigcap_{s=N}^{\infty} E_{t_s}(s)$$
which is measurable.

Now we are ready to prove the requirements of the theorem. In fact, we find from the inequalities (12.1) and (13.22) that
$$m(E \setminus K_{\epsilon, N}) = m\Big(\bigcup_{s=N}^{\infty} (E \setminus E_{t_s}(s)) \Big) \leq \sum_{s=N}^{\infty} m(E \setminus E_{t_s}(s)) < \frac{\epsilon}{2}. \tag{13.23}$$

13.3. Applications of Littlewood's Three Principles

Since $K_{\epsilon,N}$ is measurable, Remark 12.5 ensures that there is a closed set $K_\epsilon \subseteq K_{\epsilon,N}$ such that

$$m(K_{\epsilon,N} \setminus K_\epsilon) < \frac{\epsilon}{2}. \tag{13.24}$$

By combining the inequalities (13.23) and (13.24), we establish that

$$m(E \setminus K_\epsilon) = m((E \setminus K_{\epsilon,N}) \cup (K_{\epsilon,N} \setminus K)) \leq m(E \setminus K_{\epsilon,N}) + m(K_{\epsilon,N} \setminus K) < \epsilon.$$

This gives the first requirement of the theorem.

Next, we observe that $x \in K_{\epsilon,N}$ implies that $x \in E_{t_s}(s)$ for all $s \geq N$. Thus given any $\delta > 0$, we can find an $N' \geq N$ such that $\frac{1}{N'} < \delta$. Therefore, if $i \geq t_{N'}$, then the definition implies

$$|f_i(x) - f(x)| < \frac{1}{N'} < \delta \tag{13.25}$$

on $K_{\epsilon,N}$. Recall that $K_\epsilon \subseteq K_{\epsilon,N}$, so our result (13.25) is also valid on K_ϵ. This proves the second requirement of the theorem and it completes the proof of the problem. ∎

> **Problem 13.22**
>
> (⋆) Can we drop the condition $m(E) < \infty$ in Theorem 13.8 (Egorov's Theorem)?

Proof. The answer is negative. In fact, consider $E = [0, \infty)$ and $f_n = \chi_{[n,\infty)} : E \to \mathbb{R}$. Since $[n, \infty)$ is measurable, f_n is measurable by Problem 13.9. For every $x \in E$, we have

$$f_n(x) \to f(x) = 0$$

as $n \to \infty$. Assume that there was a closed set $K \subseteq E$ such that

$$m(E \setminus K) < 1 \quad \text{and} \quad f_n \to 0 \text{ uniformly on } K. \tag{13.26}$$

Since $m(E) = \infty$, it is true that $m(K) = \infty$ so that K is unbounded. Thus for every $n \in \mathbb{N}$, we have

$$K \cap [n, \infty) \neq \emptyset.$$

Let $p_n \in K \cap [n, \infty)$. Then we have

$$f_n(p_n) = \chi_{[n,\infty)}(p_n) = 1$$

for every $n \in \mathbb{N}$ which contradicts the second condition (13.26). This completes the proof of the problem. ∎

> **Problem 13.23**
>
> (⋆) Suppose that $\{f_n\}$ is a sequence of measurable functions defined on a measurable set E with $m(E) < \infty$ and $f_n \to f$ pointwise a.e. on E. Prove that there exists a sequence $\{F_k\}$ of closed sets of E such that
>
> $$m\left(E \setminus \bigcup_{k=1}^{\infty} F_k\right) = 0 \quad \text{and} \quad f_n \to f \text{ uniformly on } F_1, F_2, \ldots.$$

Proof. If $m(E) < \infty$, then Theorem 13.8 (Egorov's Theorem) implies that there is a closed set $F_1 \subseteq E$ such that
$$m(E \setminus F_1) < \frac{1}{2} \quad \text{and} \quad f_n \to f \text{ uniformly on } F_1.$$
This argument can be applied repeatedly. In fact, for every $k \in \mathbb{N}$, there is a sequence $\{F_1, F_2, \ldots, F_k\}$ of closed sets with $F_k \subseteq E \setminus \bigcup_{i=1}^{k-1} F_i$ such that
$$m\left(E \setminus \bigcup_{i=1}^{k} F_i\right) < \frac{1}{2^k} \quad \text{and} \quad f_n \to f \text{ uniformly on } F_1, F_2, \ldots, F_k. \tag{13.27}$$
Define $F = \bigcup_{i=1}^{\infty} F_i$. Then it is easy to check that
$$m(E \setminus F) = m\left(E \setminus \bigcup_{i=1}^{\infty} F_i\right) \le m\left(E \setminus \bigcup_{i=1}^{k} F_i\right)$$
for every $k \in \mathbb{N}$, so we further deduce from the inequality (13.27) that
$$m(E \setminus F) < \frac{1}{2^k} \tag{13.28}$$
for every $k \in \mathbb{N}$. Taking $k \to \infty$ in the inequality (13.28), we may conclude that
$$m(E \setminus F) = 0$$
which completes the proof of the problem. ∎

Problem 13.24

(⋆) Suppose that $\{f_n\}$ is a sequence of measurable functions defined on a measurable set E with finite measure. We say $\{f_n\}$ **converges in measure** to the measurable function f if for every $\epsilon > 0$, there corresponds an $N \in \mathbb{N}$ such that
$$m(\{x \in E \,|\, |f_n(x) - f(x)| > \epsilon\}) < \epsilon$$
for all $n > N$. Prove that if $f_n \to f$ pointwise a.e. on E, then $f_n \to f$ in measure.

Proof. Given that $\epsilon > 0$. Let $A = \{x \in E \,|\, f_n(x) \to f(x)\}$. Then we have $m(E \setminus A) = 0$. Since $m(E) < \infty$, we have $m(A) < \infty$ and we are able to apply Theorem 13.8 (Egorov's Theorem). Therefore, there is a closed set $K_\epsilon \subseteq A$ such that
$$m(A \setminus K_\epsilon) < \epsilon \quad \text{and} \quad f_n \to f \text{ uniformly on } K_\epsilon.$$
By the definition of uniform convergence, there exists an $N \in \mathbb{N}$ such that $n > N$ implies that
$$|f_n(x) - f(x)| < \epsilon$$
on K_ϵ. In other words, if $n > N$, then $|f_n(x) - f(x)| > \epsilon$ only possibly on $A \setminus K_\epsilon$ or on $E \setminus A$. Hence, for all $n > N$, we obtain
$$m(\{x \in E \,|\, |f_n(x) - f(x)| > \epsilon\}) \le m(A \setminus K_\epsilon) + m(E \setminus A) < \epsilon.$$
By the definition, $f_n \to f$ in measure and it completes the proof of the problem. ∎

13.3. Applications of Littlewood's Three Principles

Problem 13.25

(★)(★) Prove Theorem 13.9 (Lusin's Theorem) in the special case that f is a simple function defined on E.

Proof. Suppose that f takes the form

$$f(x) = \sum_{i=1}^{n} a_i \chi_{E_i}(x), \qquad (13.29)$$

where $a_1, a_2, \ldots, a_n$ are distinct and $E_1, E_2, \ldots, E_n$ are pairwise disjoint subsets of E whose union is E. By Remark 12.5, there exist closed sets $K_1, K_2, \ldots, K_n$ such that

$$K_i \subseteq E_i \quad \text{and} \quad m(E_i \setminus K_i) < \frac{\epsilon}{n},$$

where $i = 1, 2, \ldots, n$. Set $K_\epsilon = \bigcup_{i=1}^{n} K_i$ which is clearly closed in $\mathbb{R}$. Then we deduce from the inequality (12.1) that

$$m(E \setminus K_\epsilon) = m\Big(\bigcup_{i=1}^{n}(E_i \setminus K_i)\Big) \leq \sum_{i=1}^{n} m(E_i \setminus K_i) < \epsilon.$$

This gives our first assertion.

For the second assertion, we define $g : K_\epsilon \to \mathbb{R}$ by $g(x) = a_i$ for $x \in K_i$. In fact, g takes the form

$$g(x) = \sum_{i=1}^{n} a_i \chi_{K_i}(x). \qquad (13.30)$$

By comparing the two representations (13.29) and (13.30), we find that

$$f_{K_\epsilon} = g,$$

so it suffices to show that g is continuous on K_ϵ. To this end, consider $p \in K_i$ and denote $\widehat{K_i} = \bigcup_{\substack{j=1 \\ j \neq i}}^{n} K_j$. We claim that there exists a $\delta > 0$ such that

$$(p - \delta, p + \delta) \cap \widehat{K_i} = \varnothing.$$

Otherwise, for each $s \in \mathbb{N}$, we have

$$x_s \in \Big(p - \frac{1}{s}, p + \frac{1}{s}\Big) \cap \widehat{K_i}$$

so that $x_s \to p$ as $s \to \infty$. Since the set $\widehat{K_i}$ is closed in $\mathbb{R}$, it is also true that $p \in \widehat{K_i}$. This contradicts the fact that $K_1, K_2, \ldots, K_n$ are pairwise disjoint. Therefore, we have

$$(p - \delta, p + \delta) \cap K_\epsilon = (p - \delta, p + \delta) \cap K_i$$

so that
$$g(x) = a_i \tag{13.31}$$

for all $x \in (p-\delta, p+\delta) \cap K_\epsilon$. Since $(p-\delta, p+\delta) \cap K_\epsilon$ is equivalent to $|x-p| < \delta$ and $x \in K_\epsilon$. Hence the result (13.31) actually means that g is continuous at p. As K_i and then p are arbitrary, we have shown that g is continuous on K_ϵ and thus our second assertion is proven. Hence we have completed the proof of the problem. ∎

> **Problem 13.26**
>
> (★)(★) Prove Theorem 13.9 (Lusin's Theorem) with the aid of Problem 13.25.

Proof. Given $\epsilon > 0$. By Theorem 13.6 (The Simple Function Approximation Theorem), we see that there corresponds a sequence $\{s_n\}$ of simple functions defined on E converging to f pointwise on E. For each $n \in \mathbb{N}$, Problem 13.25 ensures that there exists a closed set $K_n \subseteq E$ such that
$$m(E \setminus K_n) < \frac{\epsilon}{2^{n+1}} \quad \text{and} \quad s_n \text{ is continuous on } K_n. \tag{13.32}$$

Since $s_n \to f$ pointwise on E, it follows from Theorem 13.8 (Egorov's Theorem) that there is a closed set $K_0 \subseteq E$ such that
$$m(E \setminus K_0) < \frac{\epsilon}{2} \quad \text{and} \quad s_n \to f \text{ uniformly on } K_0. \tag{13.33}$$

Now we set
$$K_\epsilon = \bigcap_{n=0}^{\infty} K_n.$$

By De Morgan's Laws and then the applications of the inequality (12.1), (13.32) and (13.33), we obtain
$$m(E \setminus K_\epsilon) = m\Big((E \setminus K_0) \cup \bigcup_{n=1}^{\infty}(E \setminus K_n)\Big) \leq m(E \setminus K_0) + \sum_{n=1}^{\infty} m(E \setminus K_n) < \frac{\epsilon}{2} + \frac{\epsilon}{2} = \epsilon.$$

Obviously, K_ϵ is closed in $\mathbb{R}$, every s_n is continuous and $s_n \to f$ uniformly on K_ϵ. Hence, by Theorem 10.6 (Uniform Convergence and Continuity), f is continuous on K_ϵ, completing the proof of the problem. ∎

> **Remark 13.4**
>
> Traditionally, Theorem 13.9 (Lusin's Theorem) is proven by using Theorems 13.6 (The Simple Function Approximation Theorem) and 13.8 (Egorov's Theorem), just as what we have done in Problem 13.26. In fact, our argument used in Problems 13.25 and 13.26 follows that of Royden [22, pp. 66, 67]. Similar proofs can be found, for examples, in [24, Theorem 2.24, pp. 55, 56], [26, Theorem 4.5, p. 34] or [32, Theorem 5.29, pp. 137, 138]. In 2004, Loeb and Talvila [15] prove Theorem 13.9 (Lusin's Theorem) without using simple functions and their proof is presented in Problem 13.27.

13.3. Applications of Littlewood's Three Principles

Problem 13.27

(★) (★) (★) Prove Theorem 13.9 (Lusin's Theorem) with the closed set K_ϵ replaced by a compact set.

Proof. Given a measurable set E with $m(E) < \infty$ and $\epsilon > 0$. We claim that there is a compact set $K \subseteq E$ such that
$$m(E \setminus K) < \epsilon.$$
To see this, we know from Remark 12.5 that there is a closed set F of $\mathbb{R}$ contained in E such that
$$m(E \setminus F) < \frac{\epsilon}{2}. \tag{13.34}$$
Since $m(E) < \infty$, we must have $m(F) < \infty$. Define $F_n = F \cap [-n, n]$ for every $n = 1, 2, \ldots$. It is evident that each F_n is measurable and compact. Furthermore, we have
$$F_1 \subseteq F_2 \subseteq \cdots \quad \text{and} \quad F = \bigcup_{n=1}^{\infty} F_n.$$
As a consequence of Theorem 12.12 (Continuity of Lebesgue Measure), we find that
$$\lim_{n \to \infty} m(F_n) = m(F).$$
Thus there exists an $N \in \mathbb{N}$ such that
$$m(F \setminus [-N, N]) = m(F \setminus F_N) < \frac{\epsilon}{2}. \tag{13.35}$$
Let $K = F_N$. Then it follows from the inequalities (13.34) and (13.35) that
$$m(E \setminus K) = m((E \setminus F) \cup (F \setminus K)) \leq m(E \setminus F) + m(F \setminus K) < \epsilon$$
which proves our claim.

Suppose that $\{V_n\}$ is an enumeration of the open intervals with rational endpoints in $\mathbb{R}$. It is clear that $\{V_n\}$ forms a basis of a topology of $\mathbb{R}$.[d] Fix an n. By Theorem 13.3, both $f^{-1}(V_n)$ and $E \setminus f^{-1}(V_n)$ are measurable. Since $f^{-1}(V_n)$ and $E \setminus f^{-1}(V_n)$ are subsets of E, they are of finite measures. Now the previous claim can be applied to conclude that there are compact sets $K_n \subseteq f^{-1}(V_n)$ and $K'_n \subseteq E \setminus f^{-1}(V_n)$ such that
$$m(E \setminus K_n) < \frac{\epsilon}{2^{n+1}} \quad \text{and} \quad m(E \setminus K'_n) < \frac{\epsilon}{2^{n+1}}$$
and they imply that
$$m(E \setminus (K_n \cup K'_n)) < \frac{\epsilon}{2^n}. \tag{13.36}$$
Set $K_\epsilon = \bigcap_{n=1}^{\infty} (K_n \cup K'_n)$. Thus the estimate (13.36) give
$$m(E \setminus K_\epsilon) = m\Big(\bigcup_{n=1}^{\infty} [E \setminus (K_n \cup K'_n)] \Big) \leq \sum_{n=1}^{\infty} m(E \setminus (K_n \cup K'_n)) < \epsilon.$$

---
[d] See the definition on [16, p. 78].

This proves the first assertion.

To show that $f|_{K_\epsilon}$ is continuous, we recall from [16, Theorem 18.1, p. 104] that it is equivalent to showing that for each $x \in K_\epsilon$ and each neighborhood W of $f(x)$, there is a neighborhood U of x in E such that
$$f|_{K_\epsilon}(U) = f(K_\epsilon \cap U) \subseteq W. \tag{13.37}$$
To verify this, given $x \in K_\epsilon$ and W a neighborhood of $f(x)$. We first notice that since $\{V_n\}$ is a basis of a topology of $\mathbb{R}$, there exists a basis element V_N such that
$$f(x) \in V_N \subseteq W.$$
Next, since $x \in f^{-1}(V_N)$ and $x \in K_\epsilon$, we gain $x \in K_N$. Let $U = E \setminus K'_N$. Since K'_N is compact, it is closed in E and then U is open in E. Recall the fact that $K_N \cap K'_N = \varnothing$, so we also have $x \in U$. Thus we establish
$$x \in U \cap K_\epsilon,$$
i.e., the set $U \cap K_\epsilon$ is a neighborhood of x in K_ϵ. Now for every $y \in U \cap K_\epsilon$, we have $y \in U$ so that $y \notin K'_N$. Since K_ϵ also contains y, the definition of K_ϵ implies that $y \in K_N$ and so
$$y \in f^{-1}(V_N),$$
i.e., $f(y) \in V_N \subseteq W$. Hence we conclude that
$$f|_{K_\epsilon}(U) = f(U \cap K_\epsilon) \subseteq V_N \subseteq W$$
which is exactly the set relation (13.37). This completes the proof of the problem. ∎

Problem 13.28

(⋆)(⋆) Suppose that E is measurable with $m(E) < \infty$ and $f : E \to \mathbb{R}$ is measurable. Given $\epsilon > 0$. Prove that there exists a step function $g : \mathbb{R} \to \mathbb{R}$ such that
$$m(\{x \in E \mid |f(x) - g(x)| \geq \epsilon\}) < \epsilon.$$

Proof. By Problem 13.27, there exists a compact set $K \subseteq E$ such that
$$m(E \setminus K) < \epsilon \quad \text{and} \quad f|_K \text{ is continuous.}$$
As a compact set, K is bounded so that we can choose an $N \in \mathbb{N}$ such that K is a *proper* subset of $[-N, N]$ and $N \notin K$. Furthermore, $f|_K$ is uniformly continuous on K. Therefore, one can choose a $\delta \in (0, \epsilon)$ such that for all $p, q \in K$ with $|p - q| < \delta$, we have
$$|f(p) - f(q)| < \epsilon. \tag{13.38}$$
Pick $n > \frac{1}{\delta}$. Let $x_k = -N + \frac{k}{n}$, where $k = 0, 1, 2, \ldots, 2nN$. Next, we denote
$$S = \{k \in \{0, 1, \ldots, 2nN - 1\} \mid [x_k, x_{k+1}) \cap K \neq \varnothing\}.$$
Now for each $k \in S$, we choose $p_k \in [x_k, x_{k+1}) \cap K$ and define $g : \mathbb{R} \to \mathbb{R}$ by
$$g(x) = \sum_{k \in S} f(p_k) \chi_{[x_k, x_{k+1})}(x)$$

13.3. Applications of Littlewood's Three Principles

which is a step function.

We observe that if $q \in K$, then there is a $t \in \{0, 1, \ldots, 2nN - 1\}$ such that $q \in [x_t, x_{t+1})$ and

$$|p_t - q| \le |x_{t+1} - x_t| = \frac{1}{n} < \delta.$$

Recall that $\chi_{[x_i, x_{i+1})}(q) = 1$ which implies that $g(q) = f(p_t)$. Using this fact and the inequality (13.38), we see that

$$|f(q) - g(q)| = |f(q) - f(p_t)| < \epsilon. \tag{13.39}$$

Since the inequality (13.39) holds for all $q \in K$, it implies that

$$m(\{x \in E \,|\, |f(x) - g(x)| \ge \epsilon\}) \le m(E \setminus K) < \epsilon$$

as required. This ends the proof of the problem. ∎

Problem 13.29

$(\star)(\star)$ Suppose that E is measurable with $m(E) < \infty$ and $f : E \to \mathbb{R}$. Prove that there corresponds a F_σ set K and a sequence $\{f_n\}$ of continuous functions on K such that

$$m(E \setminus K) = 0 \quad \text{and} \quad f_n \to f \text{ pointwise on } K.$$

Proof. For each $i \in \mathbb{N}$, we deduce from Theorem 13.9 (Lusin's Theorem) that there exists a closed set $K'_i \subseteq E$ such that

$$m(E \setminus K'_i) < \frac{1}{i} \quad \text{and} \quad f \text{ is continuous on } K'_i. \tag{13.40}$$

For every $n \in \mathbb{N}$, we define

$$K_n = \bigcup_{i=1}^{n} K'_i \tag{13.41}$$

which is clearly a closed set in $\mathbb{R}$. Now our results (13.40) imply that f is continuous on K_n. Finally, we set

$$K = \bigcup_{n=1}^{\infty} K_n$$

which is a F_σ set by the definition. By the inequality (13.40), we know that

$$m(E \setminus K_1) = m(E \setminus K'_1) < 1.$$

Besides, we have

$$\bigcap_{j=1}^{n+1}(E \setminus K_j) \subseteq \bigcap_{j=1}^{n}(E \setminus K_j)$$

for every $n = 1, 2, \ldots$, we find first from Theorem 12.12 (Continuity of Lebesgue Measure) and then using the definition (13.41) plus the inequality (13.40) to conclude that

$$m(E \setminus K) = m\Big(E \setminus \bigcup_{n=1}^{\infty} K_n\Big)$$

$$= m\Big(\bigcap_{n=1}^{\infty}(E \setminus K_n)\Big)$$
$$= \lim_{n \to \infty} m(E \setminus K_n)$$
$$= \lim_{n \to \infty} m\Big(\bigcap_{i=1}^{n}(E \setminus K_i')\Big)$$
$$\leq \lim_{n \to \infty} m(E \setminus K_n')$$
$$\leq \lim_{n \to \infty} \frac{1}{n}$$

and then $m(E \setminus K) = 0$.

Recall that f is continuous on the closed set K_n, so [23, Exercise 4.5, p. 99] implies that there exists a continuous function $f_n : \mathbb{R} \to \mathbb{R}$ such that

$$f_n(x) = f(x) \tag{13.42}$$

on K_n. By the definition (13.41), we have $K_n \subseteq K_{n'}$ if $n' \geq n$. Combining this observation and the result (13.42), we see that for every $x \in K$, there exists an $N \in \mathbb{N}$ such that $x \in K_N$ and therefore the equation (13.42) holds for all $n \geq N$. This gives the second assertion which completes the analysis of the problem. ∎

CHAPTER 14

Lebesgue Integration

14.1 Fundamental Concepts

In Chapter 13, we have defined the notion of Lebesgue measurable functions. Now we are ready to define and study the theory of Lebesgue integration. The main references that we have applied are [7, Chap. 2], [12, Chap. 6], [22, Chap. 4, 5], [23, Chap. 11], [24, Chap. 1], [26, Chap. 2] and [27, Chap. 8].

14.1.1 Integration of Nonnegative Functions

Our starting point is the integration of a simple function.

Definition 14.1 (Integration of Simple Functions). *Suppose that $s : \mathbb{R} \to [0, \infty)$ is a simple function in the form*

$$s(x) = \sum_{k=1}^{n} c_k \chi_{E_k}(x),$$

where $E_1, E_2, \ldots, E_n$ are measurable sets and $c_1, c_2, \ldots, c_k$ are distinct constants. Then the **Lebesgue integral** of s on a measurable set E is given by

$$\int_E s \, dm = \sum_{k=1}^{n} c_k m(E_k \cap E).$$

Next, with the aid of Theorem 13.6 (The Simple Function Approximation Theorem), we have the integration of a nonnegative function.

Definition 14.2 (Integration of Nonnegative Functions). *Suppose that $f : \mathbb{R} \to [0, \infty]$ is measurable. Then the **Lebesgue integral** of f on a measurable set E is given by*

$$\int_E f \, dm = \sup \int_E s \, dm,$$

where the supremum is taken over all simple measurable functions s with $0 \leq s \leq f$.

The following theorem lists some useful and important properties of the Lebesgue integration of nonnegative functions.

Theorem 14.3. Let $f, g : \mathbb{R} \to [0, \infty]$ and $E, F, E_1, E_2, \ldots \subseteq \mathbb{R}$ be measurable. Then we have

(a) If $0 \leq f \leq g$, then we have
$$\int_E f \, dm \leq \int_E g \, dm.$$

(b) If $a \leq f(x) \leq b$ for all $x \in E$, then we have
$$am(E) \leq \int_E f \, dm \leq bm(E).$$

(c) If $F \subseteq E$, then we have
$$\int_F f \, dm \leq \int_E f \, dm.$$

(d) If $A, B \in [0, \infty)$, then we have
$$\int_E (Af + Bg) \, dm = A \int_E f \, dm + B \int_E g \, dm.$$

(e) If $E = \bigcup_{k=1}^{\infty} E_k$, where $E_1, E_2, \ldots$ are pairwise disjoint, then we have
$$\int_E f \, dm = \sum_{k=1}^{\infty} \int_{E_k} f \, dm.$$

(f) If $m(E) = 0$ or $f(x) = 0$ for almost every $x \in E$, then we have
$$\int_E f \, dm = 0.$$

(g) If $\int_E f \, dm = 0$ and $m(E) > 0$, then we have $f(x) = 0$ for almost every $x \in E$.

14.1.2 Integration of Lebesgue Integrable Functions

Definition 14.4 (Lebesgue Integrable Functions). If f is a measurable function defined on the measurable set $E \subseteq \mathbb{R}$ and if
$$\int_E |f| \, dm < \infty,$$
then f is called a **Lebesgue integrable function**. The class of all Lebesgue integrable functions defined on E is denoted by $L^1(E)$.

14.1. Fundamental Concepts

> **Remark 14.1**
>
> Recall from Remark 13.3 that the functions f^+ and f^- are nonnegative and $f = f^+ - f^-$. Since $f^\pm \leq |f|$, both functions f^+ and f^- are Lebesgue integrable provided that $f \in L^1(E)$ and then we can define the **Lebesgue integral** of f by
>
> $$\int_E f \, dm = \int_E f^+ \, dm - \int_E f^- \, dm.$$

Theorem 14.5 (Properties of Integrable Functions). *Suppose that E and F are measurable sets with $F \subseteq E$. Then we have the following properties:*

(a) *If $f, g \in L^1(E)$ and $A, B \in \mathbb{R}$, then we have $Af + Bg \in L^1(E)$ and*

$$\int_E (Af + Bg) \, dm = A \int_E f \, dm + B \int_E g \, dm.$$

(b) *If $f, g \in L^1(E)$ and $f(x) \leq g(x)$ on E, then we have*

$$\int_E f \, dm \leq \int_E g \, dm.$$

(c) *If $f \in L^1(E)$, then we have $f \in L^1(F)$.*

(d) *If $f \in L^1(E)$, then $|f| \in L^1(E)$ and we have*

$$\left| \int_E f \, dm \right| \leq \int_E |f| \, dm.$$

If f is bounded by a positive constant M on the measurable set E with $m(E) < \infty$, then we have $|f| \leq M$ and Theorem 14.3(b) implies that

$$\int_E |f| \, dm \leq M m(E) < \infty.$$

In this case, it is trivial that $f \in L^1(E)$ and the inequality in Theorem 14.5(d) (Properties of Integrable Functions) also holds.

14.1.3 Fatou's Lemma and Convergence Theorems

There are several elementary but remarkable results about integration of sequences of measurable functions.

Fatou's Lemma. *If each $f_n : \mathbb{R} \to [0, \infty]$ is measurable, then we have*

$$\int_E \left(\liminf_{n \to \infty} f_n \right) dm \leq \liminf_{n \to \infty} \int_E f_n \, dm.$$

The Bounded Convergence Theorem. *Let $\{f_n\}$ be a sequence of measurable functions defined on the measurable set E with $m(E) < \infty$. If $f_n \to f$ pointwise on E and $\{f_n\}$ is uniformly bounded[a], then we have*

$$\lim_{n \to \infty} \int_E f_n \, dm = \int_E f \, dm.$$

---
[a] See Definition 10.9 (Pointwise Boundedness and Uniformly Boundedness).

The Lebesgue's Monotone Convergence Theorem. *Suppose that E is measurable and $\{f_n\}$ is a sequence of measurable functions such that*

$$0 \leq f_1(x) \leq f_2(x) \leq \cdots$$

on E. If $f_n(x) \to f(x)$ as $n \to \infty$ for almost all $x \in E$, then we have

$$\lim_{n \to \infty} \int_E f_n \, dm = \int_E f \, dm.$$

As an immediate application of the Lebesgue's Monotone Convergence Theorem, if every $f_n : \mathbb{R} \to [0, \infty]$ is measurable for $n = 1, 2, \ldots$ and

$$f(x) = \sum_{n=1}^{\infty} f_n(x)$$

exists a.e. on E, then we have

$$\int_E f \, dm = \sum_{n=1}^{\infty} \int_E f_n \, dm.$$

The Lebesgue's Dominated Convergence Theorem. *Suppose that E is measurable and $\{f_n\}$ is a sequence of measurable functions such that*

$$f(x) = \lim_{n \to \infty} f_n(x)$$

exists a.e. on E. If there exists a function $g \in L^1(E)$ such that

$$|f_n(x)| \leq g(x)$$

for all $n \in \mathbb{N}$ and a.e. on E, then we have $f \in L^1(E)$ and

$$\lim_{n \to \infty} \int_E f_n \, dm = \int_E f \, dm.$$

The series version of the Lebesgue's Dominated Convergence Theorem is the following: suppose that $\{f_n\}$ is a sequence of measurable functions defined a.e. on E and

$$\sum_{n=1}^{\infty} \int_E |f_n| \, dm < \infty.$$

Then the series

$$f(x) = \sum_{n=1}^{\infty} f_n(x)$$

converges for almost all $x \in E$, $f \in L^1(E)$ and

$$\int_E f \, dm = \sum_{n=1}^{\infty} \int_E f_n \, dm.$$

14.1.4 Connections between Riemann Integrals and Lebesgue Integrals

To distinguish Riemann integrals from Lebesgue integrals, we write the former by

$$\mathscr{R}\int_a^b f\,\mathrm{d}x.$$

Theorem 14.6. *Let $a, b \in \mathbb{R}$ and $a < b$.*

(a) *If $f \in \mathscr{R}$ on $[a,b]$, then $f \in L^1([a,b])$ and*

$$\int_{[a,b]} f\,\mathrm{d}m = \int_a^b f\,\mathrm{d}m = \mathscr{R}\int_a^b f\,\mathrm{d}x.$$

(b) *Let f be bounded on $[a,b]$. Then $f \in \mathscr{R}$ on $[a,b]$ if and only if f is continuous a.e. on $[a,b]$.*

14.2 Properties of Integrable Functions

Problem 14.1

(⋆) For every $n \in \mathbb{N}$, calculate the Lebesgue integral

$$\int_\mathbb{R} \frac{1}{n}\chi_{[0,n)}\,\mathrm{d}m.$$

Proof. Since $[0,n)$ is measurable, we obtain from Definition 14.1 (Integration of Simple Functions) directly that

$$\int_\mathbb{R} \frac{1}{n}\chi_{[0,n)}\,\mathrm{d}m = \frac{1}{n}m([0,n)\cap\mathbb{R}) = \frac{1}{n}m([0,n)) = 1$$

which completes the proof of the problem. ∎

Problem 14.2

(⋆) Let $-\infty < a < b < \infty$ and consider the function $f : [a,b] \to \mathbb{R}$ defined by

$$f(x) = \begin{cases} x, & \text{if } x \in [a,b]\setminus\mathbb{Q}; \\ 1, & \text{otherwise.} \end{cases}$$

Evaluate the Lebesgue integral of f on $[a,b]$.

Proof. Recall that $m(\mathbb{Q}) = 0$, so $m([a,b] \cap \mathbb{Q}) = 0$. Then we gain from Theorem 14.3(f) that

$$\int_{[a,b]} f \, \mathrm{d}m = \int_{[a,b]} x \, \mathrm{d}m.$$

Next, we conclude from Theorem 9.4 (The Lebesgue's Integrability Condition) that $f \in \mathscr{R}$ on $[a,b]$. Thus it follows from Theorem 14.6(a) that $f \in L^1([a,b])$ and

$$\int_{[a,b]} f \, \mathrm{d}m = \mathscr{R} \int_a^b x \, \mathrm{d}x = \frac{b-a}{2},$$

completing the proof of the problem. ∎

Problem 14.3

(⋆) *Prove that the collection $\mathscr{S}$ of simple functions defined on a set $E \subseteq \mathbb{R}$ is closed under addition.*

Proof. Let $x \in E$. Suppose that

$$s(x) = \sum_{i=1}^{n} a_i \chi_{A_i}(x) \quad \text{and} \quad t(x) = \sum_{j=1}^{m} b_j \chi_{B_j}(x), \tag{14.1}$$

where $A_1, A_2, \ldots, A_n$ and $B_1, B_2, \ldots, B_m$ are pairwise disjoint and $\bigcup_{i=1}^{n} A_i = \bigcup_{j=1}^{m} B_j = E$. Define $C_{ij} = A_i \cap B_j$ for $1 \le i \le n$ and $1 \le j \le m$. It is clear that

$$A_i = A_i \cap \bigcup_{j=1}^{m} B_j = \bigcup_{j=1}^{m} (A_i \cap B_j) = \bigcup_{j=1}^{m} C_{ij},$$

where $1 \le i \le n$. Similarly, we have

$$B_j = \bigcup_{j=1}^{n} C_{ij},$$

where $1 \le j \le m$. By the definition, it is true that

$$\bigcup_{i=1}^{n} \bigcup_{j=1}^{m} C_{ij} = E.$$

Furthermore, as $\{C_{ij}\}$ is a collection of pairwise disjoint sets, this means that

$$\chi_{A_i} = \sum_{j=1}^{m} \chi_{C_{ij}} \quad \text{and} \quad \chi_{B_j} = \sum_{i=1}^{n} \chi_{C_{ij}}. \tag{14.2}$$

By combining the expressions (14.1) and (14.2), we get

$$s(x) = \sum_{i=1}^{n} a_i \sum_{j=1}^{m} \chi_{C_{ij}}(x) \quad \text{and} \quad t(x) = \sum_{j=1}^{m} b_j \sum_{i=1}^{n} \chi_{C_{ij}}(x)$$

14.2. Properties of Integrable Functions

which imply that
$$s(x) + t(x) = \sum_{i=1}^{n} \sum_{j=1}^{m} (a_i + b_j) \chi_{C_{ij}}(x).$$

This completes the proof of the problem. ∎

> **Remark 14.2**
>
> In fact, it can be shown further that the set $\mathscr{S}$ is also closed under multiplication.

> **Problem 14.4**
>
> (⋆)(⋆) Let f be a bounded and nonnegative measurable function defined on $[0,1]$. Prove that
> $$\int_{[0,1]} f \, dm = \inf \int_{[0,1]} t \, dm, \tag{14.3}$$
> where the infimum is taken over all simple measurable functions such that $f \leq t$.

Proof. Denote S and $u = \sup S$ to be a set and its supremum. Let $-S = \{-x \mid x \in S\}$. We claim that
$$\inf(-S) = -\sup S. \tag{14.4}$$

On the one hand, we know from the definition that $x \leq u$ for all $x \in S$ so that $-x \geq -u$. Thus $-u$ is a lower bound for the set $-S$, i.e.,
$$\inf(-S) \geq -u = -\sup S. \tag{14.5}$$

On the other hand, let $v = \inf(-S)$. Then we have $v \leq -x$ for all $x \in S$. Equivalently, $-v \geq x$ for all $x \in S$. Therefore, $-v$ is an upper bound of S and this implies that
$$-\inf(-S) = -v \geq \sup S. \tag{14.6}$$

Combining the inequalities (14.5) and (14.6), we arrive at the claim (14.4).

Since f is bounded, there exists a $M > 0$ such that $0 \leq f(x) \leq M$ on $[0,1]$. Define $g : [0,1] \to \mathbb{R}$ by
$$g = M - f$$
which is a bounded (by $2M$) and nonnegative measurable function defined on $[0,1]$. Let t be a simple function such that $0 \leq f \leq t$. If we consider $t' = \min(t, M)$, then we have
$$0 \leq t' \leq M \quad \text{and} \quad f \leq t' \leq t.$$

Since $t' = \frac{1}{2}(t + M - |t - M|)$, Problem 14.3 implies that t' is simple. Note that
$$\int_{[0,1]} t' \, dm \leq \int_{[0,1]} t \, dm$$

and so
$$\inf_{f \leq t' \leq M} \int_{[0,1]} t' \, dm = \inf_{f \leq t} \int_{[0,1]} t \, dm.$$

Hence, in order to prove the representation (14.3), it suffices to consider simple functions t' such that $f \leq t' \leq M$.

Next, we know from Definition 14.2 (Integration of Nonnegative Functions) that
$$M - \int_{[0,1]} f\,dm = \int_{[0,1]} M\,dm - \int_{[0,1]} f\,dm = \int_{[0,1]} g\,dm = \sup \int_{[0,1]} s\,dm, \qquad (14.7)$$
where the supremum is taken over all simple measurable functions s with $0 \leq s \leq g$. Now $t = M - s$ is simple by Problem 14.3 and $f \leq t \leq M$. Conversely, if t is a simple function such that $f \leq t \leq M$, then $s = M - t$ is also a simple function with $0 \leq s \leq g$. Thus it follows from the formula (14.4) and the expression (14.7) that
$$\int_{[0,1]} f\,dm = \int_{[0,1]} (M - g)\,dm$$
$$= M - \sup_{0 \leq s \leq g} \int_{[0,1]} s\,dm$$
$$= \inf_{0 \leq s \leq g} \int_{[0,1]} M\,dm + \inf_{0 \leq s \leq g} \left(- \int_{[0,1]} s\,dm \right)$$
$$= \inf_{0 \leq s \leq g} \int_{[0,1]} (M - s)\,dm$$
$$= \inf_{f \leq t \leq M} \int_{[0,1]} t\,dm$$
which is our desired result. Hence, this completes the proof of the problem. ∎

Problem 14.5 (Chebyshev's Inequality)

(★) Suppose that f is a nonnegative and integrable function on a measurable set $E \subseteq \mathbb{R}$. If $\alpha > 0$, then prove that
$$m(\{x \in E \,|\, f(x) \geq \alpha\}) \leq \frac{1}{\alpha} \int_E f\,dm. \qquad (14.8)$$

Proof. Let $F = \{x \in E \,|\, f(x) \geq \alpha\}$. Then F is measurable. If $m(F) = \infty$, then $m(E) = \infty$ and Theorem 14.3(b) implies that
$$\int_E f\,dm \geq \alpha \cdot m(E) = \infty$$
which contradicts Theorem 14.5(d) (Properties of Integrable Functions). In other words, we have $m(F) < \infty$. Now it is easy to see from Theorem 14.3(e) that
$$\int_E f\,dm = \int_F f\,dm + \int_{E \setminus F} f\,dm \geq \int_F f\,dm \geq \alpha \cdot m(F)$$
which is exactly the inequality (14.8). This ends the analysis of the problem. ∎

Problem 14.6

(★) Suppose that $f : E \to [0, \infty)$ is measurable, where E is a measurable set. If $f \in L^1(E)$, prove that f is finite a.e. on E.

14.2. Properties of Integrable Functions

Proof. Let $\alpha > 0$. Denote $F_\alpha = \{x \in E \mid f(x) \geq \alpha\}$ and $F_\infty = \{x \in E \mid f(x) = \infty\}$. Obviously, we have $F_\infty \subseteq F_n$ and $F_{n+1} \subseteq F_n$ for all $n \in \mathbb{N}$. In addition, we observe from Problem 14.5 and the integrability of f that

$$m(F_1) \leq \int_E f\,dm < \infty.$$

Hence Theorem 12.12 (Continuity of Lebesgue Measure) and Problem 14.5 imply that

$$m(F_\infty) \leq \lim_{n \to \infty} m(F_n) \leq \lim_{n \to \infty} \frac{1}{n} \int_E f\,dm = 0,$$

i.e., f is finite a.e. on E which ends the proof of the problem. ∎

Problem 14.7

(⋆) Suppose that $E \subseteq \mathbb{R}$ is measurable and $f \in L^1(E)$ satisfies

$$\left| \int_F f\,dm \right| < Cm(F)$$

for every measurable subset F of E with finite measure, where C is a positive constant. Prove that

$$|f(x)| < C$$

a.e. on E.

Proof. If $m(E) = 0$, then there is nothing to prove. Thus we assume that $m(E) > 0$. Since $f \in L^1(E)$, we have $|f| \in L^1(E)$ by Theorem 14.5(d) (Properties of Integrable Functions). Now Problem 14.5 guarantees that

$$m(\{x \in E \mid |f(x)| \geq C\}) \leq \frac{1}{C} \int_E |f|\,dm < \infty. \tag{14.9}$$

Consider the sets $F_1 = \{x \in E \mid f(x) \geq C\}$ and $F_2 = \{x \in E \mid f(x) \leq -C\}$. By the estimate (14.9), both $m(F_1)$ and $m(F_2)$ are finite.

On the set F_1, the hypothesis implies that

$$Cm(F_1) \leq \int_{F_1} f\,dm = \left| \int_{F_1} f\,dm \right| < Cm(F_1)$$

so that $m(F_1) = 0$. On the set F_2, the hypothesis gives

$$Cm(F_2) \leq \left| \int_{F_2} f\,dm \right| < Cm(F_2)$$

which shows again that $m(F_2) = 0$. Equivalently, we conclude that $|f(x)| < C$ a.e. on E and this completes the proof of the problem. ∎

Problem 14.8

(⋆) Suppose that E and f are measurable and $\int_E |f|\,dm = 0$. Prove that $f(x) = 0$ a.e. on E.

Proof. For every $n = 1, 2, \ldots$, we let $E_n = \{x \in E \mid |f(x)| \geq \frac{1}{n}\}$. On the one hand, we have

$$\{x \in E \mid f(x) \neq 0\} = \bigcup_{n=1}^{\infty} E_n,$$

so we deduce from the inequality (12.1) that

$$m(\{x \in E \mid f(x) \neq 0\}) \leq \sum_{n=1}^{\infty} m(E_n). \tag{14.10}$$

On the other hand, we know from Problem 14.5 that

$$m(E_n) \leq n \int_E |f| \, \mathrm{d}m = 0. \tag{14.11}$$

Hence we conclude from the inequalities (14.10) and (14.11) that

$$m(\{x \in E \mid f(x) \neq 0\}) = 0$$

which is equivalent to the condition $f(x) = 0$ a.e. on E, completing the proof of the problem. ∎

Problem 14.9

(★) (★) Let $f : [0,1] \to \mathbb{R}$ be a measurable function. Prove that $f \in L^1([0,1])$ if and only if

$$\sum_{n=1}^{\infty} 2^n \cdot m(\{x \in [0,1] \mid |f(x)| \geq 2^n\}) < \infty. \tag{14.12}$$

Proof. Define $E_0 = \{x \in [0,1] \mid |f(x)| < 2\}$ and $g(x) = 1$ on E_0. Next, for each $n \in \mathbb{N}$, we define

$$E_n = \{x \in [0,1] \mid 2^n \leq |f(x)| < 2^{n+1}\} \quad \text{and} \quad g(x) = 2^n \text{ on } E_n. \tag{14.13}$$

It is easy to check that

$$m(E_0) \leq 1 \quad \text{and} \quad m(E_n) \leq m(\{x \in [0,1] \mid |f(x)| \geq 2^n\}) \tag{14.14}$$

for all $n \in \mathbb{N}$. Furthermore, it is also true that $0 \leq g(x) - 1 < |f(x)| \leq 2g(x)$ for all $x \in [0,1]$, thus we get from Theorem 14.3(a) that

$$\int_{[0,1]} (g-1) \, \mathrm{d}m \leq \int_{[0,1]} |f| \, \mathrm{d}m < 2 \int_{[0,1]} g \, \mathrm{d}m$$

and these mean that $f \in L^1([0,1])$ if and only if $g \in L^1([0,1])$.

We claim that $g \in L^1([0,1])$ if and only if the condition (14.12) holds. To see this, we note from the definition (14.13) that $E_1, E_2, \ldots$ are pairwise disjoint and

$$\{x \in [0,1] \mid |f(x)| \geq 2^n\} = E_n \cup E_{n+1} \cup \cdots,$$

where $n = 1, 2, \ldots$. Therefore, we follow from Theorem 12.11 (Countably Additivity) that

$$m(\{x \in [0,1] \mid |f(x)| \geq 2^n\}) = \sum_{k=n}^{\infty} m(E_k), \tag{14.15}$$

14.2. Properties of Integrable Functions

where $n = 1, 2, \ldots$. Now we establish from Theorem 14.3(e), the relations (14.14) and finally the hypothesis (14.12) that

$$\int_{[0,1]} g\, dm = \sum_{n=0}^{\infty} \int_{E_n} g\, dm$$
$$= \sum_{n=0}^{\infty} 2^n \cdot m(E_n)$$
$$\leq 1 + \sum_{n=1}^{\infty} 2^n \cdot m(\{x \in [0,1]\,|\,|f(x)| \geq 2^n\}). \tag{14.16}$$

On the other hand, we observe from the expression (14.15) that

$$\sum_{n=1}^{\infty} 2^n \cdot m(\{x \in [0,1]\,|\,|f(x)| \geq 2^n\}) = \sum_{n=1}^{\infty} 2^n \cdot \sum_{k=n}^{\infty} m(E_k)$$
$$= \sum_{k=1}^{\infty} m(E_k) \sum_{n=1}^{k} 2^n$$
$$\leq \sum_{k=1}^{\infty} 2^{k+1} m(E_k)$$
$$\leq 2 \sum_{k=0}^{\infty} 2^k m(E_k)$$
$$= 2 \int_{[0,1]} g\, dm. \tag{14.17}$$

By the inequalities (14.16) and (14.17), we see that

$$\frac{1}{2} \sum_{n=1}^{\infty} 2^n \cdot m(\{x \in [0,1]\,|\,|f(x)| \geq 2^n\}) \leq \int_{[0,1]} g\, dm \leq 1 + \sum_{n=1}^{\infty} 2^n \cdot m(\{x \in [0,1]\,|\,|f(x)| \geq 2^n\})$$

which implies that $g \in L^1([0,1])$ if and only if the hypothesis (14.12) holds. This completes the proof of the problem. ∎

Problem 14.10

(★)(★) Let E be measurable with $m(E) < \infty$ and f be nonnegative and integrable on E. Given $\epsilon > 0$, define
$$E_n = \{x \in E \,|\, n\epsilon \leq f(x) < (n+1)\epsilon\}$$
for every $n = 0, 1, 2, \ldots$ and
$$A(\epsilon) = \epsilon \sum_{n=0}^{\infty} n \cdot m(E_n).$$
Prove that
$$\int_E f\, dm = \lim_{\epsilon \to 0} A(\epsilon).$$

Proof. Define $F = \{x \in E \mid f(x) = \infty\}$. By the definition, we know that $E_1, E_2, \ldots$ are pairwise disjoint and
$$E = F \cup \bigcup_{n=0}^{\infty} E_n.$$
On the one hand, we obtain from Theorem 14.3(e) and Problem 14.6 that
$$A(\epsilon) = \epsilon \sum_{n=0}^{\infty} n \cdot m(E_n) = \sum_{n=0}^{\infty} \int_{E_n} (n\epsilon) \, dm \le \sum_{n=0}^{\infty} \int_{E_n} f \, dm = \int_{E \setminus F} f \, dm = \int_E f \, dm. \quad (14.18)$$
On the other hand, we see from Theorem 14.3(e) and Problem 14.6 again that
$$A(\epsilon) + \epsilon m(E) = \epsilon \sum_{n=0}^{\infty} n \cdot m(E_n) + \epsilon \cdot \sum_{n=0}^{\infty} m(E_n)$$
$$= \sum_{n=0}^{\infty} (n+1)\epsilon m(E_n)$$
$$= \sum_{n=0}^{\infty} \int_{E_n} (n+1)\epsilon \, dm$$
$$> \sum_{n=0}^{\infty} \int_{E_n} f \, dm$$
$$= \int_{E \setminus F} f \, dm$$
$$= \int_E f \, dm. \quad (14.19)$$
Hence we deduce from the inequalities (14.18) and (14.19) that
$$\int_E f \, dm - \epsilon m(E) < A(\epsilon) \le \int_E f \, dm$$
which implies the desired result by letting $\epsilon \to 0$. This completes the proof of the problem. ∎

Problem 14.11

(⋆) Let E be a measurable set and f be measurable on E. If there is a nonnegative function $g \in L^1(E)$ such that
$$|f(x)| \le g(x)$$
on E, prove that $f \in L^1(E)$.

Proof. By Theorem 14.3(a) and the hypothesis, we have
$$\int_E |f| \, dm \le \int_E g \, dm < \infty,$$
i.e., $f \in L^1(E)$ which completes the proof of the problem. ∎

14.2. Properties of Integrable Functions

> **Problem 14.12**
>
> (★) Let $E \in \mathscr{B}$ and f be integrable on E. Define
> $$\varphi(E) = \int_E f \, \mathrm{d}m$$
> for every $E \in \mathscr{B}$. Prove that $\varphi(E) = 0$ for all $E \in \mathscr{B}$ if and only if $f = 0$ a.e. on $\mathbb{R}$.

Proof. If $f = 0$ a.e. on $\mathbb{R}$, then it follows form Theorem 14.3(f) that

$$\varphi(E) = \int_E f \, \mathrm{d}m = 0$$

for every $E \in \mathscr{B}$. Conversely, consider $E = \{x \in \mathbb{R} \,|\, f(x) \geq 0\}$. Then we have $f^+ = f \cdot \chi_E$ so that

$$\int_{\mathbb{R}} f^+ \, \mathrm{d}m = \int_E f \, \mathrm{d}m = \varphi(E) = 0.$$

In other words, we have $f^+ = 0$ a.e. on $\mathbb{R}$ by Theorem 14.3(g). Similarly, we can show that

$$\int_{\mathbb{R}} f^- \, \mathrm{d}m = 0$$

so that $f^- = 0$ a.e. on $\mathbb{R}$. By the definition, $f = f^+ - f^- = 0$ a.e. on $\mathbb{R}$ which completes the proof of the problem. ∎

> **Problem 14.13**
>
> (★) Suppose that E is measurable and $f, g, h : E \to \mathbb{R}^*$ are measurable. If g and h are integrable on E and $g \leq f \leq h$ a.e. on E, prove that f is integrable on E.

Proof. Since h and g are integrable on E, $h - g$ is integrable on E by Theorem 14.5(a) (Properties of Integrable Functions). Obviously, we have

$$0 \leq f - g \leq h - g$$

a.e. on E, so Problem 14.11 implies that $f - g$ is also integrable on E. By Theorem 14.5(a) (Properties of Integrable Functions) again, the function

$$f = (f - g) + g$$

is integrable on E. This completes the proof of the problem. ∎

> **Problem 14.14**
>
> (★) Suppose that E is measurable, $E_1, E_2, \ldots$ are measurable subsets of E and $f : E \to \mathbb{R}^*$ is a measurable function. Let $F = \bigcup_{n=1}^{\infty} E_n$. If f is integrable on each E_n and $\sum_{n=1}^{\infty} \int_{E_n} |f| \, dm$ converges, prove that f is integrable on F and
> $$\left| \int_F f \, dm \right| \le \sum_{n=1}^{\infty} \int_{E_n} |f| \, dm.$$

Proof. Without loss of generality, we may assume that $E_1, E_2, \ldots$ are pairwise disjoint.[b] Recall from Remark 14.1 that $|f^\pm| \le |f|$ on each E_n, so we have

$$\int_{E_n} f^\pm \, dm \le \int_{E_n} |f| \, dm \tag{14.20}$$

for every $n = 1, 2, \ldots$. Since $f^\pm \ge 0$ on every E_n, we obtain from Theorem 14.3(e), the estimate (14.20) and then the hypothesis that

$$\int_F f^\pm \, dm = \sum_{n=1}^{\infty} \int_{E_n} f^\pm \, dm \le \sum_{n=1}^{\infty} \int_{E_n} |f| \, dm < \infty. \tag{14.21}$$

Recall that $|f| = f^+ + f^-$, so the estimates (14.21) give

$$\int_F |f| \, dm = \int_F f^+ \, dm + \int_F f^- \, dm < \infty.$$

By Definition 14.4 (Lebesgue Integrable Functions), the function f is integrable on F which proves the firs assertion.

For the second assertion, we notice from Theorem 14.5(d) (Properties of Integrable Functions) and then Theorem 14.3(e) that

$$\left| \int_F f \, dm \right| \le \int_F |f| \, dm = \sum_{n=1}^{\infty} \int_{E_n} |f| \, dm.$$

This completes the analysis of the problem. ∎

> **Problem 14.15**
>
> (★)(★) Suppose that E is measurable with $m(E) < \infty$ and $f, g : E \to \mathbb{R}$ are measurable. If $f, g \in L^1(E)$ and
> $$\int_F f \, dm \le \int_F g \, dm$$
> for every measurable set $F \subseteq E$. Verify that $f(x) \le g(x)$ a.e. on E.

---

[b] See the proof of Problem 12.25.

14.2. Properties of Integrable Functions

Proof. Consider
$$E_n = \left\{x \in E \,\Big|\, f(x) \geq g(x) + \frac{1}{n}\right\},$$
where $n = 1, 2, \ldots$. Then we follow from Problems 13.6 and 13.8 that each
$$E_n = \left\{x \in E \,\Big|\, f(x) > g(x) + \frac{1}{n}\right\} \cap \left\{x \in E \,\Big|\, f(x) - g(x) = \frac{1}{n}\right\} \tag{14.22}$$
is measurable. Given $n \in \mathbb{N}$ fixed, since $g \in L^1(E)$, Theorem 14.5(c) and (d) (Properties of Integrable Functions) say that
$$\left| \int_{E_n} g \, dm \right| \leq \int_{E_n} |g| \, dm < \infty. \tag{14.23}$$
Furthermore, we see immediately from the hypothesis and the expression (14.22) that
$$\int_{E_n} g \, dm \geq \int_{E_n} f \, dm \geq \int_{E_n} g \, dm + \frac{1}{n} m(E_n) \tag{14.24}$$
for every $n \in \mathbb{N}$. Consequently, we conclude from the inequalities (14.23) and (14.24) that $m(E_n) = 0$ for each $n = 1, 2, \ldots$. Thus we get from the inequality (12.1) that
$$m\left(\bigcup_{n=1}^{\infty} E_n\right) = 0. \tag{14.25}$$
Let $E' = \{x \in E \mid f(x) > g(x)\}$. If $p \in E'$, then we must have $f(p) > g(p) + \frac{1}{n}$ for some large enough $n \in \mathbb{N}$, i.e., $p \in E_n$. This means, with the aid of the result (14.25), that
$$m(E') = 0$$
and this is equivalent to $f(x) \leq g(x)$ a.e. on E. We have completed the proof of the problem. ∎

Problem 14.16

(★)(★) *Suppose that $f : [0, 1] \to [0, \infty)$ is measurable. Consider $0 < \epsilon \leq 1$ and*
$$S = \left\{ \int_E f \, dm \,\Big|\, E \text{ is a measurable subset of } [0, 1] \text{ with } m(E) \geq \epsilon \right\}.$$
Prove that $\inf(S) > 0$.

Proof. Define $E_0 = \{x \in [0, 1] \mid f(x) \geq 1\}$ and for each $n \in \mathbb{N}$, we define
$$E_n = \left\{x \in [0, 1] \,\Big|\, \frac{1}{n+1} \leq f(x) < \frac{1}{n}\right\}.$$
Then it is trivial that
$$[0, 1] = E_0 \cup \bigcup_{n=1}^{\infty} E_n. \tag{14.26}$$

Since $E_0, E_1, E_2, \ldots$ are pairwise disjoint measurable subsets of $[0,1]$, we apply Theorem 12.11 (Countably Additivity) to the representation (14.26) to get

$$\sum_{n=0}^{\infty} m(E_n) = 1$$

which implies the existence of an $N \in \mathbb{N}$ such that

$$\sum_{n=N}^{\infty} m(E_n) \leq \frac{\epsilon}{2}. \tag{14.27}$$

By Problem 12.27, there exists a measurable subset $F \subseteq [0,1]$ such that $m(F) = \epsilon$. Let

$$F_1 = F \cap \left\{ x \in [0,1] \,\Big|\, f(x) \geq \frac{1}{N+1} \right\} = F \cap \bigcup_{n=N}^{\infty} E_n \quad \text{and} \quad F_2 = F \setminus F_1.$$

Then we see from Problem 12.13 and the estimate (14.27) that

$$m(F_2) = m(F \setminus F_1) = m(F) - m(F_1) \geq \epsilon - \sum_{n=N}^{\infty} m(E_n) \geq \frac{\epsilon}{2} > 0.$$

Hence it yields from Theorem 14.3(a) that

$$\int_F f \,\mathrm{d}m \geq \int_{F_2} f \,\mathrm{d}m \geq \frac{1}{N+1} m(F_2) \geq \frac{\epsilon}{2(N+1)} > 0.$$

In other words, we have $\inf(S) > 0$ which completes the proof of the problem. ∎

Problem 14.17

(⋆) Is it true that if $f, g : \mathbb{R} \to \mathbb{R}$ are integrable, then $f \circ g : \mathbb{R} \to \mathbb{R}$ is integrable?

Proof. The answer is negative. Consider $f = \chi_{[0,1]}$ and $g = \chi_{\{1\}}$. Clearly, we have $f, g \in L^1(\mathbb{R})$. However, we note that

$$\begin{aligned} f(g(x)) &= \chi_{[0,1]}(\chi_{\{1\}}(x)) \\ &= \begin{cases} \chi_{[0,1]}(1), & \text{if } x = 1; \\ \chi_{[0,1]}(0), & \text{otherwise} \end{cases} \\ &= 1 \end{aligned}$$

which implies that $f \circ g \notin L^1(\mathbb{R})$ because

$$\int_{\mathbb{R}} |f(g(x))| \,\mathrm{d}m = m(\mathbb{R}) = \infty.$$

This completes the proof of the problem. ∎

14.2. Properties of Integrable Functions

Problem 14.18

(⋆) Suppose that E is measurable with $m(E) > 0$ and f is measurable on E with

$$\left| \int_E f \, dm \right| = \int_E |f| \, dm.$$

Prove that either $f \geq 0$ a.e. on E or $f \leq 0$ a.e. on E.

Proof. Suppose that

$$\int_E f \, dm \geq 0. \tag{14.28}$$

By the hypothesis, we have

$$\int_E (f - |f|) \, dm = 0,$$

so Theorem 14.3(g) implies that $f(x) - |f(x)| = 0$ a.e. on E and this means that

$$f(x) = |f(x)| \geq 0$$

a.e. on E.

Next, we suppose that

$$\int_E f \, dm \leq 0.$$

In this case, we consider the function $-f$ which is also measurable on E and it satisfies condition (14.28). Thus our previous analysis can be applied to $-f$ to conclude that

$$-f(x) = |-f(x)| \geq 0$$

a.e. on E and this is equivalent to

$$f(x) \leq 0$$

a.e. on E. This completes the proof of the problem. ∎

Problem 14.19

(⋆) Suppose that $f \in L^1((0, \infty))$. Prove that there exists a sequence $x_n \to \infty$ such that

$$\lim_{n \to \infty} x_n f(x_n) = 0.$$

Proof. We put

$$\alpha = \liminf_{x \to \infty} x|f(x)|.$$

If $\alpha = 0$, then since $-x|f(x)| \leq xf(x) \leq x|f(x)|$ on $(0, \infty)$, we have

$$\liminf_{x \to \infty} xf(x) = 0$$

308   Chapter 14. Lebesgue Integration

and, by the definition, such a sequence exists. Next, we suppose that $\alpha > 0$. Then there is an $N \in \mathbb{N}$ such that $x|f(x)| > \frac{\alpha}{2}$ for all $x \geq N$. Consequently, we conclude from Theorem 14.3(c) that
$$\int_{(0,\infty)} |f| \, dm \geq \int_{(N,\infty)} |f| \, dm \geq \int_N^\infty \frac{\alpha}{2x} \, dx = \frac{\alpha}{2} \ln x \Big|_N^\infty = \infty,$$
a contradiction. Hence we have completed the proof of the problem. ∎

Problem 14.20

(⋆) (⋆) Suppose that E is measurable and f is integrable on E. Given $\epsilon > 0$. Prove that there is a $\delta > 0$ such that
$$\left| \int_F f \, dm \right| < \epsilon \tag{14.29}$$
for every $F \subseteq E$ with $m(F) < \delta$.

Proof. Given $\epsilon > 0$. If f is bounded on E, then there exists a $M > 0$ such that $|f(x)| \leq M$. Since $f \in L^1(E)$, Theorem 14.5(d) (Properties of Integrable Functions) gives
$$\left| \int_F f \, dm \right| \leq M \cdot m(F). \tag{14.30}$$
If F is a measurable subset of E such that $m(F) < \frac{\epsilon}{M}$, then we conclude from the inequality (14.30) that the inequality (14.29) holds in this case.

Suppose next that f is unbounded on E. For each $n = 0, 1, 2, \ldots$, we consider the sets
$$E_n = \{x \in E \mid n \leq |f(x)| < n+1\}, \quad F_n = \bigcup_{k=0}^n E_k \quad \text{and} \quad G_n = E \setminus F_n = \bigcup_{k=N+1}^n E_k. \tag{14.31}$$
Since $|f(x)| \geq 0$ on E and $E_0, E_1, \ldots$ are pairwise disjoint subsets of E whose union is E, we know from Theorem 14.3(e) and the hypothesis that
$$\sum_{n=0}^\infty \int_{E_n} |f| \, dm = \int_E |f| \, dm < \infty.$$
Therefore, there exists an $N \in \mathbb{N}$ such that
$$\int_{G_N} |f| \, dm = \sum_{n=N+1}^\infty \int_{E_n} |f| \, dm < \frac{\epsilon}{2}. \tag{14.32}$$
Let $\delta = \frac{\epsilon}{2(N+1)}$. Thus if F is a measurable subset of E with $m(F) < \delta$, then it yields from Theorem 14.5(d) (Properties of Integrable Functions), the facts $E = F_N \cup G_N$ and $F_N \cap G_N = \varnothing$ that
$$\left| \int_F f \, dm \right| \leq \int_F |f| \, dm = \int_{F \cap F_N} |f| \, dm + \int_{F \cap G_N} |f| \, dm. \tag{14.33}$$
By applying the definition (14.31) and the inequality (14.32) to the expression (14.33), we obtain immediately that
$$\left| \int_F f \, dm \right| \leq (N+1) m(F) + \int_{G_N} |f| \, dm < (N+1) \delta + \frac{\epsilon}{2} = \frac{\epsilon}{2} + \frac{\epsilon}{2} = \epsilon.$$
This completes the proof of the problem. ∎

14.3 Applications of Fatou's Lemma

Problem 14.21

(★) Prove that we may have the strict inequality in Fatou's Lemma.

Proof. For each $n = 1, 2, \ldots$, we consider the function $f_n : \mathbb{R} \to \mathbb{R}$ defined by

$$f_n(x) = \begin{cases} 1, & \text{if } x \in [n, n+1); \\ 0, & \text{otherwise.} \end{cases} \quad (14.34)$$

If $x \in \mathbb{R}$, then it is easy to check that $x \notin [n, n+1)$ for all large enough n so that

$$\liminf_{n \to \infty} f_n(x) = 0$$

which gives

$$\int_{\mathbb{R}} \left(\liminf_{n \to \infty} f_n \right) dm = 0.$$

However, the definition of f_n implies that

$$\int_{\mathbb{R}} f_n \, dm = \int_n^{n+1} dm = 1.$$

Hence the sequence of functions (14.34) shows that the strict inequality in Fatou's Lemma can actually occur, completing the proof of the problem. ∎

Problem 14.22

(★)(★) Suppose that $f_n, g_n, f, g \in L^1(\mathbb{R})$ for every $n \in \mathbb{N}$. If $f_n \to f$, $g_n \to g$ pointwise a.e. on $\mathbb{R}$, $|f_n(x)| \le g_n(x)$ a.e. on $\mathbb{R}$ and

$$\int_{\mathbb{R}} g_n \, dm \to \int_{\mathbb{R}} g \, dm,$$

prove that

$$\int_{\mathbb{R}} f_n \, dm \to \int_{\mathbb{R}} f \, dm.$$

Proof. By the triangle inequality, we know that $|f_n - f| \le |f_n| + |f| \le g_n + |f|$ so that

$$g_n + |f| - |f_n - f| \ge 0.$$

By Fatou's Lemma, we have

$$\int_{\mathbb{R}} g \, dm + \int_{\mathbb{R}} |f| \, dm = \int_{\mathbb{R}} (g + |f|) \, dm$$
$$= \int_{\mathbb{R}} \lim_{n \to \infty} (g + |f| - |f_n - f|) \, dm$$

$$= \int_{\mathbb{R}} \liminf_{n\to\infty} (g + |f| - |f_n - f|) \, \mathrm{d}m$$

$$\leq \liminf_{n\to\infty} \int_{\mathbb{R}} (g + |f| - |f_n - f|) \, \mathrm{d}m$$

$$= \liminf_{n\to\infty} \left(\int_{\mathbb{R}} g \, \mathrm{d}m + \int_{\mathbb{R}} |f| \, \mathrm{d}m - \int_{\mathbb{R}} |f_n - f| \, \mathrm{d}m \right)$$

$$= \int_{\mathbb{R}} g \, \mathrm{d}m + \int_{\mathbb{R}} |f| \, \mathrm{d}m - \limsup_{n\to\infty} \int_{\mathbb{R}} |f_n - f|) \, \mathrm{d}m$$

so that

$$\limsup_{n\to\infty} \int_{\mathbb{R}} |f_n - f|) \, \mathrm{d}m = 0. \tag{14.35}$$

Since $f_n, f \in L^1(E)$, we deduce from Theorem 14.5(a) and (d) (Properties of Integrable Functions) that $f_n - f \in L^1(E)$ and

$$\left| \int_{\mathbb{R}} f_n \, \mathrm{d}m - \int_{\mathbb{R}} f \, \mathrm{d}m \right| = \left| \int_{\mathbb{R}} (f_n - f) \, \mathrm{d}m \right| \leq \int_{\mathbb{R}} |f_n - f| \, \mathrm{d}m. \tag{14.36}$$

Applying the result (14.35) to the inequality (14.36), we see that

$$\left| \int_{\mathbb{R}} f_n \, \mathrm{d}m - \int_{\mathbb{R}} f \, \mathrm{d}m \right| \to 0$$

as $n \to \infty$ which implies that

$$\lim_{n\to\infty} \int_{\mathbb{R}} f_n \, \mathrm{d}m = \int_{\mathbb{R}} f \, \mathrm{d}m.$$

This ends the proof of the problem. ∎

Problem 14.23

(★) Let $-\infty < a < b < \infty$. Let $\{f_n\}$ be a sequence of nonnegative measurable functions on (a, b) such that $f_n \to f$ a.e. on (a, b). Define

$$F(x) = \int_a^x f \, \mathrm{d}m \quad \text{and} \quad F_n(x) = \int_a^x f_n \, \mathrm{d}m,$$

where $n = 1, 2, \ldots$. Prove that

$$\int_{[a,b]} (f + F) \, \mathrm{d}m \leq \liminf_{n\to\infty} \int_{[a,b]} (f_n + F_n) \, \mathrm{d}m. \tag{14.37}$$

Proof. Since $\{f_n\}$ satisfies the hypothesis of Fatou's Lemma, so we know that

$$F(x) = \int_a^x f \, \mathrm{d}m = \int_a^x \liminf_{n\to\infty} f_n \, \mathrm{d}m \leq \liminf_{n\to\infty} \int_a^x f_n \, \mathrm{d}m = \liminf_{n\to\infty} F_n(x). \tag{14.38}$$

Since f_n is nonnegative, each F_n is also nonnegative. Now we apply Fatou's Lemma to the inequality (14.38) one more time, we see that

$$\int_{[a,b]} (f + F) \, \mathrm{d}m \leq \int_{[a,b]} (f + \liminf_{n\to\infty} F_n) \, \mathrm{d}m$$

14.3. Applications of Fatou's Lemma

$$= \int_{[a,b]} \liminf_{n\to\infty}(f_n + F_n)\,\mathrm{d}m$$

$$\leq \liminf_{n\to\infty}\int_{[a,b]}(f_n + F_n)\,\mathrm{d}m$$

which is exactly the expected result (14.37). It completes the proof of the problem. ∎

Problem 14.24

(★) Suppose that E is measurable and $\{f_n\}$ is a sequence of measurable functions defined on E. If g is integrable on E and $|f_n| \leq g$ on E for all $n = 1, 2, \ldots$, prove that

$$\int_E \liminf f_n\,\mathrm{d}m \leq \liminf_{n\to\infty}\int_E f_n\,\mathrm{d}m \leq \limsup_{n\to\infty}\int_E f_n\,\mathrm{d}m \leq \int_E \limsup f_n\,\mathrm{d}m. \qquad (14.39)$$

Proof. Since $|f_n| \leq g$ on E and for all $n = 1, 2, \ldots$, we must have $g - f_n \geq 0$ and $g + f_n \geq 0$ on E and for all $n = 1, 2, \ldots$. Applying Fatou's Lemma to both $g - f_n$ and $g + f_n$, we get

$$\int_E \liminf_{n\to\infty}(g + f_n)\,\mathrm{d}m \leq \liminf_{n\to\infty}\int_E (g + f_n)\,\mathrm{d}m$$

and

$$\int_E \liminf_{n\to\infty}(g - f_n)\,\mathrm{d}m \leq \liminf_{n\to\infty}\int_E (g - f_n)\,\mathrm{d}m.$$

Then they imply that

$$\int_E \liminf_{n\to\infty} f_n\,\mathrm{d}m + \int_E g\,\mathrm{d}m \leq \liminf_{n\to\infty}\int_E f_n\,\mathrm{d}m + \int_E g\,\mathrm{d}m \qquad (14.40)$$

and

$$\int_E g\,\mathrm{d}m - \int_E \limsup_{n\to\infty} f_n\,\mathrm{d}m \leq \int_E g\,\mathrm{d}m - \limsup_{n\to\infty}\int_E f_n\,\mathrm{d}m. \qquad (14.41)$$

Combining the results (14.40) and (14.41), we conclude immediately that the set of inequalities (14.39) hold and we complete the proof of the problem. ∎

Problem 14.25

(★)(★) Suppose that E is a measurable set and $\{f_n\}$ is a sequence of integrable functions defined on E that converges pointwise almost everywhere to an integrable function f defined on E. If we have

$$\lim_{n\to\infty}\int_E |f_n|\,\mathrm{d}m = \int_E |f|\,\mathrm{d}m, \qquad (14.42)$$

prove that

$$\lim_{n\to\infty}\int_F |f_n|\,\mathrm{d}m = \int_F |f|\,\mathrm{d}m$$

for every measurable subset F of E.

Proof. Let F be a measurable subset of E. Notice that

$$|f_n - f|\chi_F \leq |f_n - f| \leq |f_n| + |f|$$

on E. Thus if we set $g_n = |f_n| + |f| - |f_n - f|\chi_F$, then they are nonnegative and measurable so that we can apply Fatou's Lemma to get

$$\int_E \liminf_{n\to\infty} g_n \, dm \leq \liminf_{n\to\infty} \int_E g_n \, dm$$

$$\int_E \liminf_{n\to\infty}(|f_n| + |f| - |f_n - f|\chi_F) \, dm \leq \liminf_{n\to\infty} \int_E (|f_n| + |f| - |f_n - f|\chi_F) \, dm. \qquad (14.43)$$

Since $f_n \to f$ pointwise a.e. on E, we deduce from the hypothesis (14.42) and the inequality (14.43) that

$$2\int_E |f| \, dm - \int_E \limsup_{n\to\infty} |f_n - f|\chi_F \, dm \leq \liminf_{n\to\infty} \int_E |f_n| \, dm + \int_E |f| \, dm$$

$$- \limsup_{n\to\infty} \int_E |f_n - f|\chi_F \, dm$$

$$2\int_E |f| \, dm - 0 \leq \int_E |f| \, dm + \int_E |f| \, dm - \limsup_{n\to\infty} \int_F |f_n - f| \, dm$$

$$\limsup_{n\to\infty} \int_F |f_n - f| \, dm \leq 0$$

which implies

$$0 \leq \liminf_{n\to\infty} \int_F |f_n - f| \, dm \leq \limsup_{n\to\infty} \int_F |f_n - f| \, dm \leq 0. \qquad (14.44)$$

Consequently, the inequalities (14.44) mean that

$$\lim_{n\to\infty} \int_F |f_n - f| \, dm = 0. \qquad (14.45)$$

By Theorem 14.5(a) and (d) (Properties of Integrable Functions) and the triangle inequality, we obtain

$$\left| \int_F |f_n| \, dm - \int_F |f| \, dm \right| = \left| \int_F (|f_n| - |f|) \, dm \right| \leq \int_F ||f_n| - |f|| \, dm \leq \int_F |f_n - f| \, dm.$$

Therefore, the limit (14.45) implies that

$$\lim_{n\to\infty} \left| \int_F |f_n| \, dm - \int_F |f| \, dm \right| = 0$$

which gives the desired result. This completes the proof of the problem. ∎

Problem 14.26

(★)(★) Use Fatou's Lemma to prove the Lebesgue's Monotone Convergence Theorem.

14.3. Applications of Fatou's Lemma

Proof. Since $f_n \to f$ pointwise a.e. on E and every f_n is nonnegative, Fatou's Lemma gives

$$\int_E f \, dm \le \liminf_{n \to \infty} \int_E f_n \, dm. \tag{14.46}$$

Furthermore, we also have $f_n \le f$ a.e. on E and for all $n = 1, 2, \ldots$, so Theorem 14.3(a) implies that

$$\int_E f_n \, dm \le \int_E f \, dm$$

for all $n \in \mathbb{N}$ and then

$$\limsup_{n \to \infty} \int_E f_n \, dm \le \int_E f \, dm. \tag{14.47}$$

Hence we combine the two inequalities (14.46) and (14.47) to obtain

$$\lim_{n \to \infty} \int_E f_n \, dm = \int_E f \, dm,$$

completing the proof of the problem. ∎

Problem 14.27

(★)(★) *Use Fatou's Lemma to prove the Lebesgue's Dominated Convergence Theorem.*

Proof. By Problem 14.11, we observe that $f_n, f \in L^1(E)$ for every $n \in \mathbb{N}$. On the one hand, since $f_n + g \ge 0$ on E, Fatou's Lemma is applicable to gain

$$\int_E (f + g) \, dm = \int_E \liminf_{n \to \infty} (f + g) \, dm \le \liminf_{n \to \infty} \int_E (f_n + g) \, dm$$

or equivalently,

$$\int_E f \, dm \le \liminf_{n \to \infty} \int_E f_n \, dm. \tag{14.48}$$

On the other hand, since $g - f_n \ge 0$, Fatou's Lemma again shows that

$$\int_E (g - f) \, dm = \int_E \liminf_{n \to \infty} (g - f_n) \, dm \le \liminf_{n \to \infty} \int_E (g - f_n) \, dm = \int_E g \, dm - \limsup_{n \to \infty} \int_E f_n \, dm$$

so that

$$\int_E f \, dm \ge \limsup_{n \to \infty} \int_E f \, dm. \tag{14.49}$$

Hence our desired result follows by combining the inequalities (14.48) and (14.49), completing the proof of the problem. ∎

> **Problem 14.28**
>
> (★) Suppose that $\{f_n\}$ is a sequence of integrable functions defined on $\mathbb{R}$ and $f_n \to f$ pointwise a.e. on $\mathbb{R}$. Furthermore, for every $\epsilon > 0$, there is a measurable set $E \subseteq \mathbb{R}$ such that
> $$\int_{E^c} |f_n|\,dm \le \epsilon \qquad (14.50)$$
> for all $n \ge N$ and
> $$\int_E |f_n - f|\,dm \to 0 \qquad (14.51)$$
> as $n \to \infty$. Prove that
> $$\lim_{n \to \infty} \int_{\mathbb{R}} |f_n - f|\,dm = 0.$$

Proof. According to Fatou's Lemma and the hypothesis (14.50), we see that
$$\int_{E^c} |f|\,dm = \int_{E^c} \liminf_{n \to \infty} |f_n|\,dm \le \liminf_{n \to \infty} \int_{E^c} |f_n|\,dm \le \epsilon. \qquad (14.52)$$

Next, using the inequality (14.52), we have
$$\int_{\mathbb{R}} |f_n - f|\,dm \le \int_E |f_n - f|\,dm + \int_{E^c} |f_n - f|\,dm$$
$$\le \int_E |f_n - f|\,dm + \int_{E^c} |f_n|\,dm + \int_{E^c} |f|\,dm$$
$$\le \int_E |f_n - f|\,dm + 2\epsilon.$$

Hence we follow from the hypothesis (14.51) that
$$\limsup_{n \to \infty} \int_{\mathbb{R}} |f_n - f|\,dm \le 2\epsilon.$$

Since ϵ is arbitrary, we establish
$$\lim_{n \to \infty} \int_{\mathbb{R}} |f_n - f|\,dm = 0$$
which completes the proof of the problem. ∎

> **Problem 14.29**
>
> (★) Suppose that E is a measurable set, $\{f_n\}, \{g_n\}, \{h_n\} \subseteq L^1(E)$ and $g_n \le f_n \le h_n$ a.e. on E. Furthermore, we have $f_n \to f$, $g_n \to g$, $h_n \to h$ pointwise on E. If $g, h \in L^1(E)$ satisfy
> $$\int_E g_n\,dm \to \int_E g\,dm \quad \text{and} \quad \int_E h_n\,dm \to \int_E h\,dm$$
> as $n \to \infty$, prove that
> $$\lim_{n \to \infty} \int_E f_n\,dm = \int_E f\,dm.$$

Proof. By the hypotheses, it is clear that
$$g(x) \le f(x) \le h(x)$$
a.e. on E. Since g and h are integrable on E, Problem 14.13 ensures that $f \in L^1(E)$. Now $f_n(x) - g_n(x) \ge 0$ a.e. on E for all positive integers n, so Fatou's Lemma implies that

$$\begin{aligned}
\int_E f \, dm - \int_E g \, dm &= \int_E (f - g) \, dm \\
&= \int_E \liminf_{n \to \infty} (f_n - g_n) \, dm \\
&\le \liminf_{n \to \infty} \int_E (f_n - g_n) \, dm \\
&= \liminf_{n \to \infty} \int_E f_n \, dm - \int_E g \, dm
\end{aligned}$$

so that
$$\int_E f \, dm \le \liminf_{n \to \infty} \int_E f_n \, dm. \tag{14.53}$$

Similarly, Fatou's Lemma again shows that

$$\begin{aligned}
\int_E h \, dm - \int_E f \, dm &= \int_E (h - f) \, dm \\
&= \int_E \liminf_{n \to \infty} (h_n - f_n) \, dm \\
&\le \liminf_{n \to \infty} \int_E (h_n - f_n) \, dm \\
&= \int_E h \, dm - \limsup_{n \to \infty} \int_E f_n \, dm
\end{aligned}$$

and then
$$\int_E f \, dm \ge \limsup_{n \to \infty} \int_E f_n \, dm. \tag{14.54}$$

Hence we derive from the inequalities (14.53) and (14.54) that
$$\lim_{n \to \infty} \int_E f_n \, dm = \int_E f \, dm.$$

This completes the proof of the problem. ∎

14.4 Applications of Convergence Theorems

Problem 14.30

(★) Suppose that $f_n : E \to \mathbb{R}$ is integrable for each $n = 1, 2, \ldots$, where E is measurable with $m(E) < \infty$. Prove that if $\{f_n\}$ converges uniformly to f on E, then

$$\lim_{n \to \infty} \int_E f_n \, dm = \int_E f \, dm. \tag{14.55}$$

Proof. Since $\{f_n\}$ converges uniformly to f on E, there exists an $N \in \mathbb{N}$ such that
$$|f_n(x) - f_N(x)| < 1$$
for all $n \geq N$ and $x \in E$. Therefore, we have
$$|f_n(x)| \leq |f_N(x)| + 1$$
on E. Since $m(E) < \infty$ and $f_N \in L^1(E)$, we conclude that $|f_N| + 1 \in L^1(E)$. Hence it follows from the Lebesgue's Dominated Convergence Theorem that the limit (14.55) holds, completing the proof of the problem. ∎

Problem 14.31

(★) Prove that
$$\lim_{n \to \infty} \mathscr{R} \int_0^1 \frac{1 + nx^2}{(1+x^2)^n}\, dx = 0.$$

Proof. Let $f_n(x) = \frac{1+nx^2}{(1+x^2)^n}$ for every $n = 1, 2, \ldots$. By the binomial theorem, we know that
$$(1+x^2)^n = 1 + nx^2 + \frac{n(n-1)}{2}x^4 + \cdots + x^{2n} \geq 1 + nx^2 + \frac{n(n-1)}{2}x^4$$
on $[0,1]$ so that
$$|f_n(x)| \leq \frac{1 + nx^2}{1 + nx^2 + \frac{n(n-1)}{2}x^4} \leq 1$$
on $[0,1]$. Therefore, each f_n is bounded on $[0,1]$ and Theorem 14.6 ensures that $f_n \in L^1([0,1])$ and then
$$\int_{[0,1]} f_n\, dm = \mathscr{R} \int_0^1 f_n\, dx.$$

Now, since we have
$$\lim_{n \to \infty} \frac{1 + nx^2}{1 + nx^2 + \frac{n(n-1)}{2}x^4} = 0$$
for every $x \in (0,1]$, we have $f_n \to f = 0$ pointwise a.e. on $[0,1]$. Hence we conclude from the Lebesgue's Dominated Convergence Theorem that
$$\lim_{n \to \infty} \mathscr{R} \int_0^1 \frac{1+nx^2}{(1+x^2)^n}\, dx = \lim_{n \to \infty} \int_{[0,1]} f_n\, dm = \int_{[0,1]} \lim_{n \to \infty} f_n\, dm = 0$$
which ends the proof of the problem. ∎

Problem 14.32

(★) Let $-\infty < a < b < \infty$ and $f \in L^1([a,b])$. Prove that
$$\lim_{n \to \infty} \int_{[a,b]} n \ln\left(1 + \frac{|f(x)|^2}{n^2}\right) dm = 0.$$

14.4. Applications of Convergence Theorems

Proof. Suppose that
$$f_n(x) = n \ln\left(1 + \frac{|f(x)|^2}{n^2}\right),$$
where $n = 1, 2, \ldots$. Since $f \in L^1([a,b])$, Problem 14.6 ensures that $|f|$ is finite a.e on $[a,b]$. Now it is easy to see that
$$f_n(x) = \frac{1}{n} \ln\left(1 + \frac{|f(x)|^2}{n^2}\right)^{n^2} \le \frac{1}{n} \ln \exp(|f(x)|^2) = \frac{1}{n}|f(x)|^2 \to 0 \quad (14.56)$$
pointwise a.e. on $[a,b]$. Next, if $y \ge 0$, then we have
$$e^{2\sqrt{y}} = 1 + 2\sqrt{y} + \frac{4y}{2} + \cdots \ge 1 + y$$
so that $\ln(1+y) \le 2\sqrt{y}$. Thus we obtain
$$f_n(x) = n \ln\left(1 + \frac{|f(x)|^2}{n^2}\right) \le 2n\sqrt{\frac{|f(x)|^2}{n^2}} = 2n \cdot \frac{\sqrt{|f(x)|^2}}{n} = 2|f(x)|$$
for all $n \in \mathbb{N}$ and all $x \in [a,b]$. By the fact $f \in L^1([a,b])$ and the limit (14.56), we can apply the Lebesgue's Dominated Convergence Theorem to conclude that
$$\lim_{n\to\infty} \int_{[a,b]} n \ln\left(1 + \frac{|f(x)|^2}{n^2}\right) dm = \lim_{n\to\infty} \int_{[a,b]} f_n(x)\, dm = \int_{[a,b]} \lim_{n\to\infty} f_n(x)\, dm = 0,$$
as required. This completes the proof of the problem. ∎

Problem 14.33

(★) Suppose that $f_n, f \in L^1(\mathbb{R})$ for all $n \in \mathbb{N}$, $f_n \to f$ pointwise a.e. on $\mathbb{R}$ and
$$\lim_{n\to\infty} \int_{\mathbb{R}} |f_n|\, dm = \int_{\mathbb{R}} |f|\, dm. \quad (14.57)$$
Prove that
$$\lim_{n\to\infty} \int_E f_n\, dm = \int_E f\, dm$$
for every measurable subset E of $\mathbb{R}$.

Proof. Notice from the triangle inequality that
$$||f_n| - |f_n - f|| \le |f|,$$
so the Lebesgue's Dominated Convergence Theorem implies that
$$\lim_{n\to\infty}\left(\int_{\mathbb{R}} |f_n|\, dm - \int_{\mathbb{R}} |f_n - f|\, dm\right) = \lim_{n\to\infty} \int_{\mathbb{R}} (|f_n| - |f_n - f|)\, dm$$
$$= \int_{\mathbb{R}} \lim_{n\to\infty} (|f_n| - |f_n - f|)\, dm$$
$$= \int_{\mathbb{R}} |f|\, dm. \quad (14.58)$$

By the hypothesis (14.57) and the limit (14.58), we get

$$\int_{\mathbb{R}} |f| \, dm - \lim_{n \to \infty} \int_{\mathbb{R}} |f_n - f| \, dm = \lim_{n \to \infty} \int_{\mathbb{R}} |f_n| \, dm - \lim_{n \to \infty} \int_{\mathbb{R}} |f_n - f| \, dm$$
$$= \lim_{n \to \infty} \left(\int_{\mathbb{R}} |f_n| \, dm - \int_{\mathbb{R}} |f_n - f| \, dm \right)$$
$$= \int_{\mathbb{R}} |f| \, dm$$

which implies that

$$\lim_{n \to \infty} \int_{\mathbb{R}} |f_n - f| \, dm = 0. \tag{14.59}$$

Since $f_n - f \in L^1(\mathbb{R})$ for every positive integer n, Theorem 14.5(c) and (d) (Properties of Integrable Functions) ensures that, for every measurable subset E of $\mathbb{R}$,

$$\left| \int_E f_n \, dm - \int_E f \, dm \right| \le \int_E |f_n - f| \, dm \le \int_{\mathbb{R}} |f_n - f| \, dm. \tag{14.60}$$

Using the limit (14.59) directly to the inequality (14.60), we acquire the desired result which completes the proof of the problem. ∎

> **Problem 14.34**
>
> (★)(★)(★) *Prove the Bounded Convergence Theorem.*

Proof. It is clear that the theorem holds when $m(E) = 0$. Without loss of the generality, we may assume that $m(E) > 0$ in the following discussion.

Since each f_n is measurable on E and $f_n \to f$ pointwise on E, f is measurable on E by Theorem 13.4(c) (Properties of Measurable Functions). Since $\{f_n\}$ is uniformly bounded, there exists a $M > 0$ such that

$$|f_n(x)| \le M \tag{14.61}$$

for all $n \in \mathbb{N}$ and $x \in E$. Consequently, we also have

$$|f(x)| \le M \tag{14.62}$$

for all $x \in E$. Now since $\{f_n\}$ and f are bounded on E, each $f_n - f$ is also bounded[c] on E and the paragraph following Theorem 14.5 (Properties of Integrable Functions) guarantees that

$$\left| \int_E (f_n - f) \, dm \right| \le \int_E |f_n - f| \, dm. \tag{14.63}$$

If F is any measurable subset of E, then we have

$$\int_E f_n \, dm - \int_E f \, dm = \int_E (f_n - f) \, dm$$
$$= \int_F (f_n - f) \, dm + \int_{E \setminus F} (f_n - f) \, dm,$$

---

[c] In fact, the sequence $\{f_n - f\}$ is uniformly bounded.

14.4. Applications of Convergence Theorems

so we deduce from the bounds (14.61), (14.62) and the estimate (14.63) that

$$\left| \int_E f_n \, dm - \int_E f \, dm \right| \leq \left| \int_F (f_n - f) \, dm \right| + \left| \int_{E \setminus F} (f_n - f) \, dm \right|$$

$$\leq \int_F |f_n - f| \, dm + \int_{E \setminus F} |f_n| \, dm + \int_{E \setminus F} |f| \, dm$$

$$\leq \int_F |f_n - f| \, dm + 2M \cdot m(E \setminus F). \tag{14.64}$$

Next, given $\epsilon > 0$. Since $m(E) < \infty$, Egorov's Theorem ensures that there exists a closed set $K_\epsilon \subseteq E$ such that $m(E \setminus K_\epsilon) < \frac{\epsilon}{4M}$ and $f_n \to f$ uniformly on K_ϵ. In other words, there exists an $N \in \mathbb{N}$ such that $n \geq N$ implies

$$|f_n(x) - f(x)| < \frac{\epsilon}{2m(E)}$$

for all $x \in E$. Hence we deduce from the inequality (14.64) with F replaced by K_ϵ that

$$\left| \int_E f_n \, dm - \int_E f \, dm \right| < \frac{\epsilon}{2m(E)} \cdot m(K_\epsilon) + 2M \cdot m(E \setminus K_\epsilon) \leq \frac{\epsilon}{2} + \frac{\epsilon}{2} = \epsilon$$

for all $n \geq N$. This proves the Bounded Convergence Theorem and we have completed the proof of the problem. ∎

Problem 14.35

(⋆) Let $-\infty < a < b < \infty$. Suppose that $f_n, f : [a,b] \to \mathbb{R}$ are measurable and $f_n(x) \geq 0$ on $[a,b]$ for every $n = 1, 2, \ldots$. If $f_n \to f$ pointwise a.e. on $[a,b]$, prove that

$$\lim_{n \to \infty} \int_{[a,b]} f_n e^{-f_n} \, dm = \int_{[a,b]} f e^{-f} \, dm.$$

Proof. It is trivial to see that $f_n e^{-f_n} \to f e^{-f}$ pointwise a.e. on $[a,b]$. Since $f_n(x) \geq 0$ on $[a,b]$, we have

$$0 \leq f_n(x) e^{-f_n(x)} \leq 1$$

for all $n \in \mathbb{N}$ and $x \in [a,b]$. Since $1 \in L^1([a,b])$, it yields from the Lebesgue's Dominated Convergence Theorem that

$$\lim_{n \to \infty} \int_{[a,b]} f_n e^{-f_n} \, dm = \int_{[a,b]} \lim_{n \to \infty} f_n e^{-f_n} \, dm = \int_{[a,b]} f e^{-f} \, dm,$$

completing the proof of the problem. ∎

> **Problem 14.36**
>
> (★) Let $f : \mathbb{R} \to [0, \infty)$ be measurable. If we have
> $$\int_{\mathbb{R}} \frac{n^2 f(x)}{n^2 + x^2} \, dm \leq 1$$
> for all $n \in \mathbb{N}$, prove that
> $$\int_{\mathbb{R}} f(x) \, dm \leq 1.$$

Proof. Let $g_n(x) = \frac{n^2}{n^2+x^2}$, where $n \in \mathbb{N}$. It is evident that each g_n is measurable and
$$0 \leq g_1(x) \leq g_2(x) \leq \cdots$$
for all $x \in \mathbb{R}$. Furthermore, we have $g_n(x) \to 1$ pointwise on $\mathbb{R}$. Let $f_n = g_n f$. Then $\{f_n\}$ is an increasing sequence of nonnegative measurable functions with $f_n(x) \to f(x)$ on $\mathbb{R}$, so the Lebesgue's Monotone Convergence Theorem implies that
$$\int_{\mathbb{R}} f(x) \, dm = \int_{\mathbb{R}} \lim_{n \to \infty} f_n(x) \, dm = \lim_{n \to \infty} \int_{\mathbb{R}} f_n(x) \, dm \leq 1.$$

We have completed the proof of the problem. ∎

> **Problem 14.37**
>
> (★)(★) Prove Problem 14.20 by the Lebesgue's Monotone Convergence Theorem.

Proof. For each $n \in \mathbb{N}$, we define $A_n = \{x \in E \mid \frac{1}{n} \leq |f(x)| < n\}$. Then each A_n is measurable and $A_1 \subseteq A_2 \subseteq \cdots$. Let $A = \bigcup_{n=1}^{\infty} A_n$, $A_0 = \{x \in E \mid f(x) = 0\}$ and $A_\infty = \{x \in E \mid f(x) = \infty\}$. Notice that
$$\int_E |f| \, dm = \int_{A_0} |f| \, dm + \int_{A_\infty} |f| \, dm + \int_A |f| \, dm. \tag{14.65}$$

Since $f \in L^1(E)$, Problem 14.6 ensures that $m(A_\infty) = 0$. By Theorem 14.3(f), the first two integrals on the right-hand side of the equation (14.65) are 0, thus we have
$$\int_E |f| \, dm = \int_A |f| \, dm. \tag{14.66}$$

Next, for each $n \in \mathbb{N}$, we define $f_n = |f|\chi_{A_n} : A \to \mathbb{R}$. Then $\{f_n\}$ is an increasing sequence of nonnegative measurable functions and
$$\lim_{n \to \infty} f_n(x) = |f(x)|\chi_A(x)$$
for all $x \in A$. Now the Lebesgue's Monotone Convergence Theorem implies that
$$\lim_{n \to \infty} \int_A f_n \, dm = \int_A \lim_{n \to \infty} f_n \, dm = \int_A |f| \, dm. \tag{14.67}$$

14.4. Applications of Convergence Theorems

Now the two results (14.66) and (14.67) combine to give

$$\lim_{n\to\infty}\int_{A_n}|f|\,dm = \lim_{n\to\infty}\int_A f_n\,dm = \int_A |f|\,dm = \int_E |f|\,dm < \infty$$

or equivalently,

$$\lim_{n\to\infty}\int_{E\setminus A_n}|f|\,dm = 0.$$

Therefore, there is an $N \in \mathbb{N}$ such that

$$\int_{E\setminus A_N}|f|\,dm < \frac{\epsilon}{2}.$$

If we take $\delta = \frac{\epsilon}{2N}$ and F is any measurable subset of E with $m(F) < \delta$, then Theorem 14.5(c) and (d) (Properties of Integrable Functions) say that

$$\left|\int_F f\,dm\right| \leq \int_F |f|\,dm$$
$$\leq \int_{(E\setminus A_N)\cap F}|f|\,dm + \int_{A_N\cap F}|f|\,dm$$
$$\leq \int_{E\setminus A_N}|f|\,dm + \int_{A_N\cap F} N\,dm$$
$$< \frac{\epsilon}{2} + N\cdot m(A_N \cap F)$$
$$< \frac{\epsilon}{2} + N\cdot \frac{\epsilon}{2N}$$
$$= \epsilon.$$

This completes the proof of the problem. ∎

Problem 14.38

(⋆) Does the Lebesgue's Monotone Convergence Theorem hold for decreasing sequences of measurable functions?

Proof. The answer is negative. In fact, for each $n \in \mathbb{N}$, we consider $f_n = \chi_{[n,\infty)}$ defined on $\mathbb{R}$. Then $\{f_n\}$ is clearly a decreasing sequence of nonnegative measurable functions and

$$f_n(x) \to f = 0$$

pointwise on $\mathbb{R}$. However, we note that

$$\int_{\mathbb{R}} f_n\,dm = \int_n^\infty dx = \infty \neq 0 = \int_{\mathbb{R}} f\,dm,$$

so the Lebesgue's Monotone Convergence Theorem fails in this case. We complete the proof of the problem. ∎

> **Problem 14.39**
>
> (★)(★) Suppose that $f \in L^1(\mathbb{R})$. Prove that
> $$\lim_{n\to\infty} \int_{\mathbb{R}} \frac{xf(x-n)}{1+|x|}\,dm = \int_{\mathbb{R}} f\,dm.$$

Proof. Let n be a fixed positive integer. By the change of variable (x replaced by $x - n$), we see immediately that

$$\int_{\mathbb{R}} \frac{xf(x-n)}{1+|x|}\,dm = \int_{-n}^{\infty} \frac{(x+n)f(x)}{1+(x+n)}\,dm + \int_{-\infty}^{-n} \frac{(x+n)f(x)}{1-(x+n)}\,dm. \tag{14.68}$$

Thus we are going to compute the two integrals on the right-hand side of the equation (14.68) when $n \to \infty$.

To compute the first integral, we notice that if $x \geq -n$, then we have

$$0 \leq \frac{x+n}{1+(x+n)} < 1 \quad \text{and} \quad \lim_{n\to\infty} \frac{x+n}{1+(x+n)} = 1.$$

Therefore, we get

$$0 \leq \left|\frac{(x+n)f(x)}{1+(x+n)}\right| \leq |f(x)| \tag{14.69}$$

on $[-n, \infty)$. Define $g_n : \mathbb{R} \to \mathbb{R}$ by

$$g_n(x) = \frac{(x+n)f(x)}{1+(x+n)}\chi_{[-n,\infty)}(x),$$

so the inequalities (14.69) show that

$$|g_n| \leq |f| \quad \text{and} \quad g_n \to f \text{ pointwise a.e. on } \mathbb{R}.$$

Since $|f| \in L^1(\mathbb{R})$ by Theorem 14.5(d) (Properties of Integrable Functions), it follows from the Lebesgue's Dominated Convergence Theorem that

$$\lim_{n\to\infty} \int_{-n}^{\infty} \frac{(x+n)f(x)}{1+(x+n)}\,dm = \lim_{n\to\infty} \int_{\mathbb{R}} g_n\,dm = \int_{\mathbb{R}} \lim_{n\to\infty} g_n\,dm = \int_{\mathbb{R}} f\,dm. \tag{14.70}$$

Next, we define $h_n : \mathbb{R} \to \mathbb{R}$ by

$$h_n(x) = \frac{(x+n)f(x)}{1-(x+n)}\chi_{(-\infty,-n]}(x).$$

Now if $x \leq -n$, then we have

$$0 \leq \left|\frac{(x+n)}{1-(x+n)}\right| < 1 \quad \text{and} \quad \lim_{n\to\infty} \frac{(x+n)}{1-(x+n)} = -1.$$

Therefore, they yield that

$$|h_n| \leq |f| \quad \text{and} \quad h_n \to 0 \text{ pointwise a.e. on } \mathbb{R}.$$

14.4. Applications of Convergence Theorems

Again the Lebesgue's Dominated Convergence Theorem guarantees that

$$\lim_{n\to\infty} \int_{-\infty}^{-n} \frac{(x+n)f(x)}{1-(x+n)} \, dm = \lim_{n\to\infty} \int_{\mathbb{R}} h_n \, dm = \int_{\mathbb{R}} \lim_{n\to\infty} h_n \, dm = 0. \qquad (14.71)$$

Finally, we substitute the two limits (14.70) and (14.71) back into the equation (14.68) to conclude that

$$\lim_{n\to\infty} \int_{\mathbb{R}} \frac{xf(x-n)}{1+|x|} \, dm = \lim_{n\to\infty} \int_{-n}^{\infty} \frac{(x+n)f(x)}{1+(x+n)} \, dm + \lim_{n\to\infty} \int_{-\infty}^{-n} \frac{(x+n)f(x)}{1-(x+n)} \, dm = \int_{\mathbb{R}} f \, dm.$$

This completes the proof of the problem. ∎

> **Problem 14.40**
>
> (★) Let $E \subseteq \mathbb{R}$ and $f : E \to \mathbb{R}$ be measurable and $E_n = \{x \in E \,|\, |f(x)| \geq n\}$, where n is a positive integer. If $f \in L^1(E)$. prove that
>
> $$\lim_{n\to\infty} \int_{E_n} |f| \, dm = 0.$$

Proof. Since E is measurable, every E_n is also measurable. Let $f_n = |f|\chi_{E_n} : E \to \mathbb{R}$ for each $n = 1, 2, \ldots$. Clearly, each f_n is measurable. Since $f \in L^1(E)$, Problem 14.6 ensures that f is finite a.e. on E and then we have $|f_n(x)| \leq |f(x)|$ for all $n \in \mathbb{N}$ and all $x \in E$. In addition, it is easy to check that

$$f_n(x) \to 0$$

pointwise a.e. on E. By applying the Lebesgue's Dominated Convergence Theorem, we conclude that

$$\lim_{n\to\infty} \int_{E_n} |f| \, dm = \lim_{n\to\infty} \int_E f_n \, dm = \int_E \lim_{n\to\infty} f_n \, dm = 0.$$

Therefore, it completes the proof of the problem. ∎

> **Problem 14.41**
>
> (★) Suppose that $f \in L^1(\mathbb{R})$. Prove that the series
>
> $$\sum_{n=-\infty}^{\infty} f(x+n)$$
>
> converges absolutely a.e. on $\mathbb{R}$.

Proof. We define

$$F(x) = \sum_{n=-\infty}^{\infty} |f(x+n)| \geq 0.$$

We claim that F is finite a.e. on $\mathbb{R}$. In fact, since F is periodic of period 1, it suffices to show that F is finite a.e. on $[0,1]$. Applying the Lebesgue's Monotone Convergence Theorem for Series and then Theorem 14.3(e), we get

$$\int_{[0,1]} F(x)\,dm = \int_{[0,1]} \Big(\sum_{n=-\infty}^{\infty} |f(x+n)| \Big)\,dm$$
$$= \sum_{n=-\infty}^{\infty} \int_{[0,1]} |f(x+n)|\,dm$$
$$= \sum_{n=-\infty}^{\infty} \int_{[n,n+1]} |f(x)|\,dm$$
$$= \int_{\mathbb{R}} |f|\,dm$$
$$< \infty.$$

In other words, F is integrable on $[0,1]$ and it deduces from Problem 14.6 that F is finite a.e. on $[0,1]$, as claimed. This completes the proof of the problem. ∎

Problem 14.42

(★)(★) Suppose that $E \subseteq \mathbb{R}$ is measurable and $f \in L^1(E)$. Given $\epsilon > 0$. Prove that there exists a measurable set F with finite measure such that

$$\sup_{x \in F} |f(x)| < \infty \quad \text{and} \quad \int_{F^c} |f|\,dm < \epsilon.$$

Proof. For each $n \in \mathbb{N}$, let $F_n = \{x \in E \mid \frac{1}{n} \leq |f(x)| \leq n\} \subseteq E$. Now we deduce from Problem 14.5 that

$$m(F_n) \leq m\Big(\Big\{x \in E \,\Big|\, |f(x)| \geq \frac{1}{n}\Big\}\Big) \leq n \int_{F_n} |f|\,dm \leq n \int_E |f|\,dm$$

which reduces to

$$\frac{m(F_n)}{n} \leq \int_E |f|\,dm < \infty$$

for every $n = 1, 2, \ldots$. Thus for each *fixed* positive integer n, if we set $F = F_n$, then we have

$$m(F) < \infty \quad \text{and} \quad \sup_{x \in F} |f(x)| \leq n < \infty.$$

To find the estimate of the integral, we let $f_n = f \chi_{F_n^c} : E \to \mathbb{R}$ for each $n = 1, 2, \ldots$. Then we have $|f_n| \leq |f|$ on E. By the definition, we see that

$$F_n^c = \Big\{x \in E \,\Big|\, 0 \leq |f(x)| < \tfrac{1}{n} \text{ or } |f(x)| > n\Big\}.$$

Recall from Problem 14.6 that f is finite a.e. on E, so we conclude that $f_n(x) \to 0$ pointwise a.e. on E. Consequently, the Lebesgue's Dominated Convergence Theorem is applicable and we get

$$\lim_{n \to \infty} \int_E f_n\,dm = \int_E \lim_{n \to \infty} f_n\,dm = 0.$$

14.4. Applications of Convergence Theorems

Therefore, there exists an $N \in \mathbb{N}$ such that

$$\int_E f_N \, dm < \epsilon. \tag{14.72}$$

Finally, we know from the definition of f_N that the integral (14.72) is equal to

$$\int_{F_N^c} |f| \, dm.$$

We have completed the proof of the problem. ∎

CHAPTER 15

Differential Calculus of Functions of Several Variables

15.1 Fundamental Concepts

The definition and basic properties of derivatives of $f : \mathbb{R} \to \mathbb{R}$ have been introduced and discussed in Chapter 8. This chapter extends the differential calculus theory to functions from $\mathbb{R}^n$ to $\mathbb{R}^m$ and the main references for this chapter are [3, Chap. 12], [23, Chap. 9], [27, Chap. 6] and [33, Chap. 7 & 8]. Here we assume that you are familiar with elementary linear algebra which can be found, for examples, in [13] or [17].

15.1.1 The Directional Derivatives and the Partial Derivatives

Let S be a subset of $\mathbb{R}^n$, $\mathbf{x}$ be an interior point of S, $\mathbf{v}$ be a point of S and $\mathbf{f} : S \to \mathbb{R}^m$ be a function.

Definition 15.1 (Directional Derivatives). *The **directional derivative** of $\mathbf{f}$ at $\mathbf{x}$ in the direction $\mathbf{v}$, denoted by $\mathbf{f}'_\mathbf{v}(\mathbf{x})$ or $\nabla_\mathbf{v}\mathbf{f}(\mathbf{x})$, is defined by the limit*

$$\mathbf{f}'_\mathbf{v}(\mathbf{x}) = \lim_{h \to 0} \frac{\mathbf{f}(\mathbf{x} + h\mathbf{v}) - \mathbf{f}(\mathbf{x})}{h}$$

whenever the limit exists.[a]

Let $\{\mathbf{e}_1, \mathbf{e}_2, \ldots, \mathbf{e}_n\}$ and $\{\mathbf{u}_1, \mathbf{u}_2, \ldots, \mathbf{u}_m\}$ be the standard bases of $\mathbb{R}^n$ and $\mathbb{R}^m$ respectively, x_j be the jth component of $\mathbf{x}$ with respect to the standard basis $\{\mathbf{e}_1, \mathbf{e}_2, \ldots, \mathbf{e}_n\}$ and

$$\mathbf{f} = (f_1, f_2, \ldots, f_m).$$

Then we have

$$\mathbf{f}(\mathbf{x}) = \sum_{i=1}^{m} f_i(\mathbf{x})\mathbf{u}_i \quad \text{and} \quad f_i(\mathbf{x}) = \mathbf{f}(\mathbf{x}) \cdot \mathbf{u}_i.$$

---
[a] Since $\mathbf{x} \in S^\circ$, it is true that $\mathbf{x} + h\mathbf{v} \in S$ for all small $h > 0$ so that $\mathbf{f}'_\mathbf{v}(\mathbf{x})$ is well-defined.

Besides, it is clear from Definition 15.1 (Directional Derivatives) that $\mathbf{f}'_\mathbf{v}(\mathbf{x})$ exists if and only if all $(f_i)'_\mathbf{v}(\mathbf{x})$ exist, where $i = 1, 2, \ldots, m$. In this case, we have

$$\mathbf{f}'_\mathbf{v}(\mathbf{x}) = ((f_1)'_\mathbf{v}(\mathbf{x}), (f_2)'_\mathbf{v}(\mathbf{x}), \ldots, (f_m)'_\mathbf{v}(\mathbf{x})).$$

Definition 15.2 (Partial Derivatives). *For $1 \leq i \leq m$ and $1 \leq j \leq n$, we define*

$$(D_j f_i)(\mathbf{x}) = \lim_{h \to 0} \frac{f_i(\mathbf{x} + h\mathbf{e}_j) - f_i(\mathbf{x})}{h}$$

whenever the limit exists. Then each $(D_j f_i)(\mathbf{x})$ is called the **partial derivative** *of f_i with respect to x_j.*[b]

> **Remark 15.1**
>
> It is very easy to check from the above two definitions that the existence of $\mathbf{f}'_\mathbf{v}(\mathbf{x})$ for all $\mathbf{v} \in \mathbb{R}^n$ implies the existence of $(D_j f_i)(\mathbf{x})$, where $1 \leq i \leq m$ and $1 \leq j \leq n$. Particularly, if $\mathbf{v} = \mathbf{e}_j$, then we have
>
> $$\mathbf{f}'_{\mathbf{e}_j}(\mathbf{x}) = ((D_j f_1)(\mathbf{x}), (D_j f_2)(\mathbf{x}), \ldots, (D_j f_m)(\mathbf{x})),$$
>
> where $1 \leq j \leq n$. However, the converse is false, see Problem 15.2 below.

15.1.2 Differentiation of Functions of Several Variables

Definition 15.3. *The function $\mathbf{f} : S \subseteq \mathbb{R}^n \to \mathbb{R}^m$ is said to be* **differentiable at** $\mathbf{x} \in \mathbb{R}^n$ *if there exists a* **linear transformation** $T_\mathbf{x} : \mathbb{R}^n \to \mathbb{R}^m$ *such that*

$$\lim_{\mathbf{h} \to 0} \frac{|\mathbf{f}(\mathbf{x} + \mathbf{h}) - \mathbf{f}(\mathbf{x}) - T_\mathbf{x}(\mathbf{h})|}{|\mathbf{h}|} = 0. \tag{15.1}$$

In this case, we write $\mathbf{f}'(\mathbf{x}) = T_\mathbf{x}$. If $\mathbf{f}$ is differentiable at every point of S, then we say that $\mathbf{f}$ is **differentiable in** S.[c]

The derivative $\mathbf{f}'(\mathbf{x})$ (or equivalently the transformation $T_\mathbf{x}$) is called the **differential of f at x** or the **total derivative of f at x**. Furthermore, the limit (15.1) is equivalent to the form

$$\mathbf{f}(\mathbf{x} + \mathbf{h}) - \mathbf{f}(\mathbf{x}) = \mathbf{f}'(\mathbf{x})\mathbf{h} + \mathbf{R}(\mathbf{h}), \tag{15.2}$$

where the remainder $\mathbf{R}(\mathbf{h})$ satisfies

$$\lim_{\mathbf{h} \to 0} \frac{|\mathbf{R}(\mathbf{h})|}{|\mathbf{h}|} = 0. \tag{15.3}$$

Here we have some basic properties of the total derivative $\mathbf{f}'(\mathbf{x})$.

Theorem 15.4 (Uniqueness of the Total Derivative). *Suppose that both the linear transformations $T_\mathbf{x}$ and $\widehat{T}_\mathbf{x}$ satisfy Definition 15.3. Then we have $T_\mathbf{x} = \widehat{T}_\mathbf{x}$.*

---

[b] Another notation for $D_j f_i$ is $\dfrac{\partial f_i}{\partial x_j}$.

[c] Reader should pay attention that the $\mathbf{h}$ is a vector, *not* a scalar.

15.1. Fundamental Concepts

Theorem 15.5. *If $\mathbf{f}$ is differentiable at $\mathbf{p} \in S \subseteq \mathbb{R}^n$, then $\mathbf{f}$ is continuous at $\mathbf{p}$.*

> **Remark 15.2**
>
> A consequence of Theorem 15.5 is that directional differentiability *does not* imply total differentiability, see Problem 15.3. However, they are connected in *some* way. In fact, Problem 15.4 ensures that the existence of $\mathbf{f}'(\mathbf{x})$ implies that of $\mathbf{f}'_\mathbf{v}(\mathbf{x})$ for every $\mathbf{v} \in \mathbb{R}^n$.

Theorem 15.6 (The Chain Rule). *Suppose that S is open in $\mathbb{R}^n$, $\mathbf{f} : S \to \mathbb{R}^m$ is differentiable at $\mathbf{a}$ and $\mathbf{g} : \mathbf{f}(S) \subseteq \mathbb{R}^m \to \mathbb{R}^k$ is differentiable at $\mathbf{f}(\mathbf{a})$. Then the mapping $\mathbf{F} = \mathbf{g} \circ \mathbf{f} : S \to \mathbb{R}^k$ is differentiable at $\mathbf{a}$ and*

$$\mathbf{F}'(\mathbf{a}) = \mathbf{g}'(\mathbf{f}(\mathbf{a})) \times \mathbf{f}'(\mathbf{a}). \tag{15.4}$$

We note that on the right-hand side of the equation (15.4), we have **the product of two linear transformations**.

15.1.3 The Total Derivatives and the Jacobian Matrices

Theorem 15.7. *Let $\{\mathbf{e}_1, \mathbf{e}_2, \ldots, \mathbf{e}_n\}$ and $\{\mathbf{u}_1, \mathbf{u}_2, \ldots, \mathbf{u}_m\}$ denote the standard bases of $\mathbb{R}^n$ and $\mathbb{R}^m$ respectively. Suppose that S is open in $\mathbb{R}^n$ and $\mathbf{f} : S \subseteq \mathbb{R}^n \to \mathbb{R}^m$ is differentiable at $\mathbf{x}$. Then all the partial derivatives $(D_j f_i)(\mathbf{x})$ exist, where $1 \leq i \leq m$ and $1 \leq j \leq n$, and furthermore the total derivative $\mathbf{f}'(\mathbf{x})$ satisfies*

$$\mathbf{f}'(\mathbf{x})\mathbf{e}_j = \sum_{i=1}^{m}(D_j f_i)(\mathbf{x})\mathbf{u}_i, \tag{15.5}$$

where $1 \leq j \leq n$.

This theorem says that, once we know that $\mathbf{f}'(\mathbf{x})$ exists, we can express it in terms of the partial derivatives $(D_j f_i)(\mathbf{x})$. In addition, we recall from Definition 15.3 that $\mathbf{f}'(\mathbf{x})$ is in fact a linear transformation $T_\mathbf{x} : \mathbb{R}^n \to \mathbb{R}^m$, so elementary linear algebra tells us that there is a (unique) matrix associated with $\mathbf{f}'(\mathbf{x})$. This matrix is called the **Jacobian matrix** of $\mathbf{f}$ at $\mathbf{x}$ and it is denoted by $\mathbf{J}_\mathbf{f}(\mathbf{x})$. We see from the forms (15.5) that $\mathbf{J}_\mathbf{f}(\mathbf{x})$ is related to the partial derivatives $(D_j f_i)(\mathbf{x})$ as follows:

$$\mathbf{J}_\mathbf{f}(\mathbf{x}) = \begin{pmatrix} (D_1 f_1)(\mathbf{x}) & (D_2 f_1)(\mathbf{x}) & \cdots & (D_n f_1)(\mathbf{x}) \\ (D_1 f_2)(\mathbf{x}) & (D_2 f_2)(\mathbf{x}) & \cdots & (D_n f_2)(\mathbf{x}) \\ \vdots & \vdots & \ddots & \vdots \\ (D_1 f_m)(\mathbf{x}) & (D_2 f_m)(\mathbf{x}) & \cdots & (D_n f_m)(\mathbf{x}) \end{pmatrix}_{m \times n}. \tag{15.6}$$

In terms of the Jacobian matrices, the matrix form of Theorem 15.6 (The Chain Rule) can be written as

$$\mathbf{J}_\mathbf{F}(\mathbf{x}) = \mathbf{J}_\mathbf{g}(\mathbf{f}(\mathbf{x})) \times \mathbf{J}_\mathbf{f}(\mathbf{x}), \tag{15.7}$$

where this time the product on the right-hand side of the equation (15.7) refers to matrix product.

In addition, if $m = 1$, then $\mathbf{f} = f$ is real-valued so the formula (15.5) gives

$$f'(\mathbf{x})\mathbf{v} = v_1 f'(\mathbf{x})\mathbf{e}_1 + v_2 f'(\mathbf{x})\mathbf{e}_2 + \ldots + v_n f'(\mathbf{x})\mathbf{e}_n$$

$$= v_1(D_1f)(\mathbf{x}) + v_2(D_2f)(\mathbf{x}) + \cdots + v_n(D_nf)(\mathbf{x})$$
$$= \nabla f(\mathbf{x}) \cdot \mathbf{v},$$

where
$$\nabla f(\mathbf{x}) = ((D_1f)(\mathbf{x}), (D_2f)(\mathbf{x}), \ldots, (D_nf)(\mathbf{x})) \qquad (15.8)$$

is called the **gradient of f at x**.

15.1.4 The Mean Value Theorem for Differentiable Functions

The Mean Value Theorem for Differentiable Functions. *Suppose that S is open in $\mathbb{R}^n$ and $\mathbf{f} : S \to \mathbb{R}^m$ is differentiable in S. If $\lambda\mathbf{x} + (1-\lambda)\mathbf{y} \in S$, where $0 \leq \lambda \leq 1$, then for every $\mathbf{a} \in \mathbb{R}^m$, there exists a point $\mathbf{z}$ lying on the line segment of $\mathbf{x}$ and $\mathbf{y}$ such that*

$$\mathbf{a} \cdot [\mathbf{f}(\mathbf{y}) - \mathbf{f}(\mathbf{x})] = \mathbf{a} \cdot [\mathbf{f}'(\mathbf{z})(\mathbf{y} - \mathbf{x})].$$

Theorem 15.8. *Suppose that the open set S in the previous theorem is convex and there is a constant M such that*

$$\sup_{|\mathbf{y}| \leq 1} \{|\mathbf{f}'(\mathbf{x})\mathbf{y}|\} \leq M$$

for every $\mathbf{x} \in S$.[d] *Then we have*

$$|\mathbf{f}(\mathbf{b}) - \mathbf{f}(\mathbf{a})| \leq M|\mathbf{b} - \mathbf{a}|$$

for every $\mathbf{a}, \mathbf{b} \in S$.

Theorem 15.9. *Let $S \subseteq \mathbb{R}^n$ be open and convex. If $\mathbf{f} : S \to \mathbb{R}^m$ is differentiable in S and $\mathbf{f}'(\mathbf{x}) = \mathbf{0}$ for all $\mathbf{x} \in S$, then $\mathbf{f}$ is a constant on S.*

15.1.5 Continuously Differentiable Functions

Definition 15.10 (Continuously Differentiable Functions). Denote $L(\mathbb{R}^n, \mathbb{R}^m)$ to be the set of all linear transformations of $\mathbb{R}^n$ into $\mathbb{R}^m$. Let $S \subseteq \mathbb{R}^n$ be open and $\mathbf{f} : S \to \mathbb{R}^m$ be differentiable. We say that $\mathbf{f}$ is **continuously differentiable** in S if

$$\mathbf{f}' : S \to L(\mathbb{R}^n, \mathbb{R}^m)$$

is a continuous mapping. The class of all continuously differentiable functions in S is denoted by $\mathscr{C}'(S)$.

Now the following theorem tells us a necessary and sufficient condition for $\mathbf{f} \in \mathscr{C}'(S)$.

Theorem 15.11. *Suppose that S is open in $\mathbb{R}^n$. Then $\mathbf{f} : S \to \mathbb{R}^m$ belongs to $\mathscr{C}'(S)$ if and only if all $D_j f_i$ exist and are continuous on S for all $1 \leq i \leq m$ and $1 \leq j \leq n$.*

---

[d] Note that $\mathbf{f}'(\mathbf{x})\mathbf{y}$ is a vector in $\mathbb{R}^m$.

15.1.6 The Inverse Function Theorem and the Implicit Function Theorem

The Inverse Function Theorem. *Let E be an open subset of $\mathbb{R}^n$ and $\mathbf{f} : E \to \mathbb{R}^n$ be a **continuously differentiable** function on E. Let $\mathbf{a} \in E$ be a point such that the linear transformation $\mathbf{f}'(\mathbf{a})$ is invertible, i.e., $\det \mathbf{J_f}(\mathbf{a}) \neq 0$. Then there exists an open set $U \subseteq E$ containing $\mathbf{a}$ and an open set $V \subseteq \mathbb{R}^n$ containing $\mathbf{b} = \mathbf{f}(\mathbf{a})$ such that $\mathbf{f} : U \to V$ is a bijection. Furthermore, the inverse mapping $\mathbf{f}^{-1} : V \to U$ is differentiable at $\mathbf{x}$ and*

$$(\mathbf{f}^{-1})'(\mathbf{y}) = \frac{1}{\mathbf{f}'(\mathbf{x})}$$

for every $\mathbf{x} \in U$ and $\mathbf{y} = \mathbf{f}(\mathbf{x})$.

See Figure 15.1 for the idea of this theorem.

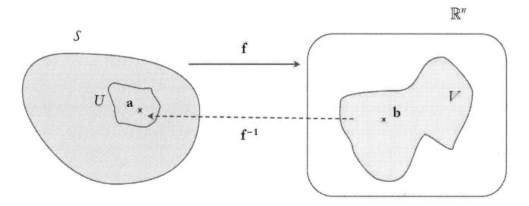

Figure 15.1: The Inverse Function Theorem.

The Implicit Function Theorem. *Let E be an open subset of $\mathbb{R}^{n+m}$ and $\mathbf{f} : E \to \mathbb{R}^m$ be a **continuously differentiable** function on E. We write $(\mathbf{x}, \mathbf{y}) = (x_1, x_2, \ldots, x_n, y_1, y_2, \ldots, y_m)$. Let $(\mathbf{a}, \mathbf{b}) \in E$ be a point such that $\mathbf{f}(\mathbf{a}, \mathbf{b}) = \mathbf{0}$, where $\mathbf{a} \in \mathbb{R}^n$ and $\mathbf{b} \in \mathbb{R}^m$. Suppose that*

$$\det \begin{pmatrix} \frac{\partial f_1}{\partial y_1}(\mathbf{a},\mathbf{b}) & \frac{\partial f_1}{\partial y_2}(\mathbf{a},\mathbf{b}) & \cdots & \frac{\partial f_1}{\partial y_m}(\mathbf{a},\mathbf{b}) \\ \frac{\partial f_2}{\partial y_1}(\mathbf{a},\mathbf{b}) & \frac{\partial f_2}{\partial y_2}(\mathbf{a},\mathbf{b}) & \cdots & \frac{\partial f_2}{\partial y_m}(\mathbf{a},\mathbf{b}) \\ \vdots & \vdots & \ddots & \vdots \\ \frac{\partial f_m}{\partial y_1}(\mathbf{a},\mathbf{b}) & \frac{\partial f_m}{\partial y_2}(\mathbf{a},\mathbf{b}) & \cdots & \frac{\partial f_m}{\partial y_m}(\mathbf{a},\mathbf{b}) \end{pmatrix}_{m \times m} \neq 0.$$

Then there exists an open set $U \subseteq \mathbb{R}^n$ containing $\mathbf{a}$ such that there exists a unique continuously differentiable function $\mathbf{g} : U \to \mathbb{R}^m$ satisfying

$$\mathbf{g}(\mathbf{a}) = \mathbf{b} \quad \text{and} \quad \mathbf{f}(\mathbf{x}, \mathbf{g}(\mathbf{x})) = \mathbf{0}$$

for all $\mathbf{x} \in U$. Furthermore, we have

$$\mathbf{g}'(\mathbf{x}) = - \begin{pmatrix} \dfrac{\partial f_1}{\partial y_1}(\mathbf{x}, \mathbf{g}(\mathbf{x})) & \cdots & \dfrac{\partial f_1}{\partial y_m}(\mathbf{x}, \mathbf{g}(\mathbf{x})) \\ \vdots & \ddots & \vdots \\ \dfrac{\partial f_m}{\partial y_1}(\mathbf{x}, \mathbf{g}(\mathbf{x})) & \cdots & \dfrac{\partial f_m}{\partial y_m}(\mathbf{x}, \mathbf{g}(\mathbf{x})) \end{pmatrix}^{-1}_{m \times m}$$

$$\times \begin{pmatrix} \dfrac{\partial f_1}{\partial x_1}(\mathbf{x}, \mathbf{g}(\mathbf{x})) & \cdots & \dfrac{\partial f_1}{\partial x_n}(\mathbf{x}, \mathbf{g}(\mathbf{x})) \\ \vdots & \ddots & \vdots \\ \dfrac{\partial f_m}{\partial x_1}(\mathbf{x}, \mathbf{g}(\mathbf{x})) & \cdots & \dfrac{\partial f_m}{\partial x_n}(\mathbf{x}, \mathbf{g}(\mathbf{x})) \end{pmatrix}_{m \times n}$$

for all $\mathbf{x} \in U$.

> **Remark 15.3**
>
> Classically, mathematicians prove first the Inverse Function Theorem and then derive from it the Implicit Function Theorem. In addition, the usual ingredients of proofs of the two theorems consist of compactness of balls in $\mathbb{R}^n$ and the Contraction Principle (see [23, Theorem 9.23, p. 220] or [27, Theorem 6.6.4, p. 150]). In 2013, Oliveira [18] employed Dini's approach to prove first the Implicit Function Theorem by induction and then he derived from it the Inverse Function Theorem. The tools he used are the Intermediate Value Theorem and the Mean Value Theorem for Derivatives. For an extensive treatment of the Implicit Function Theorem, we refer the reader to the book [11]. -

15.1.7 Higher Order Derivatives

The partial derivatives $D_1\mathbf{f}, D_2\mathbf{f}, \ldots, D_n\mathbf{f}$ of a function $\mathbf{f} : \mathbb{R}^n \to \mathbb{R}^m$ are also functions from $\mathbb{R}^n$ to $\mathbb{R}^m$. If the partial derivatives $D_j\mathbf{f}$ are differentiable, then they are called the **second-order partial derivatives of f** which are denoted by

$$D_{ij}\mathbf{f} = D_i(D_j\mathbf{f}) = \frac{\partial}{\partial x_i}\left(\frac{\partial \mathbf{f}}{\partial x_j}\right) = \frac{\partial^2 \mathbf{f}}{\partial x_i \partial x_j},$$

where $1 \leq i, j \leq n$.

Theorem 15.12 (Clairaut's Theorem). *If both partial derivatives $D_i\mathbf{f}$ and $D_j\mathbf{f}$ exist in an open set $S \subseteq \mathbb{R}^n$ and both $D_{ij}\mathbf{f}$ and $D_{ji}\mathbf{f}$ are continuous at $\mathbf{p} \in S$, then we have*

$$(D_{ij}\mathbf{f})(\mathbf{p}) = (D_{ji}\mathbf{f})(\mathbf{p}).$$

15.2 Differentiation of Functions of Several Variables

Problem 15.1

(★) Consider the function $f : \mathbb{R}^2 \to \mathbb{R}$ defined by
$$f(x,y) = \begin{cases} 1, & \text{if } xy = 0; \\ 0, & \text{otherwise.} \end{cases}$$
Show that $(D_1 f)(0,0) = (D_2 f)(0,0)$, but f is discontinuous at $(0,0)$.

Proof. We note that
$$(D_1 f)(0,0) = \lim_{h \to 0} \frac{f(h,0) - f(0,0)}{h} = \lim_{h \to 0} \frac{1-1}{h} = 0$$
and
$$(D_2 f)(0,0) = \lim_{h \to 0} \frac{f(0,h) - f(0,0)}{h} = \lim_{h \to 0} \frac{1-1}{h} = 0.$$
Thus $(D_1 f)(0,0) = (D_2 f)(0,0)$. However, we notice that
$$\lim_{x \to 0} f(x,x) = 0 \neq 1 = f(0,0)$$
which means that f is discontinuous at $(0,0)$. This completes the proof of the problem. ∎

Remark 15.4

Problem 15.1 asserts that the existence of the partial derivatives $D_1 f, D_2 f, \ldots, D_n f$ at $\mathbf{p}$ *does not* necessarily imply the continuity of f at $\mathbf{p}$. Fortunately, this really happens when all partial derivatives are bounded in an open set $E \subseteq \mathbb{R}^n$, see [23, Exercise 7, p. 239].

Problem 15.2

(★) Consider $f : \mathbb{R}^2 \to \mathbb{R}$ which is given by
$$f(x,y) = \begin{cases} -x - y, & \text{if } xy = 0; \\ 1, & \text{otherwise.} \end{cases}$$
Prove that both $(D_1 f)(\mathbf{0})$ and $(D_2 f)(\mathbf{0})$ exist, but not $f_\mathbf{v}(\mathbf{0})$ for any $\mathbf{v} = (a,b)$ with $ab \neq 0$.

Proof. By Definition 15.2 (Partial Derivatives), we have
$$(D_1 f)(\mathbf{0}) = \lim_{h \to 0} \frac{f(h,0) - f(0,0)}{h} = \lim_{h \to 0} \frac{-h}{h} = -1$$

and
$$(D_2 f)(\mathbf{0}) = \lim_{h \to 0} \frac{f(0, h) - f(0, 0)}{h} = \lim_{h \to 0} \frac{-h}{h} = -1.$$

However, if $\mathbf{v} = (a, b)$, where $ab \neq 0$, then $a \neq 0$ and $b \neq 0$. Therefore, we obtain from Definition 15.1 (Directional Derivatives) that

$$f'_{\mathbf{v}}(\mathbf{0}) = \lim_{h \to 0} \frac{f(ha, hb) - f(0, 0)}{h} = \lim_{h \to 0} \frac{1}{h}$$

which does not exist. This completes the proof of the problem. ∎

Problem 15.3

($\star$) Prove that the function $f : \mathbb{R}^2 \to \mathbb{R}$ defined by

$$f(x, y) = \begin{cases} \dfrac{xy^2}{x^2 + y^6}, & \text{if } x \neq 0; \\ 0, & \text{otherwise.} \end{cases}$$

has well-defined $f'_{\mathbf{v}}(\mathbf{0})$ for every $\mathbf{v}$ but f is discontinuous at $\mathbf{0}$.

Proof. Suppose that $\mathbf{v} = (a, b)$. By Definition 15.1 (Directional Derivatives), we have

$$\begin{aligned} f'_{\mathbf{v}}(\mathbf{0}) &= \lim_{h \to 0} \frac{f(\mathbf{0} + h\mathbf{v}) - f(\mathbf{0})}{h} \\ &= \lim_{h \to 0} \frac{f(ha, hb)}{h} \\ &= \lim_{h \to 0} \frac{ha \cdot h^2 b^2}{h(h^2 a^2 + h^6 b^6)} \\ &= \lim_{h \to 0} \frac{ab^2}{a^2 + h^4 b^6} \\ &= \begin{cases} \dfrac{b^2}{a}, & \text{if } a \neq 0; \\ 0, & \text{otherwise.} \end{cases} \end{aligned}$$

Thus $f'_{\mathbf{v}}(\mathbf{0})$ is well-defined for every $\mathbf{v}$.

To show that f is discontinuous at $\mathbf{0}$, we first note that $f(0, 0) = 0$. Next, if $y \neq 0$, then we have

$$f(y^2, y) = \frac{y^4}{y^4 + y^6} = \frac{1}{1 + y^2}$$

which implies that

$$\lim_{y \to 0} f(y^2, y) = 1 \neq f(0, 0).$$

Hence f is discontinuous at $\mathbf{0}$ and we complete the proof of the problem. ∎

15.2. Differentiation of Functions of Several Variables

> **Problem 15.4**
>
> (★) Suppose that $\mathbf{x}$ is an interior point of $S \subseteq \mathbb{R}^n$, $\mathbf{f} : S \subseteq \mathbb{R}^n \to \mathbb{R}^m$ is a function and $\mathbf{v} \in \mathbb{R}^n$. If $\mathbf{f}$ is differentiable at $\mathbf{x}$, prove that
>
> $$\mathbf{f}'_{\mathbf{v}}(\mathbf{x}) = \mathbf{f}'(\mathbf{x})\mathbf{v}. \tag{15.9}$$

Proof. Fix $\mathbf{v}$ and consider $\mathbf{h} = t\mathbf{v}$ for some sufficiently small real t. Since $\mathbf{f}$ is differentiable at $\mathbf{x}$, it follows from the approximation (15.2) that

$$\begin{aligned}
\mathbf{f}(\mathbf{x}+t\mathbf{v}) - \mathbf{f}(\mathbf{x}) - t\mathbf{f}'(\mathbf{x})\mathbf{v} &= \mathbf{f}(\mathbf{x}+t\mathbf{v}) - \mathbf{f}(\mathbf{x}) - \mathbf{f}'(\mathbf{x})(t\mathbf{v}) \\
&= \mathbf{f}(\mathbf{x}+\mathbf{h}) - \mathbf{f}(\mathbf{x}) - \mathbf{f}'(\mathbf{x})\mathbf{h} \\
&= \mathbf{R}(\mathbf{h}) \\
&= \mathbf{R}(t\mathbf{v}).
\end{aligned}$$

By the limit (15.3), we see that

$$\lim_{t \to 0} \frac{|\mathbf{R}(t\mathbf{v})|}{|t|} = 0.$$

Thus we get

$$\mathbf{f}'_{\mathbf{v}}(\mathbf{x}) - \mathbf{f}'(\mathbf{x})\mathbf{v} = \lim_{t \to 0} \frac{\mathbf{f}(\mathbf{x}+t\mathbf{v}) - \mathbf{f}(\mathbf{x}) - t\mathbf{f}'(\mathbf{x})\mathbf{v}}{t} = \lim_{t \to 0} \frac{\mathbf{R}(t\mathbf{v})}{t} = \mathbf{0},$$

i.e., $\mathbf{f}'_{\mathbf{v}}(\mathbf{x}) = \mathbf{f}'(\mathbf{x})\mathbf{v}$. This completes the proof of the problem. ∎

> **Remark 15.5**
>
> We notice that the left-hand side of the formula (15.9) is a limit, while its right-hand side is a matrix product. In other words, this gives an easier way (matrix product) to compute the directional derivative $\mathbf{f}'_{\mathbf{v}}(\mathbf{x})$ whenever $\mathbf{f}$ is differentiable at $\mathbf{x}$.

> **Problem 15.5**
>
> (★) Suppose that $\mathbf{f}'_{\mathbf{v}}(\mathbf{x})$ and $\mathbf{g}'_{\mathbf{v}}(\mathbf{x})$ exist. Prove that $(\mathbf{f}+\mathbf{g})'_{\mathbf{v}}(\mathbf{x}) = \mathbf{f}'_{\mathbf{v}}(\mathbf{x}) + \mathbf{g}'_{\mathbf{v}}(\mathbf{x})$.

Proof. By Definition 15.1 (Directional Derivatives), we have

$$\begin{aligned}
(\mathbf{f}+\mathbf{g})'_{\mathbf{v}}(\mathbf{x}) &= \lim_{h \to 0} \frac{(\mathbf{f}+\mathbf{g})(\mathbf{x}+h\mathbf{v}) - (\mathbf{f}+\mathbf{g})(\mathbf{x})}{h} \\
&= \lim_{h \to 0} \frac{\mathbf{f}(\mathbf{x}+h\mathbf{v}) - \mathbf{f}(\mathbf{x})}{h} + \lim_{h \to 0} \frac{\mathbf{g}(\mathbf{x}+h\mathbf{v}) - \mathbf{g}(\mathbf{x})}{h} \\
&= \mathbf{f}'_{\mathbf{v}}(\mathbf{x}) + \mathbf{g}'_{\mathbf{v}}(\mathbf{x}),
\end{aligned}$$

completing the proof of the problem. ∎

336   Chapter 15. Differential Calculus of Functions of Several Variables

> **Problem 15.6**
>
> (⋆) Prove Theorem 15.5.

Proof. Since $\mathbf{f}'(\mathbf{p}) = T_{\mathbf{p}}$, we rewrite the formula (15.2) as
$$\mathbf{f}(\mathbf{p}+\mathbf{h}) - \mathbf{f}(\mathbf{p}) = T_{\mathbf{p}}(\mathbf{h}) + \mathbf{R}(\mathbf{h}). \tag{15.10}$$
Let $\mathbf{h} = h_1 \mathbf{e}_1 + h_2 \mathbf{e}_2 + \cdots + h_n \mathbf{e}_n$. Then $\mathbf{h} \to \mathbf{0}$ if and only if $h_i \to 0$ for all $i = 1, 2, \ldots, n$. Since $T_{\mathbf{p}}$ is linear, we have
$$T_{\mathbf{p}}(\mathbf{h}) = h_1 T_{\mathbf{p}}(\mathbf{e}_1) + h_2 T_{\mathbf{p}}(\mathbf{e}_2) + \cdots + h_n T_{\mathbf{p}}(\mathbf{e}_n)$$
so that $T_{\mathbf{p}}(\mathbf{h}) \to \mathbf{0}$ as $\mathbf{h} \to \mathbf{0}$. By the limit (15.3), we see that $\mathbf{R}(\mathbf{h}) \to \mathbf{0}$ as $\mathbf{h} \to \mathbf{0}$. Hence we conclude from the formula (15.10) that
$$\lim_{\mathbf{h} \to \mathbf{0}} \mathbf{f}(\mathbf{p}+\mathbf{h}) = \mathbf{f}(\mathbf{p}),$$
i.e., $\mathbf{f}$ is continuous at $\mathbf{p}$. This completes the proof of the problem. ∎

> **Problem 15.7**
>
> (⋆) Prove Theorem 15.7.

Proof. By Problem 15.4 and then Remark 15.1, we obtain
$$\mathbf{f}'(\mathbf{x})\mathbf{e}_j = \mathbf{f}'_{\mathbf{e}_j}(\mathbf{x}) = ((D_j f_1)(\mathbf{x}), (D_j f_2)(\mathbf{x}), \ldots, (D_j f_m)(\mathbf{x})) = \sum_{i=1}^{m} (D_j f_i)(\mathbf{x}) \mathbf{u}_i.$$

We have completed the proof of the problem. ∎

> **Problem 15.8**
>
> (⋆)(⋆) Let $S \subseteq \mathbb{R}^n$ and $\mathbf{x}$ be an interior point of S. Suppose that $\mathbf{f} : S \to \mathbb{R}^m$ is given by $\mathbf{f} = (f_1, f_2, \ldots, f_m)$. Use Definition 15.3 to prove that $\mathbf{f}$ is differentiable at $\mathbf{x}$ if and only if every $f_i : S \to \mathbb{R}$ is differentiable at $\mathbf{x}$.

Proof. We notice that for any $\mathbf{y} = (y_1, y_2, \ldots, y_m) \in \mathbb{R}^m$, we have
$$|y_i| \le |\mathbf{y}| \le \sqrt{m} \max_{1 \le i \le m} |y_i| \tag{15.11}$$
for all $i = 1, 2, \ldots, m$. Now the Jacobian matrices for the component functions f_i, where $i = 1, 2, \ldots, m$, at $\mathbf{x}$ are given by
$$\mathbf{J}_{f_i}(\mathbf{x}) = \begin{pmatrix} (D_1 f_i)(\mathbf{x}) & (D_2 f_i)(\mathbf{x}) & \cdots & (D_n f_i)(\mathbf{x}) \end{pmatrix}.$$

15.2. Differentiation of Functions of Several Variables

If we compare these with the matrix (15.6), then we assert that

$$\mathbf{J_f}(\mathbf{x}) = \begin{pmatrix} \mathbf{J}_{f_1}(\mathbf{x}) \\ \mathbf{J}_{f_2}(\mathbf{x}) \\ \vdots \\ \mathbf{J}_{f_m}(\mathbf{x}) \end{pmatrix} \quad \text{or} \quad \mathbf{f}'(\mathbf{x}) = \begin{pmatrix} f_1'(\mathbf{x}) \\ f_2'(\mathbf{x}) \\ \vdots \\ f_m'(\mathbf{x}) \end{pmatrix}.$$

Therefore, the ith component function of $\mathbf{f}(\mathbf{x} + \mathbf{h}) - \mathbf{f}(\mathbf{x}) - \mathbf{f}'(\mathbf{x})\mathbf{h}$ is exactly

$$f_i(\mathbf{x} + \mathbf{h}) - f_i(\mathbf{x}) - f_i'(\mathbf{x})\mathbf{h}$$

and then we follow from the inequalities (15.11) that

$$\frac{|f_i(\mathbf{x} + \mathbf{h}) - f_i(\mathbf{x}) - f_i'(\mathbf{x})\mathbf{h}|}{|\mathbf{h}|} \leq \frac{|\mathbf{f}(\mathbf{x} + \mathbf{h}) - \mathbf{f}(\mathbf{x}) - \mathbf{f}'(\mathbf{x})\mathbf{h}|}{|\mathbf{h}|}$$
$$\leq \sqrt{m} \times \frac{|f_i(\mathbf{x} + \mathbf{h}) - f_i(\mathbf{x}) - f_i'(\mathbf{x})\mathbf{h}|}{|\mathbf{h}|}. \quad (15.12)$$

Hence, by taking $\mathbf{h} \to \mathbf{0}$ to the inequalities (15.12), we get the desired result. This completes the proof of the problem. ∎

Problem 15.9

(★) Let $T : \mathbb{R}^n \to \mathbb{R}^m$ be linear. Prove that T is differentiable everywhere on $\mathbb{R}^n$ and $T'(\mathbf{x}) = T$.

Proof. Since T is linear, we have

$$T(\mathbf{x} + \mathbf{h}) - T(\mathbf{x}) - T(\mathbf{h}) = \mathbf{0}$$

which is in the form (15.2). Next Theorem 15.4 (Uniqueness of the Total Derivative) implies that

$$T'(\mathbf{x}) = T$$

which ends the proof of the problem. ∎

Problem 15.10

(★) Use Theorem 15.6 (The Chain Rule) and Problem 15.9 to prove Problem 15.8 again.

Proof. Suppose that $\mathbf{f}$ is differentiable at $\mathbf{x}$. Let $\pi_i : \mathbb{R}^m \to \mathbb{R}$ be the projection

$$\pi_i(\mathbf{y}) = \pi_i(y_1, y_2, \ldots, y_m) = y_i,$$

where $1 \leq i \leq m$. Since each π_i is linear, every π_i is differentiable everywhere on $\mathbb{R}^m$ by Problem 15.9. Obviously, we have $f_i = \pi_i \circ \mathbf{f}$, so Theorem 15.6 (The Chain Rule) implies that f_i is also differentiable at $\mathbf{x}$.

Conversely, we suppose that every f_i is differentiable at $\mathbf{x} \in S$. Define $\phi_i : \mathbb{R} \to \mathbb{R}^m$ by

$$\phi_i(y) = (0, \ldots, 0, y, 0, \ldots, 0),$$

where the y is the ith coordinate. It is obvious that each ϕ_i is linear so that it is differentiable everywhere on $\mathbb{R}$ by Problem 15.9. By Theorem 15.6 (The Chain Rule), the composite function $\phi_i \circ f_i : S \to \mathbb{R}^m$ is differentiable at $\mathbf{x}$. Since

$$\mathbf{f} = (f_1, f_2, \ldots, f_m) = \sum_{i=1}^{m} \phi_i \circ f_i,$$

we have the desired result that $\mathbf{f}$ is differentiable at $\mathbf{x}$. This completes the proof of the problem. ∎

Problem 15.11

(★) Suppose that $\mathbf{f}, \mathbf{g} : \mathbb{R}^n \to \mathbb{R}^m$ are functions such that $\mathbf{f}$ is differentiable at $\mathbf{p}$, $\mathbf{f}(\mathbf{p}) = \mathbf{0}$ and $\mathbf{g}$ is continuous at $\mathbf{p}$. If $\mathbf{h}(\mathbf{x}) = \mathbf{g}(\mathbf{x}) \cdot \mathbf{f}(\mathbf{x})$, prove that

$$\mathbf{h}'(\mathbf{p})\mathbf{v} = \mathbf{g}(\mathbf{p}) \cdot [\mathbf{f}'(\mathbf{p})\mathbf{v}], \tag{15.13}$$

where $\mathbf{v} \in \mathbb{R}^n$.

Proof. Since $\mathbf{f}$ is differentiable at $\mathbf{p}$ and $\mathbf{f}(\mathbf{p}) = \mathbf{0}$, we get from the formula (15.2) that

$$\mathbf{f}(\mathbf{p} + \mathbf{h}) = \mathbf{f}'(\mathbf{p})\mathbf{h} + \mathbf{R}(\mathbf{h})$$

which gives

$$\begin{aligned}
\mathbf{h}(\mathbf{p} + \mathbf{h}) - \mathbf{h}(\mathbf{p}) &= \mathbf{g}(\mathbf{p} + \mathbf{h}) \cdot \mathbf{f}(\mathbf{p} + \mathbf{h}) - \mathbf{g}(\mathbf{p}) \cdot \mathbf{f}(\mathbf{p}) \\
&= \mathbf{g}(\mathbf{p} + \mathbf{h}) \cdot [\mathbf{f}'(\mathbf{p})\mathbf{h} + \mathbf{R}(\mathbf{h})] \\
&= \mathbf{g}(\mathbf{p} + \mathbf{h}) \cdot [\mathbf{f}'(\mathbf{p})\mathbf{h} + \mathbf{R}(\mathbf{h})] + \mathbf{g}(\mathbf{p}) \cdot \mathbf{f}'(\mathbf{p})\mathbf{h} - \mathbf{g}(\mathbf{p}) \cdot \mathbf{f}'(\mathbf{p})\mathbf{h} \\
&= \mathbf{g}(\mathbf{p}) \cdot \mathbf{f}'(\mathbf{p})\mathbf{h} + \mathbf{g}(\mathbf{p} + \mathbf{h}) \cdot \mathbf{R}(\mathbf{h}) + [\mathbf{g}(\mathbf{p} + \mathbf{h}) - \mathbf{g}(\mathbf{p})] \cdot \mathbf{f}'(\mathbf{p})\mathbf{h}.
\end{aligned} \tag{15.14}$$

Since $\mathbf{g}$ is continuous at $\mathbf{p}$, the limit (15.3) implies

$$\lim_{\mathbf{h} \to \mathbf{0}} \frac{|\mathbf{g}(\mathbf{p} + \mathbf{h}) \cdot \mathbf{R}(\mathbf{h})|}{|\mathbf{h}|} = |\mathbf{g}(\mathbf{p})| \lim_{\mathbf{h} \to \mathbf{0}} \frac{|\mathbf{R}(\mathbf{h})|}{|\mathbf{h}|} = 0.$$

Furthermore, we see that

$$\begin{aligned}
\lim_{\mathbf{h} \to \mathbf{0}} \frac{[\mathbf{g}(\mathbf{p} + \mathbf{h}) - \mathbf{g}(\mathbf{p})] \cdot \mathbf{f}'(\mathbf{p})\mathbf{h}}{|\mathbf{h}|} &= \lim_{\mathbf{h} \to \mathbf{0}} [\mathbf{g}(\mathbf{p} + \mathbf{h}) - \mathbf{g}(\mathbf{p})] \cdot \mathbf{f}'(\mathbf{p})\Big(\frac{\mathbf{h}}{|\mathbf{h}|}\Big) \\
&= \mathbf{0} \cdot \lim_{\mathbf{h} \to \mathbf{0}} \mathbf{f}'(\mathbf{p})\Big(\frac{\mathbf{h}}{|\mathbf{h}|}\Big) \\
&= 0.
\end{aligned}$$

Thus the equation (15.14) can be expressed as

$$\mathbf{h}(\mathbf{p} + \mathbf{h}) - \mathbf{h}(\mathbf{p}) = \mathbf{g}(\mathbf{p}) \cdot \mathbf{f}'(\mathbf{p})\mathbf{h} + \widehat{R}(\mathbf{h}),$$

15.2. Differentiation of Functions of Several Variables

where
$$\lim_{\mathbf{h} \to 0} \frac{|\widehat{\mathbf{R}}(\mathbf{h})|}{|\mathbf{h}|} = 0.$$

By Definition 15.3, we have the expected result (15.13), completing the proof of the problem. ∎

Problem 15.12

(★) Suppose that $\mathbf{f} : \mathbb{R}^3 \to \mathbb{R}^2$ is given by
$$\mathbf{f}(x, y, z) = (x^2 - y + z, e^{-xy} \sin z + xz).$$
Prove that $\mathbf{f}$ is differentiable at $\mathbf{x} \in \mathbb{R}^3$ and find $\mathbf{J_f}(\mathbf{p})$.

Proof. We notice that the component functions f_1 and f_2 of $\mathbf{f}$ are given by
$$f_1(x, y, z) = x^2 - y + z \quad \text{and} \quad f_2(x, y, z) = e^{-xy} \sin z + xz$$
so that
$$\begin{aligned} D_1 f_1 &= 2x, \quad D_2 f_1 = -1, \quad D_3 f_1 = 1, \\ D_1 f_2 &= -y e^{-xy} \sin z + z, \quad D_2 f_2 = -x e^{-xy} \sin z, \quad D_3 f_2 = e^{-xy} \cos z + x. \end{aligned} \quad (15.15)$$

Since all $D_j f_i$ exist and are continuous on $\mathbb{R}^3$, we deduce from Theorem 15.11 that $\mathbf{f}$ is differentiable in $\mathbb{R}^3$. Furthermore, we gain immediately from the partial derivatives (15.15) that
$$\mathbf{J_f}(\mathbf{x}) = \begin{pmatrix} 2x & -1 & 1 \\ -y e^{-xy} \sin z + z & -x e^{-xy} \sin z & e^{-xy} \cos z + x \end{pmatrix}.$$

This completes the analysis of the problem. ∎

Problem 15.13

(★) Suppose that $f, g, h : \mathbb{R}^2 \to \mathbb{R}$ are differentiable such that $z = f(x, y), x = g(r, \theta)$ and $y = h(r, \theta)$. Verify that
$$\frac{\partial z}{\partial r} = \frac{\partial z}{\partial x} \cdot \frac{\partial x}{\partial r} + \frac{\partial z}{\partial y} \cdot \frac{\partial y}{\partial r} \quad \text{and} \quad \frac{\partial z}{\partial \theta} = \frac{\partial z}{\partial x} \cdot \frac{\partial x}{\partial \theta} + \frac{\partial z}{\partial y} \cdot \frac{\partial y}{\partial \theta}. \quad (15.16)$$

Proof. Let $z = \psi(r, \theta) = f(g(r, \theta), h(r, \theta))$ and $\phi = (g, h)$. Since $z = f(x, y)$, the matrix form (15.6) or the gradient form (15.8) shows that
$$\mathbf{J}_f = \begin{pmatrix} \dfrac{\partial f}{\partial x} & \dfrac{\partial f}{\partial y} \end{pmatrix} = \begin{pmatrix} \dfrac{\partial z}{\partial x} & \dfrac{\partial z}{\partial y} \end{pmatrix}.$$

Similarly, we have
$$\mathbf{J}_\psi = \begin{pmatrix} \dfrac{\partial z}{\partial r} & \dfrac{\partial z}{\partial \theta} \end{pmatrix} \quad \text{and} \quad \mathbf{J}_\phi = \begin{pmatrix} \dfrac{\partial x}{\partial r} & \dfrac{\partial x}{\partial \theta} \\ \dfrac{\partial y}{\partial r} & \dfrac{\partial y}{\partial \theta} \end{pmatrix}.$$

Since $\psi = f \circ \phi$, the matrix form (15.7) gives $\mathbf{J}_\psi = \mathbf{J}_f \times \mathbf{J}_\phi$ or explicitly,

$$\begin{pmatrix} \dfrac{\partial z}{\partial r} & \dfrac{\partial z}{\partial \theta} \end{pmatrix} = \begin{pmatrix} \dfrac{\partial z}{\partial x} & \dfrac{\partial z}{\partial y} \end{pmatrix} \times \begin{pmatrix} \dfrac{\partial x}{\partial r} & \dfrac{\partial x}{\partial \theta} \\ \dfrac{\partial y}{\partial r} & \dfrac{\partial y}{\partial \theta} \end{pmatrix}$$

$$= \begin{pmatrix} \dfrac{\partial z}{\partial x} \cdot \dfrac{\partial x}{\partial r} + \dfrac{\partial z}{\partial y} \cdot \dfrac{\partial y}{\partial r} & \dfrac{\partial z}{\partial x} \cdot \dfrac{\partial x}{\partial \theta} + \dfrac{\partial z}{\partial y} \cdot \dfrac{\partial y}{\partial \theta} \end{pmatrix}$$

which is exactly the formulas (15.16), completing the analysis of the problem. ■

Problem 15.14

(⋆) Let $\mathbf{p} \in S \subseteq \mathbb{R}^n$. Prove that there is no function $f : S \to \mathbb{R}$ such that $f'_\mathbf{v}(\mathbf{p}) > 0$ for every $\mathbf{v} \in \mathbb{R}^n \setminus \{\mathbf{0}\}$.

Proof. Assume that there was a function $f : S \to \mathbb{R}$ such that $f'_\mathbf{v}(\mathbf{p}) > 0$ for every $\mathbf{v} \in \mathbb{R}^n \setminus \{\mathbf{0}\}$. Let $\alpha \in \mathbb{R} \setminus \{0\}$. Then we notice from Definition 15.1 (Directional Derivatives) that

$$f'_{\alpha\mathbf{v}}(\mathbf{p}) = \lim_{h \to 0} \frac{f(\mathbf{p} + h\alpha\mathbf{v}) - f(\mathbf{p})}{h} = \alpha \lim_{\alpha h \to 0} \frac{f(\mathbf{p} + h\alpha\mathbf{v}) - f(\mathbf{p})}{\alpha h} = \alpha f'_\mathbf{v}(\mathbf{p}).$$

In particular, we have

$$f'_{-\mathbf{v}}(\mathbf{p}) = -f'_\mathbf{v}(\mathbf{p}) < 0$$

for every $\mathbf{v} \in \mathbb{R}^n \setminus \{\mathbf{0}\}$, a contradiction. Hence we end the proof of the problem. ■

Problem 15.15

(⋆) Suppose that $f, g, h, k : S \subseteq \mathbb{R}^n \to \mathbb{R}$ are functions such that $h = fg$ and $k = \frac{f}{g}$. Find ∇h and ∇k.

Proof. By the definition (15.8), we have

$$\nabla h = (D_1 h, D_2 h, \ldots, D_n h). \tag{15.17}$$

For $1 \le i \le n$, since $D_i h = D_i(fg) = (D_i f)g + f(D_i g)$, we obtain from the expression (15.17) that

$$\nabla h = ((D_1 f)g + f(D_1 g), (D_2 f)g + f(D_2 g), \ldots, (D_n f)g + f(D_n g))$$
$$= (D_1 f, D_2 f, \ldots, D_n f)g + f(D_1 g, D_2 g, \ldots, D_n g)$$
$$= (\nabla f)g + f(\nabla g). \tag{15.18}$$

Write $k = fg^{-1}$, so the formula (15.18) gives

$$\nabla k = (\nabla f)g^{-1} + f[\nabla(g^{-1})]. \tag{15.19}$$

By the definition (15.8) again, we know that

$$\nabla(g^{-1}) = (D_1(g^{-1}), D_2(g^{-1}), \ldots, D_n(g^{-1})) = \Big(\frac{-D_1 g}{g^2}, \frac{-D_2 g}{g^2}, \ldots, \frac{-D_n g}{g^2}\Big) = -\frac{\nabla g}{g^2}. \tag{15.20}$$

15.3. The Mean Value Theorem for Differentiable Functions

If we substitute the expression (15.20) into the formula (15.19), then we assert that

$$\nabla k = \frac{\nabla f}{g} - \frac{f \nabla g}{g^2} = \frac{1}{g^2}(g \nabla f - f \nabla g).$$

This completes the analysis of the problem. ∎

Problem 15.16

(★) *Suppose that $f : \mathbb{R} \to \mathbb{R}$ is differentiable in $\mathbb{R}$ and $g : \mathbb{R}^3 \to \mathbb{R}$ is defined by*

$$g(x, y, z) = x^k + y^k + z^k,$$

where $k \in \mathbb{N}$. Denote $h = f \circ g$. Prove that

$$|\nabla h(x,y,z)|^2 = k^2[x^{2(k-1)} + y^{2(k-1)} + z^{2(k-1)}] \times [f'(g(x,y,z))]^2. \qquad (15.21)$$

Proof. We have $h(x, y, z) = f(x^k + y^k + z^k)$, so Theorem 15.6 (The Chain Rule) implies

$$D_1 h = kx^{k-1} f'(x^k + y^k + z^k).$$

Similarly, we have

$$D_2 h = ky^{k-1} f'(x^k + y^k + z^k) \quad \text{and} \quad D_3 h = kz^{k-1} f'(x^k + y^k + z^k).$$

Thus it follows from the formula (15.8) that

$$\nabla h(x, y, z) = (kx^{k-1} f'(x^k + y^k + z^k), ky^{k-1} f'(x^k + y^k + z^k), kz^{k-1} f'(x^k + y^k + z^k))$$

and then

$$\begin{aligned}|\nabla h(x,y,z)|^2 &= k^2[x^{2(k-1)} + y^{2(k-1)} + z^{2(k-1)}] \times [f'(x^k + y^k + z^k)]^2 \\ &= k^2[x^{2(k-1)} + y^{2(k-1)} + z^{2(k-1)}] \times [f'(g(x,y,z)]^2\end{aligned}$$

which is the desired result (15.21). This ends the proof of the problem. ∎

15.3 The Mean Value Theorem for Differentiable Functions

Problem 15.17

(★)(★) *Prove the Mean Value Theorem for Differentiable Functions.*

Proof. Fix $\mathbf{a} \in \mathbb{R}^m$. Define the function $F : [0, 1] \to \mathbb{R}$ by

$$F(t) = \mathbf{a} \cdot \mathbf{f}(\mathbf{x} + t(\mathbf{y} - \mathbf{x})). \qquad (15.22)$$

The hypothesis shows that

$$\mathbf{x} + t(\mathbf{y} - \mathbf{x}) = (1-t)\mathbf{x} + t\mathbf{y} \in S$$

for all $t \in [0,1]$ so that the function F is well-defined. Since $\mathbf{f}$ is differentiable in S, Theorem 15.5 guarantees that $\mathbf{f}$ is continuous on S and then F must be continuous on $[0,1]$. Furthermore, we apply Theorem 15.6 (The Chain Rule) to the definition (15.22) to get

$$F'(t) = \mathbf{a} \cdot \mathbf{f}'(\mathbf{x} + t(\mathbf{y} - \mathbf{x}))(\mathbf{y} - \mathbf{x}).$$

Now the Mean Value Theorem for Derivatives implies the existence of a $\xi \in (0,1)$ such that

$$F(1) - F(0) = F'(\xi) = \mathbf{a} \cdot [\mathbf{f}'(\mathbf{x} + \xi(\mathbf{y} - \mathbf{x}))(\mathbf{y} - \mathbf{x})]. \tag{15.23}$$

We know that $F(1) = \mathbf{a} \cdot \mathbf{f}(\mathbf{y})$ and $F(0) = \mathbf{a} \cdot \mathbf{f}(\mathbf{x})$, so if we let $\mathbf{z} = \mathbf{x} + \xi(\mathbf{y} - \mathbf{x}) \in S$, then the formula (15.23) can be rewritten as

$$\mathbf{a} \cdot [\mathbf{f}(\mathbf{y}) - \mathbf{f}(\mathbf{x})] = \mathbf{a} \cdot [\mathbf{f}'(\mathbf{z})(\mathbf{y} - \mathbf{x})].$$

This completes the proof of the problem. ∎

> **Problem 15.18**
>
> (★) Let S be open and convex in $\mathbb{R}^n$ and $f : S \to \mathbb{R}$. Let f be differentiable in S and $\mathbf{a}, \mathbf{a} + \mathbf{h} \in S$. Prove that there exists a $\lambda \in (0,1)$ such that
>
> $$f(\mathbf{a} + \mathbf{h}) - f(\mathbf{a}) = f'(\mathbf{a} + \lambda\mathbf{h})\mathbf{h}.$$

Proof. Let $c \neq 0$. Since S is convex, we have $\mathbf{a} + \lambda\mathbf{h} = \lambda(\mathbf{a} + \mathbf{h}) + (1-\lambda)\mathbf{a} \in S$, where $0 \leq \lambda \leq 1$. By the Mean Value Theorem for Differentiable Functions, we get

$$c[f(\mathbf{a} + \mathbf{h}) - f(\mathbf{a})] = c[f'(\mathbf{z})(\mathbf{a}\mathbf{h} - \mathbf{a})] = c[f'(\mathbf{z})\mathbf{h}] \tag{15.24}$$

for some $\mathbf{z} = \mathbf{a} + \lambda\mathbf{h}$. Hence our result follows from dividing the expression (15.24) by the nonzero constant c and we have completed the proof of the problem. ∎

> **Problem 15.19**
>
> (★) Prove Theorem 15.9.

Proof. Since $\mathbf{f}'(\mathbf{x}) = \mathbf{0}$ on S, we have $|\mathbf{f}'(\mathbf{x})\mathbf{y}| = 0$ for every $|\mathbf{y}| \leq 1$. Thus we may take $M = 0$ in Theorem 15.8 to conclude that

$$|\mathbf{f}(\mathbf{b}) - \mathbf{f}(\mathbf{a})| = 0$$

for all $\mathbf{a}, \mathbf{b} \in S$. This means that $\mathbf{f}$ is constant on S, completing the proof of the problem. ∎

> **Problem 15.20**
>
> (★)(★) Let $\mathbf{a} \in S \subseteq \mathbb{R}^n$ and $B_r(\mathbf{a}) = \{\mathbf{x} \in S \mid |\mathbf{x} - \mathbf{a}| < r\}$, where $r > 0$. Suppose that $f : S \to \mathbb{R}$ satisfies $f'_{\mathbf{v}}(\mathbf{x}) = 0$ for every $\mathbf{x} \in B_r(\mathbf{a})$ and every $\mathbf{v} \in \mathbb{R}^n$. Prove that f is constant on $B_r(\mathbf{a})$.

15.4. The Inverse Function Theorem and the Implicit Function Theorem

Proof. We first claim that if the directional derivative $f'_{\mathbf{v}}(\mathbf{a}+t\mathbf{v})$ exists for each $t \in [0,1]$, then there exists a $\lambda \in (0,1)$ such that

$$f(\mathbf{a}+\mathbf{v}) - f(\mathbf{a}) = f'_{\mathbf{v}}(\mathbf{a}+\lambda\mathbf{v}). \tag{15.25}$$

To this end, we define $g : [0,1] \to \mathbb{R}$ by

$$g(t) = f(\mathbf{a}+t\mathbf{v}). \tag{15.26}$$

By Definition 15.1 (Directional Derivatives), we see that

$$g'(t) = \lim_{h \to 0} \frac{g(t+h)-g(t)}{h} = \lim_{h \to 0} \frac{f(\mathbf{a}+t\mathbf{v}+h\mathbf{v}) - f(\mathbf{a}+t\mathbf{v})}{h} = f'_{\mathbf{v}}(\mathbf{a}+t\mathbf{v}), \tag{15.27}$$

where $t \in [0,1]$. Therefore, the Mean Value Theorem for Derivatives shows that there exists a $\lambda \in (0,1)$ such that

$$g(1) - g(0) = g'(\lambda)(1-0) = g'(\lambda). \tag{15.28}$$

By the definition (15.26), we have $g(1) = f(\mathbf{a}+\mathbf{v})$ and $g(0) = f(\mathbf{a})$. Hence our desired result (15.25) follows directly from the comparison of the expressions (15.27) and (15.28).

Next, for every $\mathbf{x} \in B_r(\mathbf{a})$, let $\mathbf{v} = \mathbf{x} - \mathbf{a}$ so that $|\mathbf{v}| < r$. Since $|\mathbf{a}+\lambda\mathbf{v}-\mathbf{a}| = \lambda|\mathbf{v}| < \lambda r < r$, the hypothesis shows that $f'_{\mathbf{v}}(\mathbf{a}+\lambda\mathbf{v}) = 0$ and then

$$f(\mathbf{x}) = f(\mathbf{a}+\mathbf{v}) = f(\mathbf{a}).$$

In other words, f is constant on $B_r(\mathbf{a})$ which completes the proof of the problem. ∎

15.4 The Inverse Function Theorem and the Implicit Function Theorem

Problem 15.21

(★) Suppose that $E = \{(x,y) \in \mathbb{R}^2 \,|\, x > y\}$. Define $\mathbf{f} : E \to \mathbb{R}^2$ by

$$\mathbf{f}(x,y) = (f_1(x,y), f_2(x,y)) = (x+y, x^2+y^2).$$

Prove that $\mathbf{f}$ is locally bijective.

Proof. Since $D_1 f_1 = 1$, $D_2 f_1 = 1$, $D_1 f_2 = 2x$ and $D_2 f_2 = 2y$ are all continuous on E, Theorem 15.11 implies that $\mathbf{f} \in \mathscr{C}'(E)$. Furthermore, we have

$$\mathbf{J_f}(x,y) = \begin{pmatrix} 1 & 1 \\ 2x & 2y \end{pmatrix}.$$

Since $x > y$ if $(x,y) \in E$, it is true that $\det \mathbf{J_f}(x,y) = 2(y-x) \neq 0$. Hence it deduces from the Inverse Function Theorem that $\mathbf{f}$ is locally bijective. This completes the analysis of the problem. ∎

Problem 15.22

(★) Suppose that $\mathbf{f}: B_r(\mathbf{a}) \to \mathbb{R}^n$ is differentiable at $\mathbf{a}$ and its inverse $\mathbf{f}^{-1}$ exists and is differentiable at $\mathbf{f}(\mathbf{a})$. Prove that $\det \mathbf{J_f}(\mathbf{a}) \neq 0$.

Proof. Assume that $\det \mathbf{J_f}(\mathbf{a}) = 0$. By Problem 15.9 and Theorem 15.6 (The Chain Rule), we can establish
$$I_n = (\mathbf{f}^{-1} \circ \mathbf{f})(\mathbf{a}) = (\mathbf{f}^{-1} \circ \mathbf{f})'(\mathbf{a}) = (\mathbf{f}^{-1})'(\mathbf{f}(\mathbf{a})) \times \mathbf{f}'(\mathbf{a})$$
so that
$$1 = \det I_n = \det \mathbf{J_{f^{-1}}}(\mathbf{f}(\mathbf{a})) \times \det \mathbf{J_f}(\mathbf{a}) = 0,$$
a contradiction. Hence we must have $\det \mathbf{J_f}(\mathbf{a}) \neq 0$ and we complete the proof of the problem. ∎

Remark 15.6

In other words, Problem 15.22 says that the hypothesis $\det \mathbf{J_f}(\mathbf{a}) \neq 0$ cannot be omitted in the Inverse Function Theorem.

Problem 15.23

(★) Show, by a counterexample, that the hypothesis $\mathbf{f} \in \mathscr{C}'(E)$ cannot be dropped in the Inverse Function Theorem.

Proof. We consider the function $f : \mathbb{R} \to \mathbb{R}$ defined by
$$f(x) = \begin{cases} x + 2x^2 \sin \dfrac{1}{x}, & \text{if } x \neq 0; \\ 0, & \text{otherwise.} \end{cases}$$
Obviously, f is differentiable in $(-1, 1)$ and
$$f'(0) = \lim_{h \to 0} \frac{f(h) - f(0)}{h} = \lim_{h \to 0} \left(1 + 2h \sin \frac{1}{h}\right) = 1 \neq 0.$$
However, for $x \neq 0$, we have
$$f'(x) = 1 + 4x \sin \frac{1}{x} - 2 \cos \frac{1}{x}$$
so that $\lim_{x \to 0} f'(x)$ does not exist. In other words, $f \notin \mathscr{C}'((-1, 1))$.

Let V be any neighborhood of 0. Then there exists an $N \in \mathbb{N}$ such that $\frac{2}{(4N-3)\pi} \in V$. It is clear that V also contains $\frac{2}{(4N-1)\pi}$ and $\frac{2}{(4N+1)\pi}$ because
$$\frac{2}{(4N+1)\pi} < \frac{2}{(4N-1)\pi} < \frac{2}{(4N-3)\pi}.$$
By direct computation, we see that
$$f\left(\frac{2}{(4N-1)\pi}\right) < f\left(\frac{2}{(4N+1)\pi}\right) < f\left(\frac{2}{(4N-3)\pi}\right). \tag{15.29}$$

15.4. The Inverse Function Theorem and the Implicit Function Theorem

Now the inequalities (15.29) and the continuity of f imply that f is *not* injective in V. Thus the hypothesis $\mathbf{f} \in \mathscr{C}'(E)$ cannot be dropped in the Inverse Function Theorem and it completes the proof of the problem. ∎

Problem 15.24

(★)(★) Suppose that $\mathbf{f} : \mathbb{R}^m \to \mathbb{R}^m$ is an element of $\mathscr{C}'(\mathbb{R}^m)$ and $\det \mathbf{J_f}(\mathbf{x}) \neq 0$ for every $\mathbf{x} \in \mathbb{R}^n$. If $\mathbf{f}^{-1}(K)$ is compact whenever K is compact, prove that $\mathbf{f}(\mathbb{R}^m) = \mathbb{R}^m$.

Proof. Since $\mathbf{f}$ is continuous on $\mathbb{R}^m$ and $\mathbb{R}^m$ is connected, $\mathbf{f}(\mathbb{R}^m)$ is also connected. We claim that $\mathbf{f}(\mathbb{R}^m)$ is both open and closed in $\mathbb{R}^m$. To this end, let $\mathbf{x} \in \mathbb{R}^m$ and $\mathbf{y} = \mathbf{f}(\mathbf{x}) \in \mathbf{f}(\mathbb{R}^m)$. Since $\det \mathbf{J_f}(\mathbf{x}) \neq 0$, the Inverse Function Theorem implies that there exist open sets $V_\mathbf{x}$ and $V_\mathbf{y}$ containing $\mathbf{x}$ and $\mathbf{y}$ respectively such that $\mathbf{f} : V_\mathbf{x} \to V_\mathbf{y}$ is a bijection which gives

$$\mathbf{y} \in V_\mathbf{y} = \mathbf{f}(V_\mathbf{x}) \subseteq \mathbf{f}(\mathbb{R}^m).$$

In other words, $\mathbf{f}(\mathbb{R}^m)$ is open in $\mathbb{R}^m$.

Next, we prove that $\mathbf{f}(\mathbb{R}^m)$ is closed in $\mathbb{R}^m$. Let $\{\mathbf{y}_n\}$ be a sequence of the set $\mathbf{f}(\mathbb{R}^m)$ and $\mathbf{y}_n \to \mathbf{y} \in \mathbb{R}^m$. Then there is a corresponding sequence $\{\mathbf{x}_n\}$ of $\mathbb{R}^m$ such that

$$\mathbf{f}(\mathbf{x}_n) = \mathbf{y}_n \tag{15.30}$$

for all $n \in \mathbb{N}$. Now the set $K = \{\mathbf{y}_n\} \cup \{\mathbf{y}\}$ is compact, so the hypothesis guarantees that $\mathbf{f}^{-1}(K)$ is also compact and then it is bounded by the Heine-Borel Theorem. By the definition, we have $\{\mathbf{x}_n\} \subseteq \mathbf{f}^{-1}(K)$, so $\{\mathbf{x}_n\}$ is a bounded sequence and the Bolzano-Weierstrass Theorem (Problem 5.25) ensures that $\{\mathbf{x}_n\}$ contains a convergent subsequence, namely $\{\mathbf{x}_{n_k}\}$ and $\mathbf{x}_{n_k} \to \mathbf{x} \in \mathbb{R}^m$. Recall that $\mathbf{f}$ is continuous on $\mathbb{R}^m$, so we use the relations (15.30) to conclude that

$$\mathbf{f}(\mathbf{x}) = \lim_{k\to\infty} \mathbf{f}(\mathbf{x}_{n_k}) = \lim_{k\to\infty} \mathbf{y}_{n_k} = \mathbf{y}.$$

Therefore, $\mathbf{f}(\mathbb{R}^m)$ is also closed in $\mathbb{R}^m$ and we have the claim. Since $\det \mathbf{J_f}(\mathbf{x}) \neq 0$, we have $\mathbf{f}(\mathbb{R}^m) \neq \varnothing$ and thus $\mathbf{f}(\mathbb{R}^m) = \mathbb{R}^m$. We complete the proof of the problem. ∎

Problem 15.25

(★) Suppose that $f, g : \mathbb{R} \to \mathbb{R}$ are continuously differentiable with $f(0) = 0$ and $f'(0) \neq 0$. Consider the equation

$$f(x) = tg(x), \tag{15.31}$$

where $t \in \mathbb{R}$. Prove that there exists a $\delta > 0$ such that in $(-\delta, \delta)$, there is a unique continuous function $x(t)$ satisfying the equation (15.31) and $x(0) = 0$.

Proof. We define $F : \mathbb{R}^2 \to \mathbb{R}$ by

$$F(x, t) = f(x) - tg(x).$$

Since $f, g, t \in \mathscr{C}'(\mathbb{R})$, we have $F \in \mathscr{C}'(\mathbb{R}^2)$. In addition, we have $F(0,0) = 0$ and

$$D_1 F(0,0) = f'(0) - 0 \times g'(0) = f'(0) \neq 0.$$

Hence we follow from the Implicit Function Theorem that one can find a $\delta > 0$ and a unique continuously function $x : (-\delta, \delta) \to \mathbb{R}$ such that
$$x(0) = 0 \quad \text{and} \quad F(x(t), t) = 0$$
which mean that $x(t)$ is a solution of the equation (15.31). Thus we complete the proof of the problem. ∎

> **Problem 15.26**
>
> (★) Consider the system of equations
> $$\begin{aligned} x^2 + 3y^2 + z^2 - w &= 9, \\ x^3 + 4y^2 + z + w^2 &= 22. \end{aligned}$$
> Prove that z and w can be written as differentiable functions of x and y around $(1, 1, 1, -4)$.

Proof. We write
$$f_1(x, y, z, w) = x^2 + 3y^2 + z^2 - w - 9 \quad \text{and} \quad f_2(x, y, z, w) = x^3 + 4y^2 + z + w^2 - 22.$$
Next, we define $F : \mathbb{R}^4 \to \mathbb{R}^2$ by
$$\begin{aligned} F(x, y, z, w) &= (f_1(x, y, z, w), f_2(x, y, z, w)) \\ &= (x^2 + 3y^2 + z^2 - w - 9, x^3 + 4y^2 + z + w^2 - 22). \end{aligned}$$
By direct computation, we have $F(1, 1, 1, -4) = (0, 0)$ and
$$D_1 f_1 = 2x, \quad D_2 f_1 = 6y, \quad D_3 f_1 = 2z, \quad D_4 f_1 = -1,$$
$$D_1 f_2 = 3x^2, \quad D_2 f_2 = 8y, \quad D_3 f_2 = 1, \quad D_4 f_2 = 2w.$$
Since all the partial derivatives exist and continuous on $\mathbb{R}^4$, we deduce from Theorem 15.11 that $F \in \mathscr{C}'(\mathbb{R}^4)$. Besides, we know that
$$\det \begin{pmatrix} D_3 f_1(1, 1, 1, -4) & D_4 f_1(1, 1, 1, -4) \\ D_3 f_2(1, 1, 1, -4) & D_4 f_2(1, 1, 1, -4) \end{pmatrix} = \det \begin{pmatrix} 2 & -1 \\ 1 & -8 \end{pmatrix} = -15 \neq 0.$$
Hence it follows from the Implicit Function Theorem that z and w can be written as differentiable functions of x and y around $(1, 1, 1, -4)$. This completes the proof of the problem. ∎

> **Problem 15.27**
>
> (★) Show that there exist functions $f, g : \mathbb{R}^4 \to \mathbb{R}$ which are functions of x, y, z, w, continuously differentiable in $B_\delta(2, 1, 1, -2)$ for some $\delta > 0$ such that $f(2, 1, 1, -2) = 4$, $g(2, 1, 1, -2) = 3$ and the equations
> $$f^2 + g^2 + w^2 = 29 \quad \text{and} \quad \frac{f^2}{x^2} + \frac{g^2}{y^2} + \frac{w^2}{z^2} = 17$$
> hold on $B_\delta(2, 1, 1, -2)$.

15.5. Higher Order Derivatives

Proof. Let $F_1(f, g, x, y, z, w) = f^2 + g^2 + w^2 - 29$ and $F_2(f, g, x, y, z, w) = \frac{f^2}{x^2} + \frac{g^2}{y^2} + \frac{w^2}{z^2} - 17$. Define $\mathbf{F} : \mathbb{R}^6 \to \mathbb{R}^2$ by

$$\mathbf{F}(f, g, x, y, z, w) = (F_1(f, g, x, y, z, w), F_2(f, g, x, y, z, w))$$
$$= \left(f^2 + g^2 + w^2 - 29, \frac{f^2}{x^2} + \frac{g^2}{y^2} + \frac{w^2}{z^2} - 17\right).$$

It is clear that $\mathbf{F}(4, 3, 2, 1, 1, -2) = (0, 0)$. Furthermore, we have

$$D_1 F_1 = 2f, \quad D_2 F_1 = 2g, \quad D_3 F_1 = 0, \quad D_4 F_1 = 0, \quad D_5 F_1 = 0, \quad D_6 F_1 = 2w,$$

$$D_1 F_2 = \frac{2f}{x^2}, \quad D_2 F_2 = \frac{2g}{y^2}, \quad D_3 F_2 = -\frac{2f^2}{x^3}, \quad D_4 F_2 = -\frac{2g^2}{y^3},$$

$$D_5 F_2 = -\frac{2w^2}{z^3}, \quad D_6 F_2 = \frac{2w}{z^2}.$$

If we take $\delta = \frac{1}{2}$, then any point (x, y, z, w) in $B_{\frac{1}{2}}(2, 1, 1, -2)$ satisfy $xyzw \neq 0$. Otherwise, assume for example that $y = 0$, then we obtain

$$(x - 2)^2 + (0 - 1)^2 + (z - 1)^2 + (w + 2)^2 < \frac{1}{4}$$

which is a contradiction. In other words, all partial derivatives $D_i F_j$ exist and continuous on $B_{\frac{1}{2}}(2, 1, 1, -2)$ so that Theorem 15.11 implies $\mathbf{F} \in \mathscr{C}'(B_{\frac{1}{2}}(2, 1, 1, -2))$. Finally, since

$$\det \begin{pmatrix} \frac{\partial F_1}{\partial f} & \frac{\partial F_1}{\partial g} \\ \frac{\partial F_2}{\partial f} & \frac{\partial F_2}{\partial g} \end{pmatrix} = \begin{pmatrix} 2f & 2g \\ \frac{2f}{x^2} & \frac{2g}{y^2} \end{pmatrix} = 4fg\left(\frac{1}{y^2} - \frac{1}{x^2}\right)$$

which is nonzero when $f = 4$, $g = 3$, $x = 2$ and $y = 1$, the Implicit Function Theorem ensures the existence of such functions f and g, completing the proof of the problem. ∎

15.5 Higher Order Derivatives

Problem 15.28

(★) Define $f : \mathbb{R}^2 \to \mathbb{R}$ by

$$f(x, y) = \begin{cases} xy \cdot \dfrac{x^2 - y^2}{x^2 + y^2}, & \text{if } (x, y) \neq (0, 0); \\ 0, & \text{otherwise.} \end{cases}$$

Prove that $(D_{12} f)(0, 0) \neq (D_{21} f)(0, 0)$.

Proof. Now direct computation gives

$$(D_1 f)(x,y) = \begin{cases} y \cdot \dfrac{x^2 - y^2}{x^2 + y^2} + \dfrac{4x^2 y^3}{(x^2 + y^2)^2}, & \text{if } (x,y) \neq (0,0); \\ 0, & \text{otherwise} \end{cases}$$

and

$$(D_2 f)(x,y) = \begin{cases} x \cdot \dfrac{x^2 - y^2}{x^2 + y^2} - \dfrac{4x^3 y^2}{(x^2 + y^2)^2}, & \text{if } (x,y) \neq (0,0); \\ 0, & \text{otherwise.} \end{cases}$$

Finally, we have

$$(D_{12} f)(0,0) = \lim_{h \to 0} \frac{(D_2 f)(h,0) - (D_2 f)(0,0)}{h} = \lim_{h \to 0} \frac{1}{h} \times \left(h \cdot \frac{h^2 - 0^2}{h^2 + 0^2} \right) = 1$$

and

$$(D_{21} f)(0,0) = \lim_{h \to 0} \frac{(D_1 f)(0,h) - (D_1 f)(0,0)}{h} = \lim_{h \to 0} \frac{1}{h} \times \left(h \cdot \frac{0^2 - h^2}{0^2 + h^2} \right) = -1$$

which imply that

$$(D_{12} f)(0,0) \neq (D_{21} f)(0,0),$$

completing the analysis of the problem. ∎

Problem 15.29

(⋆) *Consider the function* $f : \mathbb{R}^2 \to \mathbb{R}$ *given by*

$$f(x,y) = \begin{cases} \dfrac{x^2 y^2}{x^2 + y^2}, & \text{if } (x,y) \neq (0,0); \\ 0, & \text{otherwise.} \end{cases}$$

Prove that $(D_{12} f)(0,0) = (D_{21} f)(0,0)$ *but* $D_{21} f$ *is discontinuous at* $(0,0)$.

Proof. By the definition, we have

$$(D_1 f)(x,y) = \begin{cases} \dfrac{2xy^4}{(x^2 + y^2)^2}, & \text{if } (x,y) \neq (0,0); \\ 0, & \text{otherwise} \end{cases} \qquad (15.32)$$

and

$$(D_2 f)(x,y) = \begin{cases} \dfrac{2x^4 y}{(x^2 + y^2)^2}, & \text{if } (x,y) \neq (0,0); \\ 0, & \text{otherwise.} \end{cases}$$

15.5. Higher Order Derivatives

Furthermore, we have
$$(D_{12}f)(0,0) = \lim_{h \to 0} \frac{(D_2 f)(h,0) - (D_2 f)(0,0)}{h} = 0$$
and
$$(D_{21}f)(0,0) = \lim_{h \to 0} \frac{(D_1 f)(0,h) - (D_1 f)(0,0)}{h} = 0$$
so that
$$(D_{12}f)(0,0) = (D_{21}f)(0,0).$$

However, we get from the definition (15.32) that
$$(D_{21}f)(x,y) = \begin{cases} \dfrac{8x^3 y^3}{(x^2 + y^2)^3}, & \text{if } (x,y) \neq (0,0); \\ 0, & \text{otherwise.} \end{cases}$$

Therefore, when $x = y = h$, we have
$$\lim_{h \to 0} (D_{21}f)(h,h) = \lim_{h \to 0} \frac{8h^6}{(2h^2)^3} = 1 \neq 0 = (D_{21}f)(0,0).$$

Consequently, $(D_{21}f)(x,y)$ is not continuous at $(0,0)$, completing the proof of the problem. ∎

Remark 15.7

(a) Problem 15.28 tells us that it is *not* always true that $(D_{12}f)(\mathbf{p}) = (D_{21}f)(\mathbf{p})$.

(b) Problem 15.29 is a counterexample that the converse of Theorem 15.12 (Clairaut's Theorem) is false.

Problem 15.30

(⋆) Suppose that the third order partial derivatives of the function $f : \mathbb{R}^2 \to \mathbb{R}$ are continuous on $\mathbb{R}^2$. Prove that
$$D_{122}f = D_{212}f = D_{221}f$$
on $\mathbb{R}^2$.

Proof. If the third order partial derivatives are all continuous on $\mathbb{R}^2$, then so are the second order ones. Hence, by repeated applications of Theorem 15.12 (Clairaut's Theorem), we conclude that
$$D_{221}f = D_2(D_{21}f) = D_2(D_{12}f) = D_{212}f = D_{21}(D_2 f) = D_{12}(D_2 f) = D_{122}f.$$

This completes the proof of the problem. ∎

Problem 15.31

(⋆)(⋆) Prove Theorem 15.12 (Clairaut's Theorem).

350 Chapter 15. Differential Calculus of Functions of Several Variables

Proof. If the result holds for each component of **f**, then it is true for the function **f**. Therefore, we may assume that we are working with $f : \mathbb{R}^n \to \mathbb{R}$. Furthermore, we prove the theorem for $\mathbf{p} = \mathbf{0}$. Otherwise, we can replace the function $f(\mathbf{x})$ by $f(\mathbf{x} + \mathbf{p})$. The case is trivial if $i = j$, so we assume that $i \neq j$. Suppose that

$$A = (D_{ij}f)(\mathbf{0}) \quad \text{and} \quad B = (D_{ji}f)(\mathbf{0}).$$

Given $\epsilon > 0$. Since $D_{ij}f$ and $D_{ji}f$ are continuous at $\mathbf{0}$, there exists a $\delta > 0$ such that

$$|(D_{ij}f)(\mathbf{x}) - A| < \frac{\epsilon}{2} \quad \text{and} \quad |(D_{ji}f)(\mathbf{x}) - B| < \frac{\epsilon}{2} \tag{15.33}$$

whenever $|\mathbf{x}| < 2\delta$.

Next, we consider

$$J = f(\delta \mathbf{e}_i + \delta \mathbf{e}_j) - f(\delta \mathbf{e}_i) - [f(\delta \mathbf{e}_j) - f(\mathbf{0})]. \tag{15.34}$$

Applying the Second Fundamental Theorem of Calculus to the variable in $\mathbf{e}_i$, we see that

$$f(\delta \mathbf{e}_i + \delta \mathbf{e}_j) - f(\delta \mathbf{e}_j) = \int_0^\delta (D_i f)(x_i \mathbf{e}_i + \delta \mathbf{e}_j)\, dx_i \tag{15.35}$$

and

$$f(\delta \mathbf{e}_i) - f(\mathbf{0}) = \int_0^\delta (D_i f)(x_i \mathbf{e}_i)\, dx_i. \tag{15.36}$$

Now we substitute the expressions (15.35) and (15.36) into the definition (15.34) to get

$$J = \int_0^\delta \left[(D_i f)(x_i \mathbf{e}_i + \delta \mathbf{e}_j) - (D_i f)(x_i \mathbf{e}_i)\right] dx_i. \tag{15.37}$$

By applying the Mean Value Theorem for Derivatives to the differentiable function $D_i f$ with respect to the variable in $\mathbf{e}_j$, we obtain

$$(D_i f)(x_i \mathbf{e}_i + \delta \mathbf{e}_j) - (D_i f)(x_i \mathbf{e}_i) = (\delta - 0)[D_j(D_i f)](x_i \mathbf{e}_i + x_j \mathbf{e}_j)$$
$$= \delta (D_{ji} f)(x_i \mathbf{e}_i + x_j \mathbf{e}_j)$$

for some $x_j \in (0, \delta)$. Since $x_i, x_j \in (0, \delta)$, we know that $|x_i \mathbf{e}_i + x_j \mathbf{e}_j| < 2\delta$ so that the first inequality (15.33) gives

$$|(D_i f)(x_i \mathbf{e}_i + \delta \mathbf{e}_j) - (D_i f)(x_i \mathbf{e}_i) - \delta A| = |\delta(D_{ji} f)(x_i \mathbf{e}_i + x_j \mathbf{e}_j) - \delta A|$$
$$= \delta |(D_{ji} f)(x_i \mathbf{e}_i + x_j \mathbf{e}_j) - A|$$
$$< \frac{\delta \epsilon}{2}$$

or equivalently

$$-\frac{\delta \epsilon}{2} < (D_i f)(x_i \mathbf{e}_i + \delta \mathbf{e}_j) - (D_i f)(x_i \mathbf{e}_i) - \delta A < \frac{\delta \epsilon}{2}. \tag{15.38}$$

Here we integrable each part of the inequalities (15.38) from 0 to δ and then use the representation (15.37) to get

$$|J - \delta^2 A| < \frac{\delta^2 \epsilon}{2}. \tag{15.39}$$

Similarly, we can verify that

$$|J - \delta^2 B| < \frac{\delta^2 \epsilon}{2}. \tag{15.40}$$

15.5. Higher Order Derivatives

Finally, we deduce immediately from the inequalities (15.39) and (15.40) that

$$|\delta^2 A - \delta^2 B| \leq |\delta^2 A - J| + |J - \delta^2 B| < \frac{\delta^2 \epsilon}{2} + \frac{\delta^2 \epsilon}{2} = \delta^2 \epsilon$$

which implies that
$$|A - B| < \epsilon.$$

Since ϵ is arbitrary, we conclude that
$$A = B$$

which completes the proof of the problem. ∎

CHAPTER **16**

Integral Calculus of Functions of Several Variables

16.1 Fundamental Concepts

In this chapter, we study the Riemann integral of real-valued functions of several real variables. This is a direct generalization of the Riemann integral of functions of one variable which we have investigated in Chapter 9. The main references of this chapter are [3, Chap. 14], [26, Chap. 1], [28, §1.1], [29, Chap. 12] and [34, Chap. 11].

16.1.1 The Measure of Intervals in $\mathbb{R}^n$

Definition 16.1 (Intervals in $\mathbb{R}^n$). For $1 \leq k \leq n$, consider $-\infty < a_k \leq b_k < \infty$. Then each $I_k = [a_k, b_k]$ is an (closed) interval in $\mathbb{R}$. We call I an (closed) n-**dimensional interval in** $\mathbb{R}^n$ if I has the form

$$I = I_1 \times I_2 \times \cdots \times I_n = \{(x_1, \ldots, x_n) \,|\, x_k \in I_k \text{ for } 1 \leq k \leq n\}. \tag{16.1}$$

Definition 16.2 (The Measure of Intervals in $\mathbb{R}^n$). The **measure** or **volume** of the interval I in the form (16.1), denoted by $\mu(I)$, is defined to be

$$\mu(I) = \mu(I_1) \times \mu(I_2) \times \cdots \times \mu(I_n) = \prod_{k=1}^{n}(b_k - a_k). \tag{16.2}$$

When $n = 1, 2$ and 3, the value (16.2) corresponds to the **length**, the **area** and the **volume of** I.[a]

Theorem 16.3. *Suppose that* $I, \mathcal{I}_1, \mathcal{I}_2, \ldots, \mathcal{I}_m$ *are intervals in* $\mathbb{R}^n$ *having the form (16.1). Then the measure of the interval* I *in* $\mathbb{R}^n$ *has the following properties:*

---

[a] It is easily seen that $\mu(I) = 0$ if $\mu(I_k) = 0$ for some k. Furthermore, if each I_k is an open interval in $\mathbb{R}$, then the form (16.1) gives us an open interval I in $\mathbb{R}^n$ and it has the *same* measure as its corresponding closed interval.

(a) If $I \subseteq \mathcal{I}_1 \cup \mathcal{I}_2 \cup \cdots \cup \mathcal{I}_m$, then we have

$$\mu(I) \leq \sum_{k=1}^{m} \mu(\mathcal{I}_k).$$

(b) If $I = \bigcup_{k=1}^{m} \mathcal{I}_k$ and no two of $\mathcal{I}_1, \mathcal{I}_2, \ldots, \mathcal{I}_m$ have a common interior point (i.e., $\mathcal{I}_s^\circ \cap \mathcal{I}_t^\circ = \varnothing$ if $s \neq t$), then we have

$$\mu(I) = \sum_{k=1}^{m} \mu(\mathcal{I}_k).$$

Readers should discover the similarity between Theorem 16.3 and the inequality (12.1) and Theorem 12.11 (Countably Additivity).

16.1.2 The Riemann Integral in $\mathbb{R}^n$

Definition 16.4 (Partitions of I). Suppose that $I = I_1 \times \cdots \times I_n$ is an interval in $\mathbb{R}^n$. Let P_k be a partition of I_k.[b] Then the product

$$P = P_1 \times P_2 \times \cdots \times P_n \qquad (16.3)$$

is called a **partition of** I.[c] A partition P' of I is said to be **finer than** the partition P of I if $P \subseteq P'$.

It is clear that if P_k makes I_k into m_k subintervals, then the partition (16.3) decomposes I into a union of $(m_1 \times m_2 \times \cdots \times m_n)$ **subintervals in** $\mathbb{R}^n$. See Figure 16.1 for an example of a collection of subintervals in $\mathbb{R}^2$, where each small rectangle is a subinterval of the large rectangle.

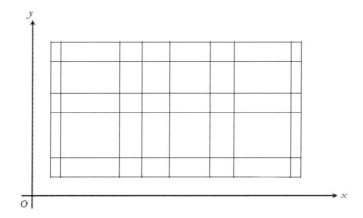

Figure 16.1: The subinterval in $\mathbb{R}^2$.

---

[b] For the definition of a partition of $[a,b]$.
[c] Some books call it a **grid of** I.

16.1. Fundamental Concepts

Definition 16.5 (Riemann Sum). Suppose that I is an interval in $\mathbb{R}^n$ and $f : I \to \mathbb{R}$ is a bounded function. Let P be a partition of I into m subintervals $I_1, I_2, \ldots, I_m$ and $\mathbf{t}_k \in I_k$. Then the sum

$$S(P, f) = \sum_{k=1}^{m} f(\mathbf{t}_k) \mu(I_k)$$

is called a **Riemann sum**.

Definition 16.6 (Riemann Integrable on I). The function $f : I \subseteq \mathbb{R}^n \to \mathbb{R}$ is called **Riemann integrable on** I, denoted by $f \in \mathscr{R}$ on I, if there corresponds a **unique** real number A whenever for every $\epsilon > 0$, there exists a partition P_ϵ of I such that

$$|S(P, f) - A| < \epsilon$$

for every partition P finer than P_ϵ. When this number exists, we write

$$A = \int_I f\, d\mathbf{x} = \int_I f(x_1, x_2, \ldots, x_n)\, dx_1\, dx_2 \cdots dx_n. \tag{16.4}$$

Remark 16.1

When $n = 2$ and $n = 3$ in the integral (16.4), it is our usual **double** and **triple integral** and they are denoted by

$$\iint_I f(x, y)\, dx\, dy \quad \text{and} \quad \iiint_I f(x, y, z)\, dx\, dy\, dz.$$

respectively.

Definition 16.7 (Lower and Upper Riemann Integrals). Suppose that I is an interval in $\mathbb{R}^n$ and $f : I \to \mathbb{R}$ is a bounded function. Let P be a partition of I into p subintervals $I_1, I_2, \ldots, I_p$ and let

$$m_k(f) = \inf\{f(\mathbf{x}) \,|\, \mathbf{x} \in I_k\} \quad \text{and} \quad M_k(f) = \sup\{f(\mathbf{x}) \,|\, \mathbf{x} \in I_k\},$$

where $k = 1, 2, \ldots, p$. The two sums

$$L(P, f) = \sum_{k=1}^{p} m_k(f) \mu(I_k) \quad \text{and} \quad U(P, f) = \sum_{k=1}^{p} M_k(f) \mu(I_k)$$

are called the **lower** and **upper Riemann sums** respectively. Then the **lower** and **upper Riemann integrals** of f on I are defined by

$$\underline{\int_I} f\, d\mathbf{x} = \sup\{L(P, f) \,|\, P \text{ is a partition of } I\}$$

and

$$\overline{\int_I} f\, d\mathbf{x} = \inf\{U(P, f) \,|\, P \text{ is a partition of } I\}$$

respectively.

16.1.3 Criteria for Integrability on Intervals

Theorem 16.8 (The Riemann Integrability Condition). We have $f \in \mathscr{R}$ on the interval I in $\mathbb{R}^n$ if and only if for every $\epsilon > 0$, there exists a partition P_ϵ such that

$$U(P_\epsilon, f) - L(P_\epsilon, f) < \epsilon. \tag{16.5}$$

In addition, the inequality (16.5) holds for every refinement of P_ϵ and the inequality (16.5) is also equivalent to the condition that

$$\underline{\int_I} f \, d\mathbf{x} = \overline{\int_I} f \, d\mathbf{x} = \int_I f \, d\mathbf{x}.$$

Definition 16.9 (n-measure zero). A subset $S \subseteq \mathbb{R}^n$ is of n-**measure zero**, i.e., $\mu(S) = 0$, provided for every $\epsilon > 0$, there exists a countable collection of n-dimensional intervals $I_1, I_2, \ldots$ such that

$$S \subseteq \bigcup_{k=1}^\infty I_k \quad \text{and} \quad \sum_{k=1}^\infty \mu(I_k) < \epsilon.$$

Theorem 16.10 (The Lebesgue's Integrability Condition). Suppose that I is an interval in $\mathbb{R}^n$ and $f : I \to \mathbb{R}$ is a bounded function. Then we have $f \in \mathscr{R}$ on I if and only if the set of discontinuities of f in I has n-measure zero. Particularly, we have $f \in \mathscr{R}$ on I for every continuous function $f : I \to \mathbb{R}$.

> **Remark 16.2**
>
> (a) In particular, if $\mu(S_k) = 0$ for $k = 1, 2, \ldots$ and $S \subseteq S_1 \cup S_2 \cup \cdots$, then we have $\mu(S) = 0$.
>
> (b) It is easy to see from Definition 16.7 (Lower and Upper Riemann Integrals) that if $f(\mathbf{x}) = 0$ on I, then $L(P, f) = U(P, f) = 0$ for every partition P of I. Hence Theorem 16.8 (The Riemann Integrability Condition) implies that $f \in \mathscr{R}$ on I and
>
> $$\int_I f \, d\mathbf{x} = \int_I 0 \, d\mathbf{x} = 0. \tag{16.6}$$
>
> (c) We suggest the reader to compare our Theorems 16.8 (The Riemann Integrability Condition) and 16.10 (The Lebesgue's Integrability Condition) with Theorems 9.2 (The Riemann Integrability Condition) and 9.4 (The Lebesgue's Integrability Condition) respectively.

16.1.4 Jordan Measurable Sets in $\mathbb{R}^n$

Roughly speaking, the **Jordan measure** (or **Jordan content**) in $\mathbb{R}^n$ is an extension of the notion of size (length, area and volume) to more complicated shapes other than triangles or disks.

Definition 16.11 (Jordan Measure). Suppose that $E \subseteq I$, where I is an interval in the form (16.1). We define the **outer Jordan measure** of E to be

$$J^*(E) = \inf \sum_{k=1}^N \mu(I_k),$$

16.1. Fundamental Concepts

where the infimum runs through over all *finite* coverings $E \subseteq \bigcup_{k=1}^{N} I_k$ by intervals in $\mathbb{R}^n$. Similarly, the **inner Jordan measure** of E is defined to be

$$J_*(E) = \sup \sum_{k=1}^{N} \mu(I_k),$$

where the supremum runs through over all unions of *finite* intervals in $\mathbb{R}^n$ which are contained in E. In the case that

$$J^*(E) = J_*(E),$$

the set E is said to be **Jordan measurable** and we denote this common value by $J(E)$ which is called the **Jordan measure** or **Jordan content** of E.

For examples, every finite subset of $\mathbb{R}^n$ is Jordan measurable and all open and closed balls in $\mathbb{R}^n$ are Jordan measurable too. In Figure 16.2, the sum of the areas of the orange rectangles is an *inside* approximation of the area of the Jordan measurable set in $\mathbb{R}^2$. Similarly, the sum of the areas of the green rectangles is an *outside* approximation of the area of the Jordan measurable set in $\mathbb{R}^2$.

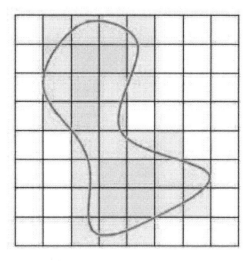

Figure 16.2: The outer and the inner Jordan measures.

Remark 16.3

Readers should compare the similarities between Definition 16.11 (Jordan Measure) and Definition 12.2 (Lebesgue Outer Measure).

Definition 16.12 (Boundary Points). *We call that a point $\mathbf{x} \in E \subseteq \mathbb{R}^n$ is a **boundary point** of E if*

$$N_r(\mathbf{x}) \cap E \neq \varnothing \quad \text{and} \quad N_r(\mathbf{x}) \cap (\mathbb{R}^n \setminus E) \neq \varnothing$$

for every $r > 0$. The set of all boundary points of E is called the **boundary** of E and it is denoted by ∂E.

Theorem 16.13. *Suppose that E is a bounded set of $\mathbb{R}^n$. Then we have*

$$J^*(\partial E) = J^*(E) - J_*(E).$$

In addition, E is Jordan measurable if and only if $J(\partial E) = 0$.

It is easy to see that if $J(E) = 0$, then $\mu(E) = 0$. However, the converse is not true. See Problem 16.2 for proofs of these. The following theorem tells us some basic properties of Jordan measurable sets.

Theorem 16.14 (Properties of Jordan Measurable Sets). *Suppose that E and F are Jordan measurable. Then $E \cup F$, $E \cap F$ and $E \setminus F$ are also Jordan measurable. Furthermore, we have*

(a) $J(E \cup F) = J(E) + J(F) - J(E \cap F)$.

(b) *if* $E \cap F = \emptyset$, *then* $J(E \cup F) = J(E) + J(F)$.

(c) *if* $F \subseteq E$, *then* $J(E \setminus F) = J(E) - J(F)$.

16.1.5 Integration on Jordan Measurable Sets

Definition 16.15 (Riemann Integrable on E). *Suppose that E is a Jordan measurable set of $\mathbb{R}^n$, f is bounded on E and I is an interval in the form (16.1) containing E. Define $g : I \to \mathbb{R}$ by*

$$g(\mathbf{x}) = \begin{cases} f(\mathbf{x}), & \text{if } \mathbf{x} \in E; \\ 0, & \text{if } \mathbf{x} \in I \setminus E. \end{cases}$$

We say that $f \in \mathscr{R}$ on E if and only if $g \in \mathscr{R}$ on I and in this case, we have

$$\int_E f \, d\mathbf{x} = \int_I g \, d\mathbf{x}.$$

Theorem 16.16. *Suppose that E is a Jordan measurable set of $\mathbb{R}^n$. Then we have $f \in \mathscr{R}$ on E if and only if the set of discontinuous points of f in E has n-measure 0. In particular, $f \in \mathscr{R}$ on E for every continuous function $f : E \to \mathbb{R}$.*

Theorem 16.17 (Properties of Integration on E). *Suppose that E is a Jordan measurable set in $\mathbb{R}^n$, $f, g : E \to \mathbb{R}$ and $\alpha \in \mathbb{R}$.*

(a) *If $f, g \in \mathscr{R}$ on E, then so are $f + g$ and αf. In fact, we have*

$$\int_E (f + g) \, d\mathbf{x} = \int_E f \, d\mathbf{x} + \int_E g \, d\mathbf{x}$$

and

$$\int_E \alpha f \, d\mathbf{x} = \alpha \int_E f \, d\mathbf{x}.$$

16.1. Fundamental Concepts

(b) If $f, g \in \mathscr{R}$ on E and $f(\mathbf{x}) \leq g(\mathbf{x})$ for all $\mathbf{x} \in E$, then we have
$$\int_E f \, d\mathbf{x} \leq \int_E g \, d\mathbf{x}.$$

(c) If A and B are Jordan measurable sets satisfying $A \cap B = \varnothing$ and $E = A \cup B$ and $f \in \mathscr{R}$ on E, then we have $f \in \mathscr{R}$ on A and on B and furthermore,
$$\int_E f \, d\mathbf{x} = \int_A f \, d\mathbf{x} + \int_B f \, d\mathbf{x}.$$

The following famous theorem suggests a practical way to evaluate double integrals on a rectangle in terms of singles integrals. For multiple integrals on an interval in $\mathbb{R}^n$, please read Problem 16.11.

Fubini's Theorem. *Suppose that $R = [a, b] \times [c, d]$ and $f : R \to \mathbb{R}$ is well-defined. Furthermore, suppose that for each $x \in [a, b]$, $g(y) = f(x, \cdot) \in \mathscr{R}$ on $[c, d]$ and for each $y \in [c, d]$, $h(x) = f(\cdot, y) \in \mathscr{R}$ on $[a, b]$. If $f \in \mathscr{R}$ on R, then we have*
$$\int_R f(x, y) \, d(x, y) = \iint_R f(x, y) \, dx \, dy = \int_a^b \Big(\int_c^d f(x, y) \, dy \Big) \, dx = \int_c^d \Big(\int_a^b f(x, y) \, dx \Big) \, dy.$$

For evaluation of integrals on a general Jordan measurable set in terms of iterated integrals, we consider the concept of a projectable region first. We call a set $E \subseteq \mathbb{R}^n$ a **projectable region** if there exists a closed Jordan measurable set $F \subseteq \mathbb{R}^{n-1}$, $k \in \{1, 2, \ldots, n\}$ and continuous functions $\phi, \varphi : F \subseteq \mathbb{R}^{n-1} \to \mathbb{R}$ such that
$$E = \{\mathbf{x} \in \mathbb{R}^n \, | \, \widehat{\mathbf{x}}_k \in F \text{ and } \phi(\widehat{\mathbf{x}}_k) \leq x_k \leq \varphi(\widehat{\mathbf{x}}_k)\}, \tag{16.7}$$
where $\widehat{\mathbf{x}}_k = (x_1, \ldots, x_{k-1}, x_{k+1}, \ldots, x_n)$. Then we have

Theorem 16.18. *Suppose that E is a projectable region in $\mathbb{R}^n$ with k, F, ϕ and φ as given in the definition (16.7). Then E is a Jordan measurable set and if $f \in \mathscr{R}$ on E, then*
$$\int_E f \, d\mathbf{x} = \int_F \Big(\int_{\phi(\widehat{\mathbf{x}}_k)}^{\varphi(\widehat{\mathbf{x}}_k)} f(\widehat{\mathbf{x}}_k) \, dx_k \Big) \, d\widehat{\mathbf{x}}_k.$$

16.1.6 Two Important Theorems

In Chapter 9, we study the Mean Value Theorems for Integrals and the Change of Variables Theorem in the case of a single variable. Here we present their multi-dimensional versions as follows:

The Mean Value Theorem for Multiple Integrals. *Suppose that E is a Jordan measurable set of $\mathbb{R}^n$. Let $f, g \in \mathscr{R}$ on E and $g(\mathbf{x}) \geq 0$ on E. Denote*
$$M = \sup\{f(\mathbf{x}) \, | \, \mathbf{x} \in E\} \quad \text{and} \quad m = \inf\{f(\mathbf{x}) \, | \, \mathbf{x} \in E\}.$$
Then there is a number $\lambda \in [m, M]$ such that
$$\int_E fg \, d\mathbf{x} = \lambda \int_E g \, d\mathbf{x}.$$
Particularly, we have
$$mJ(E) \leq \int_E f \, d\mathbf{x} \leq MJ(E).$$

The Change of Variables Theorem. Suppose that V is open in $\mathbb{R}^n$, $\phi : V \to \mathbb{R}^n$ is continuously differentiable and injective on V. If $\det \mathbf{J}_\phi(\mathbf{x}) \neq 0$ for all $\mathbf{x} \in V$, E is a Jordan measurable set with $\overline{E} \subseteq V$, $f \circ \phi \in \mathscr{R}$ on E and $f \in \mathscr{R}$ on $\phi(E)$, then we have

$$\int_{\phi(E)} f(\mathbf{y})\,\mathrm{d}\mathbf{y} = \int_E f(\phi(\mathbf{x})) \cdot |\det \mathbf{J}_\phi(\mathbf{x})|\,\mathrm{d}\mathbf{x}. \tag{16.8}$$

Recall that **polar coordinates** in $\mathbb{R}^2$ have the form

$$x = r\cos\theta \quad \text{and} \quad y = r\sin\theta,$$

where r is the distance between (x,y) and the origin and θ is the angle between the positive x-axis and the line connecting (x,y) and $(0,0)$. Set $\phi(r,\theta) = (r\cos\theta, r\sin\theta)$ so that

$$\det \mathbf{J}_\phi((r,\theta)) = \det\begin{pmatrix} \cos\theta & -r\sin\theta \\ \sin\theta & r\cos\theta \end{pmatrix} = r.$$

Obviously, ϕ is continuously differentiable on any open set V in $\mathbb{R}^2$ by Theorem 15.11. Furthermore, ϕ is injective and $\det \mathbf{J}_\phi(r,\theta) \neq 0$ on V if V does not intersect the set $\{(0,\theta)\,|\,\theta \in \mathbb{R}\}$. Hence the formula (16.8) becomes

$$\int_{\phi(E)} f(x,y)\,\mathrm{d}x\,\mathrm{d}y = \int_E f(r\cos\theta, r\sin\theta)\, r\,\mathrm{d}r\,\mathrm{d}\theta. \tag{16.9}$$

Remark 16.4

(a) There are two common, helpful and practical changes of variables in $\mathbb{R}^3$. They are the **cylindrical coordinates** and the **spherical coordinates** which are given by

$$x = r\cos\theta, \quad y = r\sin\theta, \quad z = z$$

and

$$x = r\sin\varphi\cos\theta, \quad y = r\sin\varphi\sin\theta, \quad z = r\cos\theta$$

respectively.

(b) In [29, p. 426], Wade points out that the formula (16.9) holds even though ϕ is not injective or $\det \mathbf{J}_\phi((r,\theta)) = 0$ in the whole $r\theta$-plane. Similar situations happen for the cylindrical coordinates and the spherical coordinates.

(c) Wade calls this Change of Variables Theorem a global version which can be shown by a local version of the theorem, see [29, Lemma 12.45, p. 424].

16.2 Jordan Measurable Sets

Problem 16.1

(★) Prove that $J^*(E) = J^*(\overline{E})$ for every $E \subseteq \mathbb{R}$.

16.2. Jordan Measurable Sets

Proof. Since $E \subseteq \overline{E}$, we always have

$$E \subseteq \overline{E} \subseteq \bigcup_{k=1}^{N} I_k$$

so that $J^*(E) \leq J^*(\overline{E})$. To prove the reverse direction, it suffices to show that if $E \subseteq \bigcup_{k=1}^{N} I_k$, then $\overline{E} \subseteq \bigcup_{k=1}^{N} I_k$. To this end, let $I = \bigcup_{k=1}^{N} I_k$. Assume that $p \in \overline{E}$ but $p \notin I$. Particularly, $p \notin E$. Since $\overline{E} = E \cup E'$, we have $p \in E'$ which implies that $(p-\epsilon, p+\epsilon) \cap E \neq \emptyset$ for every $\epsilon > 0$ and then

$$(p-\epsilon, p+\epsilon) \cap I \neq \emptyset$$

for every $\epsilon > 0$. Thus p is a limit point of I. By Definition 16.1 (Intervals in $\mathbb{R}^n$), I is closed in $\mathbb{R}^n$, so $p \in I$ which is a contradiction. Therefore, it is true that $\overline{E} \subseteq I$ and then $J^*(\overline{E}) \leq J^*(E)$. Consequently, we get the desired result that

$$J^*(E) = J^*(\overline{E}).$$

This completes the proof of the problem. ∎

Problem 16.2

(⋆) *Prove that if $J(E) = 0$, then $\mu(E) = 0$. Show also that the converse is not true.*

Proof. Suppose that $J(E) = 0$. Given $\epsilon > 0$. By Definition 16.11 (Jordan Measure), there is a *finite* collection of intervals $\{I_1, I_2, \ldots, I_N\}$ whose union covers E such that

$$\sum_{k=1}^{N} \mu(I_k) < \epsilon. \tag{16.10}$$

If we consider the countable collection $\{I_1, I_2, \ldots, I_N, I_{N+1}, \ldots\}$, where $I_n = \emptyset$ for all $n \geq N+1$, then its union also covers E. Now the estimate (16.10) and the fact $\mu(\emptyset) = 0$ certainly give

$$\sum_{k=1}^{\infty} \mu(I_k) < \epsilon.$$

By Theorem 16.3(a), we see that $\mu(E) = 0$.

Let $F = \mathbb{Q} \cap [0,1]$. We know from Problem 12.3 and Theorem 12.5(a) (Properties of Measurable Sets) that $m(F) = 0$. Since F is dense in $[0,1]$, we observe from Problem 16.1 that

$$J^*(F) = J^*(\overline{F}) = J^*([0,1]) = 1.$$

However, the density of irrationals in $[0,1]$ implies that F has *only* empty interval so that $J_*(F) = 0$. Since $J_*(F) \neq J^*(F)$, Definition 16.11 (Jordan Measure) shows that F is not Jordan measurable. This completes the proof of the problem. ∎

> **Remark 16.5**
>
> (a) By Problem 16.2, the same result (16.6) holds for any bounded f if $J(E) = 0$.
>
> (b) In addition, we recall that a rational $q \in [0,1]$ is Jordan measurable because $\{q\}$ is a finite set, but their union is $F = \mathbb{Q} \cap [0,1]$ in Problem 16.2 is not Jordan measurable. In other words, this counterexample shows that countable union of Jordan measurable sets is *not* necessarily Jordan measurable.

> **Problem 16.3**
>
> (★) If $E \subseteq F \subseteq \mathbb{R}^n$, prove that $J^*(E) \leq J^*(F)$.

Proof. Suppose that $\{I_1, I_2, \ldots, I_N\}$ is a finite collection of intervals in $\mathbb{R}^n$ covering F. Then we have

$$E \subseteq F \subseteq \bigcup_{k=1}^{N} I_k. \tag{16.11}$$

By Definition 16.11 (Jordan Measure), we have

$$J^*(E) \leq \sum_{k=1}^{N} \mu(I_k)$$

for every finite collection of intervals satisfying the set relations (16.11). By Definition 16.11 (Jordan Measure) again, we conclude that $J^*(E) \leq J^*(F)$ which completes the proof of the problem. ∎

> **Problem 16.4**
>
> (★) Suppose that E is a Jordan measurable subset of $\mathbb{R}^n$ and $J(E) = 0$. If $F \subseteq E$, prove that F is also Jordan measurable and $J(F) = 0$.

Proof. Since $J(E) = 0$, we have $J^*(E) = 0$ by Definition 16.11 (Jordan Measure). By Problem 16.3, we know that $J^*(F) = 0$. By the definition again, we always have

$$J_*(F) \leq J^*(F).$$

Thus we get $J_*(F) = J^*(F) = 0$ and the definition again shows that F is Jordan measurable and $J(F) = 0$. We complete the proof of the problem. ∎

> **Problem 16.5**
>
> (★) Let $E \subseteq \mathbb{R}$, prove that
> $$J^*(E) = \inf J^*(V),$$
> where the infimum takes over all open sets V containing E.

16.2. Jordan Measurable Sets

Proof. Let $S = \{J^*(V) \,|\, V \text{ is an open set containing } E\}$. If $J^*(E) = \infty$, then Problem 16.4 implies that $J^*(V) = \infty$ and our result follows. Therefore, we may assume that $J^*(E) < \infty$. By Problem 16.4 again, $J^*(E)$ is a lower bound of S. Now we want to show that for every $\epsilon > 0$, there is an open sets V containing E such that

$$J^*(V) \leq J^*(E) + \epsilon. \tag{16.12}$$

To see this, given $\epsilon > 0$, then the definition of the infimum implies that there exists a finite collection of intervals $\{I_1, I_2, \ldots, I_N\}$ in $\mathbb{R}^n$ covering E such that

$$\sum_{k=1}^{N} \mu(I_k) \leq J^*(E) + \frac{\epsilon}{2}. \tag{16.13}$$

Recall from Definition 16.1 (Intervals in $\mathbb{R}^n$) that there are $a_{k1}, b_{k1}, a_{k2}, b_{k2}, \ldots, a_{kn}, b_{kn}$ such that

$$I_k = [a_{k1}, b_{k1}] \times [a_{k2}, b_{k2}] \times \cdots \times [a_{kn}, b_{kn}],$$

where $k = 1, 2, \ldots, N$. Let $\delta > 0$ and we consider the open intervals

$$V_k = (a_{k1} - \delta, b_{k1} + \delta) \times (a_{k2} - \delta, b_{k2} + \delta) \times \cdots \times (a_{kn} - \delta, b_{kn} + \delta).$$

Then we have $I_k \subseteq V_k$ and

$$\mu(V_k) = \prod_{j=1}^{N} (b_{kj} - a_{kj} + 2\delta)$$

by the definition (16.2). Now we can make δ as small as possible so that

$$\mu(V_k) \leq \mu(I_k) + \frac{\epsilon}{2^{k+1}}. \tag{16.14}$$

If we define $V = \bigcup_{k=1}^{N} V_k$, then it is open in $\mathbb{R}^n$ and

$$E \subseteq \bigcup_{k=1}^{N} I_k \subseteq V,$$

so we follow from Definition 16.11 (Jordan Measure) and the inequalities (16.14) that

$$J^*(V) \leq \sum_{k=1}^{N} \mu(V_k) \leq \sum_{k=1}^{N} \left[\mu(I_k) + \frac{\epsilon}{2^{k+1}}\right] = \sum_{k=1}^{N} \mu(I_k) + \frac{\epsilon}{2}. \tag{16.15}$$

Combining the inequalities (16.13) and (16.15), we gain

$$J^*(V) \leq J^*(E) + \epsilon$$

which is exactly the required result (16.12). Hence we obtain $J^*(E) = \inf J^*(V)$ and this completes the proof of the problem. ∎

Problem 16.6

(⋆) Prove that there exists a non-Jordan measurable subset of $S = [0,1] \times [0,1]$.

Proof. We consider the subset
$$E = \{(x,y) \,|\, x, y \in \mathbb{Q} \cap [0,1]\}.$$

We note that an interval in $\mathbb{R}^2$ is actually a rectangle. On the one hand, it is clear that there is *no* rectangle contained in E because a rectangle must contain a point with irrational coordinates. Thus it means that
$$J_*(E) = 0.$$

On the other hand, let R be an interval contained in S. Then the density of $\mathbb{Q}$ shows that
$$R \cap E \neq \varnothing. \tag{16.16}$$

This observation means that if $E \subseteq \bigcup_{k=1}^{N} I_k$, then we have
$$\bigcup_{k=1}^{N} I_k = S.$$

Otherwise, since all I_k are rectangles, one can find a rectangle R such that
$$R \subseteq S \setminus \bigcup_{k=1}^{N} I_k$$

and then $R \cap E = \varnothing$ which contradicts the observation (16.16). Therefore, we have
$$J^*(E) = \mu(S) = 1.$$

Consequently, $J^*(E) \neq J_*(E)$ and it follows from Definition 16.11 (Jordan Measure) that E is not Jordan measurable set. This ends the analysis of the problem. ■

16.3 Integration on $\mathbb{R}^n$

Problem 16.7

(★) Suppose that E is a compact Jordan measurable set in $\mathbb{R}^n$. Prove that
$$J(E) = \int_E d\mathbf{x}.$$

Proof. Let I be an interval containing E in the form (16.1). By the definition, we have
$$\chi_E(\mathbf{x}) = \begin{cases} 1, & \text{if } \mathbf{x} \in E; \\ 0, & \text{if } \mathbf{x} \in I \setminus E. \end{cases}$$

Clearly, the set of discontinues points of χ_E in I are exactly ∂E. Since E is Jordan measurable, it follows from Theorem 16.13 that $J(\partial E) = 0$. By Problem 16.2, we have $\mu(\partial E) = 0$ and

16.3. Integration on $\mathbb{R}^n$

then Theorem 16.10 (The Lebegus's Integrability Condition) ensures that $\chi_E \in \mathscr{R}$ on I. By Definition 16.15 (Riemann Integrable on E), we conclude that

$$\int_E \mathrm{d}\mathbf{x}$$

exists and then Theorem 16.8 (The Riemann Integrability Condition) guarantees that

$$\overline{\int}_E \mathrm{d}\mathbf{x} = \underline{\int}_E \mathrm{d}\mathbf{x} = \int_E \mathrm{d}\mathbf{x}. \tag{16.17}$$

Next, we let P be a partition of I into subintervals $I_1, I_2, \ldots, I_p$ and

$$S = \{k \in \{1, 2, \ldots, p\} \mid I_k \cap E \neq \varnothing\}.$$

Therefore, we obtain

$$E \subseteq \bigcup_{k \in S} I_k$$

and

$$M_k(\chi_E) = \sup\{\chi_E(\mathbf{x}) \mid \mathbf{x} \in I_k\} = \begin{cases} 1, & \text{if } k \in S; \\ 0, & \text{if } k \notin S. \end{cases}$$

By Definition 16.7 (Lower and Upper Riemann Integrals) and then Definition 16.11 (Jordan Measure), we see that

$$U(P, \chi_E) = \sum_{k=1}^{p} M_k(\chi_E) \mu(I_k) = \sum_{k \in S} \mu(I_k)$$

which implies

$$\overline{\int}_I \chi_E \, \mathrm{d}\mathbf{x} = \inf\{U(P, \chi_E) \mid P \text{ is a partition of } I\}$$
$$= \inf\left\{\sum_{k \in S} \mu(I_k) \,\Big|\, E \subseteq \bigcup_{k \in S} I_k\right\}$$
$$= J^*(E). \tag{16.18}$$

Since E is Jordan measurable, the expression (16.18) reduces to

$$\overline{\int}_I \chi_E \, \mathrm{d}\mathbf{x} = J(E). \tag{16.19}$$

By combining the expressions (16.17) and (16.19), we establish that

$$J(E) = \overline{\int}_I \chi_E \, \mathrm{d}\mathbf{x} = \overline{\int}_E \mathrm{d}\mathbf{x} = \int_E \mathrm{d}\mathbf{x},$$

completing the proof of the problem. ∎

Remark 16.6

In fact, the condition that E is compact can be dropped in Problem 16.7.

> **Problem 16.8**
>
> (⋆) Prove Theorem 16.16.

Proof. Let I be an interval containing E. Define
$$g(\mathbf{x}) = \begin{cases} f(\mathbf{x}), & \text{if } \mathbf{x} \in E; \\ 0, & \text{if } \mathbf{x} \in I \setminus E. \end{cases}$$

Let $D_f(E)$ be the set of all discontinuities of f on E. By Theorem 16.10 (The Lebesgue's Integrability Condition) or Theorem 16.16, $g \in \mathscr{R}$ on I if and only if $\mu(D_g(I)) = 0$.

We notice that the discontinuities of f are also discontinuities of g and g may have *more* discontinuities on ∂E, so we obtain
$$D_g(I) = D_f(E) \cup D_g(\partial E).$$

Since $E^\circ \cap (\partial E)^\circ = \varnothing$, Theorem 16.3(b) implies that
$$\mu(D_g(I)) = \mu(D_f(E)) + \mu(D_g(\partial E)). \tag{16.20}$$

Next, it is obvious that $D_g(\partial E) \subseteq \partial E$. Since E is Jordan measurable, Theorem 16.13 shows that $J(\partial E) = 0$. Applying Problems 16.2 and 16.3, we conclude that $\mu(D_g(\partial E)) = 0$. Hence it follows from the expression (16.20) that $\mu(D_g(I)) = 0$ if and only if $\mu(D_f(E)) = 0$. Hence our desired result follows directly from Definition 16.15 (Riemann Integrable on E). We have completed the proof of the problem. ∎

> **Problem 16.9**
>
> (⋆) Prove Theorem 16.17(c).

Proof. We remark that $\chi_E = \chi_{A \cup B} = \chi_A + \chi_B - \chi_{A \cap B}$, so we have
$$\int_E f \, d\mathbf{x} = \int_{E \subseteq I} f\chi_E \, d\mathbf{x} = \int_I f\chi_A \, d\mathbf{x} + \int_I f\chi_B \, d\mathbf{x} - \int_I f\chi_{A \cap B} \, d\mathbf{x}. \tag{16.21}$$

By Remark 16.2(b), since $A \cap B = \varnothing$, $f\chi_{A \cap B} = 0$ on I and then
$$\int_I f\chi_{A \cap B} \, d\mathbf{x} = 0.$$

By this, the expression (16.21) reduces to
$$\int_E f \, d\mathbf{x} = \int_I f\chi_A \, d\mathbf{x} + \int_I f\chi_B \, d\mathbf{x}$$
which is our required result, completing the proof of the problem. ∎

16.3. Integration on $\mathbb{R}^n$

> **Problem 16.10**
>
> (★) Let $S = [0,1] \times [0,1]$. Prove that
> $$\int_S y^3 e^{xy^2} \, d(x,y) = \frac{e-2}{2}.$$

Proof. It is evident that f is Riemann integrable with respect to each variable. Furthermore, since $f(x,y) = y^3 e^{xy^2}$ is continuous on S, Theorem 16.10 (The Lebesgue's Integrability Condition) ensures that f satisfies the hypotheses of Fubini's Theorem. Hence we obtain

$$\int_S y^3 e^{xy^2} \, d(x,y) = \int_0^1 \int_0^1 y^3 e^{xy^2} \, dx \, dy$$
$$= \int_0^1 y^3 \left(\int_0^1 e^{xy^2} \, dx \right) dy$$
$$= \int_0^1 y(e^{y^2} - 1) \, dy$$
$$= \frac{e-2}{2}.$$

This completes the proof of the problem. ∎

> **Problem 16.11**
>
> (★) Suppose that $f_k \in \mathscr{R}$ on $I_k = [a_k, b_k]$, where $k = 1, 2, \ldots, n$. Verify that
> $$\int_I f_1(x_1) \cdots f_n(x_n) \, dx_1 \cdots dx_n = \left(\int_{a_1}^{b_1} f_1(x_1) \, dx_1 \right) \times \cdots \times \left(\int_{a_n}^{b_n} f_n(x_n) \, dx_n \right),$$
> where $I = I_1 \times \cdots \times I_n$.

Proof. For each $k = 1, 2, \ldots, n$, let D_k be the set of discontinuities of f_k on I_k. Let D be the set of discontinuities of the function
$$f = f_1 f_2 \cdots f_n$$
on I. By Theorem 16.10 (The Lebesgue's Integrability Condition), we see that $\mu(D_k) = 0$, where $k = 1, 2, \ldots, n$. It is evident that

$$D \subseteq \bigcup_{k=1}^n I_1 \times \cdots \times I_{k-1} \times D_k \times I_{k+1} \times \cdots \times I_n.$$

Thus we know from Theorem 16.3(a) and then Definition 16.2 (The Measure of Intervals in $\mathbb{R}^n$) that

$$\mu(D) \leq \sum_{k=1}^n \mu(I_1 \times \cdots \times I_{k-1} \times D_k \times I_{k+1} \times \cdots \times I_n)$$
$$= \sum_{k=1}^n \mu(I_1) \times \cdots \times \mu(I_{k-1}) \times \mu(D_k) \times \mu(I_{k+1}) \times \cdots \times \mu(I_n)$$

$$= 0.$$

By Theorem 16.10 (The Lebesgue's Integrability Condition), it yields that $f \in \mathscr{R}$ on I. By repeated use of Fubini's Theorem, we establish that

$$\int_I f_1(x_1) \cdots f_n(x_n) \, \mathrm{d}x_1 \cdots \mathrm{d}x_n = \int_{a_1}^{b_1} \Big(\int_{I_2 \times \cdots \times I_k} f_1(x_1) \cdots f_n(x_n) \, \mathrm{d}x_2 \cdots \mathrm{d}x_n \Big) \mathrm{d}x_1$$

$$= \Big(\int_{a_1}^{b_1} f_1(x_1) \, \mathrm{d}x_1 \Big) \times \Big(\int_{I_2 \times \cdots \times I_k} f_2(x_2) \cdots f_n(x_n) \, \mathrm{d}x_2 \cdots \mathrm{d}x_n \Big)$$

$$= \Big(\int_{a_1}^{b_1} f_1(x_1) \, \mathrm{d}x_1 \Big) \times \Big(\int_{a_2}^{b_2} f_2(x_2) \, \mathrm{d}x_2 \Big)$$

$$\times \Big(\int_{I_3 \times \cdots \times I_k} f_3(x_3) \cdots f_n(x_n) \, \mathrm{d}x_3 \cdots \mathrm{d}x_n \Big)$$

$$= \cdots$$

$$= \Big(\int_{a_1}^{b_1} f_1(x_1) \, \mathrm{d}x_1 \Big) \Big(\int_{a_2}^{b_2} f_2(x_2) \, \mathrm{d}x_2 \Big) \cdots \Big(\int_{a_n}^{b_n} f_n(x_n) \, \mathrm{d}x_n \Big).$$

We have completed the proof of the problem. ∎

Problem 16.12

(⋆) Suppose that $Q = [0,1] \times \cdots \times [0,1]$ and $\mathbf{y} = (1, 1, \ldots, 1)$. Prove that

$$\int_Q e^{-\mathbf{x} \cdot \mathbf{y}} \, \mathrm{d}\mathbf{x} = \Big(\frac{e-1}{e} \Big)^n.$$

Proof. If $\mathbf{x} = (x_1, x_2, \ldots, x_n)$, then we have $-\mathbf{x} \cdot \mathbf{y} = -(x_1 + x_2 + \cdots + x_n)$ and thus

$$e^{-\mathbf{x} \cdot \mathbf{y}} = e^{-x_1} e^{-x_2} \cdots e^{-x_n}.$$

Since $e^{-x_k} \in \mathscr{R}$ on $[0,1]$, it deduces from Problem 16.11 that

$$\int_Q e^{-\mathbf{x} \cdot \mathbf{y}} \, \mathrm{d}\mathbf{x} = \Big(\int_0^1 e^{-x_1} \, \mathrm{d}x_1 \Big) \times \cdots \times \Big(\int_0^1 e^{-x_n} \, \mathrm{d}x_n \Big) = \Big(\frac{e-1}{e} \Big)^n,$$

completing the proof of the problem. ∎

Problem 16.13

(⋆) Let $a < A$ and $b < B$. Suppose that $f(x,y) = \dfrac{\partial^2}{\partial x \partial y} F(x,y)$ is continuous on $Q = [a, A] \times [b, B]$ and

$$I = \int_Q f(x,y) \, \mathrm{d}(x,y).$$

Show that

$$I = F(A, B) - F(a, B) - F(A, b) + F(a, b).$$

16.3. Integration on $\mathbb{R}^n$

Proof. Since f is continuous on Q, it satisfies all the hypotheses of Fubini's Theorem. Therefore, we have
$$I = \int_b^B \left(\int_a^A f(x,y)\,\mathrm{d}x \right) \mathrm{d}y. \tag{16.22}$$
By the Second Fundamental Theorem of Calculus, we know that
$$\int_a^A f(x,y)\,\mathrm{d}x = \int_a^A \frac{\partial^2}{\partial x \partial y} F(x,y)\,\mathrm{d}x = \frac{\partial}{\partial y} F(x,y)\Big|_a^A = \frac{\partial}{\partial y} F(A,y) - \frac{\partial}{\partial y} F(a,y). \tag{16.23}$$
Substituting the result (16.23) into the integral (16.22), we get
$$I = \int_b^B \left[\frac{\partial}{\partial y} F(A,y) - \frac{\partial}{\partial y} F(a,y) \right] \mathrm{d}y = \int_b^B \frac{\partial}{\partial y} F(A,y)\,\mathrm{d}y - \int_b^B \frac{\partial}{\partial y} F(a,y)\,\mathrm{d}y.$$
Applying the Second Fundamental Theorem of Calculus again, we obtain immediately that
$$I = F(A,y)\Big|_b^B - F(a,y)\Big|_b^B = F(A,B) - F(A,b) - F(a,B) + F(a,b).$$
This completes the proof of the problem. ∎

Problem 16.14

(⋆) *Suppose that $S = [0,1] \times [0,1]$ and*
$$f(x,y) = \begin{cases} x + y - 1, & \text{if } x + y \leq 1; \\ 0, & \text{otherwise.} \end{cases} \tag{16.24}$$
Prove that
$$\int_S f\,\mathrm{d}(x,y) = -\frac{1}{6}.$$

Proof. Since f is continuous on S, it satisfies all the requirements of Fubini's Theorem. Thus we have
$$\int_S f\,\mathrm{d}(x,y) = \int_0^1 \left(\int_0^1 f(x,y)\,\mathrm{d}y \right) \mathrm{d}x. \tag{16.25}$$
By the definition (16.24), the integral (16.25) reduces to
$$\begin{aligned}
\int_S f\,\mathrm{d}(x,y) &= \int_0^1 \left(\int_0^{1-x} (x+y-1)\,\mathrm{d}y \right) \mathrm{d}x \\
&= \int_0^1 \left(xy + \frac{y^2}{2} - y \right)\Big|_0^{1-x}\,\mathrm{d}x \\
&= \int_0^1 \left[x(1-x) + \frac{(1-x)^2}{2} - (1-x) \right] \mathrm{d}x \\
&= -\frac{1}{2} \int_0^1 (1-x)^2\,\mathrm{d}x \\
&= -\frac{1}{6},
\end{aligned}$$
completing the proof of the problem. ∎

> **Problem 16.15**
>
> $(\star)(\star)$ Denote $S = [0,1] \times [0,1]$. Define $f : S \to \mathbb{R}$ by
>
> $$f(x,y) = \begin{cases} 0, & \text{if at least one of } x \text{ or } y \text{ is irrational;} \\ \frac{1}{n}, & \text{if } x, y \in \mathbb{Q} \text{ and } x = \frac{m}{n}, \end{cases} \quad (16.26)$$
>
> where m and n are relatively prime and $n > 0$. Prove that
>
> $$\int_0^1 f(x,y)\,dx = \int_0^1 \left(\int_0^1 f(x,y)\,dx\right) dy = \int_S f(x,y)\,d(x,y) = 0,$$
>
> but $f(x,y) \notin \mathscr{R}$ on $[0,1]$ for every rational x.

Proof. On $([0,1] \setminus \mathbb{Q}) \times [0,1]$, we have $f(x,y)$ is continuous and zero. Given $\epsilon > 0$. Let $\{q_1, q_2, \ldots\} = \mathbb{Q} \cap [0,1]$ and $I_k = (q_k - \frac{\epsilon}{2^{k+1}}, q_k + \frac{\epsilon}{2^{k+1}})$. Then we have

$$\mathbb{Q} \cap [0,1] \subseteq \bigcup_{k=1}^\infty I_k \quad \text{and} \quad \sum_{k=1}^\infty \mu(I_k) = \sum_{k=1}^\infty \frac{\epsilon}{2^k} < \epsilon.$$

By the definition, $\mu(\mathbb{Q} \cap [0,1]) = 0$ and Theorem 16.10 (The Lebesgue's Integrability Condition) implies that $f(x,y) \in \mathscr{R}$ on S. If P is a partition of S into p rectangles $I_1, I_2, \ldots, I_p$. Since each rectangle must contain a point with irrational coordinates, the definition (16.26) shows that $L(P, f) = 0$ and then

$$\int_S f(x,y)\,d(x,y) = 0 \quad (16.27)$$

by Theorem 16.8 (The Riemann Integrability Condition).

Next, if y is irrational, then $f(x,y) = 0$ for all $x \in [0,1]$. Therefore, Remark 16.2(b) shows that

$$\int_0^1 f(x,y)\,dx = 0. \quad (16.28)$$

If $y \in \mathbb{Q} \cap [0,1]$, then we know from [23, Exercise 18, p. 100] that the function[d] $f(x,y)$ is continuous at every irrational x in $[0,1]$. Thus it follows from Theorem 9.4 (The Lebesgue's Integrability Condition) that $f(x,y) \in \mathscr{R}$ on $[0,1]$ for every rational $y \in \mathbb{Q} \cap [0,1]$. By Theorem 14.6(a), we see that

$$\mathscr{R}\int_0^1 f(x,y)\,dx = \int_{[0,1]} f(x,y)\,dm = \int_{\mathbb{Q}\cap[0,1]} f(x,y)\,dm + \int_{[0,1]\setminus\mathbb{Q}} f(x,y)\,dm. \quad (16.29)$$

Since $m(\mathbb{Q} \cap [0,1]) = 0$ and $f(x,y) = 0$ on $[0,1] \setminus \mathbb{Q}$ by the definition (16.26), Theorem 14.3(f) ensures that the two Lebesgue integrals on the right-hand side of the equation (16.29) are zero. In other words, we have

$$\int_0^1 f(x,y)\,dx = 0 \quad (16.30)$$

---

[d] It is, in fact, the **Riemann function**, see Problems 7.5 and 9.8.

16.3. Integration on $\mathbb{R}^n$

if $y \in \mathbb{Q} \cap [0,1]$. Now we combine the results (16.28) and (16.30) to conclude that

$$\int_0^1 f(x,y)\,\mathrm{d}x = 0 \tag{16.31}$$

for every $y \in [0,1]$ and so

$$\int_0^1 \Big(\int_0^1 f(x,y)\,\mathrm{d}x\Big)\,\mathrm{d}y = 0. \tag{16.32}$$

Hence our desired results follows immediately from the results (16.27), (16.31) and (16.32).

However, suppose that $x \in \mathbb{Q} \cap [0,1]$. Then we have $x = \frac{m}{n}$, where m and n are relatively prime and $n > 0$. Thus we get from the definition (16.26) that

$$f(x,y) = \begin{cases} 0, & \text{if } y \in [0,1] \setminus \mathbb{Q}; \\ \frac{1}{n}, & \text{if } y \in \mathbb{Q} \cap [0,1]. \end{cases} \tag{16.33}$$

Obviously, the function (16.33) is nowhere continuous because it is a multiple of the **Dirichlet function** $D(y)$.[e] Consequently, $f(x,y) \notin \mathscr{R}$ on $[0,1]$ for every fixed $x \in \mathbb{Q} \cap [0,1]$. This completes the proof of the problem. ∎

Remark 16.7

Problem 16.15 shows that the condition $f(\cdot, y) \in \mathscr{R}$ on $[c,d]$ for every $x \in [a,b]$ in Fubini's Theorem cannot be relaxed. In fact, one can find counterexamples to show that the other hypotheses cannot be omitted.

Problem 16.16

(⋆) Suppose that $E = \{(x,y,z) \in \mathbb{R}^3 \mid 0 \leq x \leq 1,\, 0 \leq y \leq 1-x \text{ and } 0 \leq z \leq 1-x-y\}$ and $f(x,y,z) = x$. Prove that

$$\int_E f\,\mathrm{d}\mathbf{x} = \frac{1}{24}.$$

Proof. We notice that if $F = \{(x,y) \in \mathbb{R}^2 \mid 0 \leq x \leq 1 \text{ and } 0 \leq y \leq 1-x\}$, $\phi(x,y) = 0$ and $\varphi(x,y) = 1-x-y$, then F is clearly a closed Jordan measurable subset of $\mathbb{R}^2$ (in fact, F is the area bounded by the straight lines $y = 1-x$, $x = 0$ and $y = 0$) and both ϕ and φ are continuous on F. Thus Theorem 16.18 implies that

$$\int_E f\,\mathrm{d}\mathbf{x} = \iiint_E f(x,y,z)\,\mathrm{d}x\,\mathrm{d}y\,\mathrm{d}z = \int_F \Big(\int_0^{1-x-y} x\,\mathrm{d}z\Big)\,\mathrm{d}(x,y). \tag{16.34}$$

Similarly, if $I = [0,1]$, $\phi(x) = 0$ and $\varphi(x) = 1-x$, then I is a closed Jordan measurable subset of $\mathbb{R}$ and both $\phi(x) = 0$ and $\varphi(x) = 1-x$ are continuous on I. Therefore, we apply Theorem 16.18 again to the integral on the right-hand side of (16.34) to obtain

$$\int_E f\,\mathrm{d}\mathbf{x} = \int_F \Big(\int_0^{1-x-y} x\,\mathrm{d}z\Big)\,\mathrm{d}(x,y)$$

---
[e] Read Problems 7.2 and 7.11.

$$= \int_0^1 \int_0^{1-x} \int_0^{1-x-y} x \, dz \, dy \, dx$$

$$= \int_0^1 \int_0^{1-x} (x - x^2 - xy) \, dy \, dx$$

$$= \frac{1}{2} \int_0^1 (x - 2x^2 + x^3) \, dx$$

$$= \frac{1}{24},$$

completing the proof of the problem. ∎

16.4 Applications of the Mean Value Theorem

Problem 16.17

(⋆)(⋆) Suppose that $E \subseteq \mathbb{R}^n$, $f \in \mathscr{R}$ on E and

$$\int_E f \, d\mathbf{x} = 0. \tag{16.35}$$

Let $F = \{\mathbf{x} \in E \,|\, f(\mathbf{x}) < 0\}$ and $J(F) = 0$. Prove that there corresponds a set S with $\mu(S) = 0$ and $f(\mathbf{x}) = 0$ for every $\mathbf{x} \in E \setminus S$.

Proof. Since $J(F) = 0$, F is Jordan measurable and the particular case of the Mean Value Theorem for Multiple Integrals implies that

$$\int_F f \, d\mathbf{x} = 0. \tag{16.36}$$

Besides, it follows from Theorem 16.14 that $E \setminus F$ is also Jordan measurable. By the conditions (16.35) and (16.36), and Theorem 16.17(c) (Properties of Integration on E), we obtain

$$\int_{E \setminus F} f \, d\mathbf{x} = \int_E f \, d\mathbf{x} - \int_F f \, d\mathbf{x} = 0,$$

i.e., $f \in \mathscr{R}$ on $E \setminus F$. Next, if A is the set of the discontinuities of f in $E \setminus F$, then we deduce from Theorem 16.16 that $\mu(A) = 0$.

Take $\mathbf{p} \in E \setminus (F \cup A \cup \partial E)$. Since $\mathbf{p} \notin A$, f is continuous at $\mathbf{p}$. If $f(\mathbf{p}) \neq 0$, then $f(\mathbf{p}) > 0$. Since $\mathbf{p} \notin \partial E$, the continuity of f at $\mathbf{p}$ shows that there exists a $\delta > 0$ such that $N_\delta(\mathbf{p}) \subseteq E \setminus F$ and

$$f(\mathbf{x}) > \frac{f(\mathbf{p})}{2} \tag{16.37}$$

for all $\mathbf{x} \in N_\delta(\mathbf{p})$. By the definition, if $\mathbf{x} \in E \setminus F$, then we have $f(\mathbf{x}) \geq 0$. Note that $(E \setminus F) \setminus N_\delta(\mathbf{p})$ is Jordan measurable so that

$$\int_{(E \setminus F) \setminus N_\delta(\mathbf{p})} f \, d\mathbf{x} \geq 0 \tag{16.38}$$

16.4. Applications of the Mean Value Theorem

by Theorem 16.17(b) (Properties of Integration on E). In addition, we know from the inequalities (16.37) and (16.38) that

$$\int_{E\setminus F} f\,d\mathbf{x} = \int_{N_\delta(\mathbf{p})} f\,d\mathbf{x} + \int_{(E\setminus F)\setminus N_\delta(\mathbf{p})} f\,d\mathbf{x} \geq \int_{N_\delta(\mathbf{p})} f\,d\mathbf{x} > \mu(N_\delta(\mathbf{p})) \cdot \frac{f(\mathbf{p})}{2} > 0$$

which contradicts the hypothesis (16.35). In conclusion, we must have $f(\mathbf{p}) = 0$ and thus

$$f(\mathbf{x}) = 0 \tag{16.39}$$

on $E \setminus (F \cup A \cup \partial E)$.

Finally, we claim that the set

$$S = F \cup A \cup \partial E$$

satisfies the requirements. By the result (16.39), it suffices to prove that $\mu(S) = 0$. Recall that $J(F) = 0$, so $\mu(F) = 0$ by Problem 16.2. By the properties of the boundary of a set,[f] we have

$$\partial E \subseteq \partial F \cup \partial(E \setminus F) \quad \text{and} \quad \partial(E \setminus F) = \overline{E \setminus F} \cap \overline{E \setminus (E \setminus F)} = \overline{E \setminus F} \cap \overline{F} = \overline{F}.$$

Consequently, we have

$$\partial E \subseteq \partial F \cup \overline{F}.$$

By Theorem 16.13, we have $J(\partial F) = 0$, so $\mu(\partial F) = 0$ by Problem 16.2. By Problem 16.1, we know that $J(\overline{F}) = J^*(\overline{F}) = J^*(F) = J(F) = 0$ which implies $\mu(\overline{F}) = 0$ by Problem 16.2 again. Using Remark 16.2(a) twice, we get

$$\mu(\partial E) = 0 \quad \text{and} \quad \mu(S) = 0$$

which proves the claim. Hence we complete the analysis of the problem. ∎

Problem 16.18

(⋆) Let E be a Jordan measurable set of $\mathbb{R}^n$, $f \in \mathscr{R}$ on E and $g : E \to \mathbb{R}$ be bounded. If $F \subseteq E$ satisfies $J(F) = 0$ and $g(\mathbf{x}) = f(\mathbf{x})$ on $E \setminus F$, prove that $g \in \mathscr{R}$ on E and

$$\int_E f\,d\mathbf{x} = \int_E g\,d\mathbf{x}.$$

Proof. Since $f = g$ on $E \setminus F$, it follows from Theorem 16.17(c) (Properties of Integration on E) and Remark 16.5(a) that

$$\int_E g\,d\mathbf{x} = \int_{E\setminus F} g\,d\mathbf{x} + \int_F g\,d\mathbf{x} = \int_{E\setminus F} g\,d\mathbf{x} + 0 = \int_{E\setminus F} f\,d\mathbf{x} + \int_F f\,d\mathbf{x} = \int_E f\,d\mathbf{x}.$$

This completes the proof of the problem. ∎

---
[f] Refer to [3, Exercises 3.51 and 3.52, p. 69].

> **Problem 16.19**
>
> (⋆)(⋆) Suppose that E is an open connected Jordan measurable subset of $\mathbb{R}^n$, $f \in \mathscr{R}$ on E and f is continuous on E. Prove that there is a point $\mathbf{p} \in E$ such that
> $$\int_E f \, d\mathbf{x} = f(\mathbf{p}) J(E). \tag{16.40}$$

Proof. Since $f \in \mathscr{R}$ on E, f is bounded on E. Therefore, both $M = \sup\{f(\mathbf{x}) \,|\, \mathbf{x} \in E\}$ and $m = \inf\{f(\mathbf{x}) \,|\, \mathbf{x} \in E\}$ are finite and $m \le f(\mathbf{x}) \le M$ on E. By the Mean Value Theorem for Multiple Integrals, we have
$$mJ(E) \le \int_E f \, d\mathbf{x} \le MJ(E)$$
which means that there exists a $\lambda \in [m, M]$ such that
$$\int_E f \, d\mathbf{x} = \lambda J(E). \tag{16.41}$$

If $\lambda = m$, then we must have $m = M$. Otherwise, one can find a point $\mathbf{a} \in E$ such that
$$m < f(\mathbf{a}) < M. \tag{16.42}$$

Since f is continuous on E, the Sign-preserving Property[g] implies that there exists a $\delta > 0$ such that $f(\mathbf{x}) > m$ on $N_\delta(\mathbf{a})$. However, Theorem 16.17 (Properties of Integration on E) and Remark 16.6 give
$$\begin{aligned}
\int_E f \, d\mathbf{x} &= \int_{N_\delta(\mathbf{a})} f \, d\mathbf{x} + \int_{E \setminus N_\delta(\mathbf{a})} f \, d\mathbf{x} \\
&> \int_{N_\delta(\mathbf{a})} m \, d\mathbf{x} + \int_{E \setminus N_\delta(\mathbf{a})} m \, d\mathbf{x} \\
&= m \int_E d\mathbf{x} \\
&= mJ(E)
\end{aligned} \tag{16.43}$$

which contradicts the formula (16.41). In other words, f is a constant function and in fact, $f(\mathbf{x}) = m$ on E. Hence the formula (16.40) holds trivially.

Next, if $\lambda = M$, then we also have $m = M$. Otherwise, the inequality (16.42) holds for another point $\mathbf{b} \in E$. Thus the previous analysis can be repeated to obtain the same contradiction (16.43). Now we may suppose that
$$m < \lambda < M.$$

Since f is continuous on E and E is connected, $f(E)$ is a connected[h] subset of $[m, M]$. In fact, we conclude from Theorem 4.16 that $f(E)$ is an interval. By the definitions of M and m, we certainly have
$$(m, M) \subseteq f(E) \subseteq [m, M].$$

---
[g] In fact, we require its generalized version with $\mathbb{R}^n$ as the domain.
[h] See Theorem 7.12 (Continuity and Connectedness).

16.5 Applications of the Change of Variables Theorem

Problem 16.20

(★) Evaluate
$$\int_D \sin\sqrt{x^2+y^2}\,d(x,y),$$
where $D = \{(x,y)\,|\,\pi^2 \le x^2 + y^2 \le 4\pi^2\}$.

Proof. Let $E = \{(r,\theta)\,|\,\pi \le r \le 2\pi \text{ and } 0 < \theta < 2\pi\} = [\pi, 2\pi] \times (0, 2\pi)$. Then E is obviously Jordan measurable because $J(\partial E) = 0$ and $\phi(E) = D$, where
$$\phi(r,\theta) = (r\cos\theta, r\sin\theta)$$
and D is an annulus. Next, we note that if $V = (\frac{\pi}{2}, \frac{5\pi}{2}) \times (-\frac{\pi}{2}, \frac{5\pi}{2})$, then V is an open set containing $\overline{E} = [\pi, 2\pi] \times [0, 2\pi]$ and does not intersect the θ-axis. See Figure 16.3 for detials.

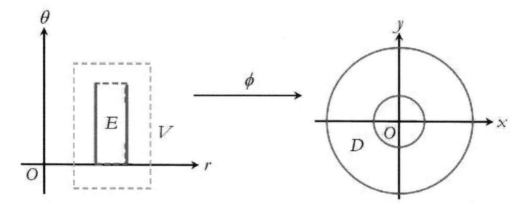

Figure 16.3: The mapping $\phi : E \to D$.

Finally, the function $f(x,y) = \sin\sqrt{x^2+y^2}$ is continuous on $\mathbb{R}^2$ and $f(r\cos\theta, r\sin\theta) = \sin r$, so we must have $f \in \mathscr{R}$ on D and $f(r\cos\theta, r\sin\theta) \in \mathscr{R}$ on E. Hence we may apply the formula (16.9) to get
$$\int_D \sin\sqrt{x^2+y^2}\,dx\,dy = \int_E f(r\cos\theta, r\sin\theta)r\,dr\,d\theta$$

$$= \int_0^{2\pi} \int_\pi^{2\pi} r \sin r \, dr \, d\theta$$
$$= -\int_0^{2\pi} \left[\left. r \cos r \right|_\pi^{2\pi} - \int_\pi^{2\pi} \cos r \, dr \right] d\theta$$
$$= -\int_0^{2\pi} 3\pi \, d\theta$$
$$= -6\pi^2.$$

This completes the analysis of the problem. ∎

Problem 16.21

(⋆) Evaluate
$$\int_R (x+y)^3 \, dx \, dy,$$
where R is the parallelogram with vertices $(1,0), (3,1), (2,2)$ and $(0,1)$.

Proof. Consider $f(x,y) = (x+y)^3$ and $\phi : \mathbb{R}^2 \to \mathbb{R}^2$ defined by
$$\phi(u,v) = \left(\frac{2u+v}{3}, \frac{u-v}{3} \right).$$

Then it is continuously differentiable and injective on $\mathbb{R}^2$. In addition, if E is the parallelogram with vertices $(1,-2), (4,-2), (4,1)$ and $(1,1)$, then it is easily shown that $\phi(E) = D$, see Figure 16.4 below:

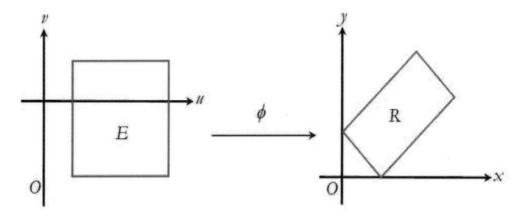

Figure 16.4: The mapping $\phi : E \to R$.

Since f is continuous on $\mathbb{R}^2$, we have $f \in \mathscr{R}$ on R by Theorem 16.16. Since $f(\phi(u,v)) = u^3$ which is continuous on $\mathbb{R}^2$, we have $f \circ \phi \in \mathscr{R}$ on E by Theorem 16.16. Finally, since E is clearly

16.5. Applications of the Change of Variables Theorem

a Jordan measurable set, $\overline{E} \subseteq \mathbb{R}^2$ and

$$\det \mathbf{J}_\phi((u,v)) = \det \begin{pmatrix} \dfrac{2}{3} & \dfrac{1}{3} \\ \dfrac{1}{3} & -\dfrac{1}{3} \end{pmatrix} = -\dfrac{1}{3} \neq 0,$$

we obtain from the formula (16.8) that

$$\int_R (x+y)^3 \, dx \, dy = \int_{\phi(E)} (x+y)^3 \, dx \, dy = \int_E u^3 \cdot \dfrac{1}{3} \, du \, dv = \dfrac{1}{3} \int_{-2}^{1} \int_{1}^{4} u^3 \, du \, dv = \dfrac{255}{3},$$

completing the proof of the problem. ∎

Problem 16.22

(★) Evaluate

$$\int_D e^{-x^2-y^2} \, d(x,y),$$

where $D = \{(x,y) \in \mathbb{R}^2 \mid x^2 + y^2 \leq 1\}$.

Proof. Let $f(x,y) = e^{-x^2-y^2}$. Since it is continuous on D, $f \in \mathcal{R}$ on D by Theorem 16.16. Let $E = \{(r,\theta) \mid 0 \leq r \leq 1 \text{ and } 0 \leq \theta < 2\pi\} = [0,1] \times [0,2\pi)$. Then E is Jordan measurable because $J(\partial E) = 0$ and $\phi(E) = D$, where

$$\phi(r,\theta) = (r\cos\theta, r\sin\theta).$$

Using the polar coordinates (with the aid of Remark 16.4(b)), we have

$$\begin{aligned}
\int_D e^{-x^2-y^2} \, d(x,y) &= \int_{\phi(E)} e^{-x^2-y^2} \, d(x,y) \\
&= \int_E r e^{-r^2} \, dr \, d\theta \\
&= \int_0^{2\pi} \int_0^1 r e^{-r^2} \, dr \, d\theta \\
&= -\dfrac{1}{2} \int_0^{2\pi} (e^{-1} - 1) \, d\theta \\
&= \pi(1 - e^{-1}).
\end{aligned}$$

This ends the proof of the problem. ∎

Problem 16.23

(★) Suppose that E is a Jordan measurable subset of $\mathbb{R}^n$ and for every $\mathbf{x} \in E$, we have $-\mathbf{x} \in E$. Let f be an odd function such that $f \in \mathcal{R}$ on E. Prove that

$$\int_E f \, d\mathbf{x} = 0.$$

Proof. Define $\phi(\mathbf{x}) = -\mathbf{x}$. Then ϕ is continuously differentiable and injective on $\mathbb{R}^n$. In addition, we have $\phi(E) = E$ so that $f \in \mathscr{R}$ on $\phi(E)$. Therefore, we get $f \circ \phi \in \mathscr{R}$ on E. Finally, since $\overline{E} \subseteq \mathbb{R}^n$ and

$$\det \mathbf{J}_\phi(\mathbf{x}) = \det \begin{pmatrix} -1 & 0 & \cdots & 0 \\ 0 & -1 & \cdots & 0 \\ \vdots & \vdots & \ddots & 0 \\ 0 & 0 & \cdots & -1 \end{pmatrix} = (-1)^n \neq 0,$$

the integral in question satisfies all the requirements of the Change of Variables Theorem. Hence we deduce from the formula (16.8) and the fact $f(\phi(\mathbf{x})) = f(-\mathbf{x}) = -f(\mathbf{x})$ that

$$\int_E f(\mathbf{y})\,d\mathbf{y} = \int_{\phi(E)} f(\mathbf{y})\,d\mathbf{y} = \int_E f(\phi(\mathbf{x})) \cdot |\det \mathbf{J}_\phi(\mathbf{x})|\,d\mathbf{x} = -\int_E f(\mathbf{x})\,d\mathbf{x}. \quad (16.44)$$

Since the variable is dummy, the formula (16.45) implies that

$$\int_E f\,d\mathbf{x} = 0$$

which is our desired result. This completes the proof of the problem. ∎

> **Problem 16.24**
>
> (★)(★) Show that
>
> $$\iint\limits_{0<x<y<\pi} \ln|\sin(x-y)|\,dx\,dy = -\frac{1}{2}\pi^2 \ln 2.$$

Proof. Let $D = \{(x,y) \in \mathbb{R}^2 \,|\, 0 < x < y < \pi\}$ and $E = \{(u,v) \in \mathbb{R}^2 \,|\, 0 < u < v < 2\pi - u\}$. Define $\phi : \mathbb{R}^2 \to \mathbb{R}^2$ by

$$\phi(u,v) = \left(\frac{-u+v}{2}, \frac{u+v}{2}\right).$$

Then it is easy to see that $\phi(E) = D$, ϕ is continuously differentiable and injective on $\mathbb{R}^2$. Furthermore, E is Jordan measurable, $\overline{E} \subseteq \mathbb{R}^2$ and

$$\det \mathbf{J}_\phi((u,v)) = \det \begin{pmatrix} -\frac{1}{2} & \frac{1}{2} \\ \frac{1}{2} & \frac{1}{2} \end{pmatrix} = -\frac{1}{2} \neq 0.$$

Now the function $f(x,y) = \ln|\sin(x-y)|$ is continuous on D, so $f \in \mathscr{R}$ on D by Theorem 16.16. Since $f(\phi(u,v)) = \ln|\sin u|$ which is continuous on $\mathbb{R}^2$, Theorem 16.16 implies that $f \circ \phi \in \mathscr{R}$ on E. Thus it follows from the formula (16.8) that

$$I = \iint_D \ln|\sin(x-y)|\,dx\,dy$$
$$= \frac{1}{2}\int_E \ln|\sin u|\,du\,dv$$
$$= \frac{1}{2}\int_0^\pi \int_u^{2\pi-u} \ln|\sin u|\,dv\,du$$

16.5. Applications of the Change of Variables Theorem

$$= \int_0^\pi (\pi - u) \ln|\sin u|\, du. \tag{16.45}$$

Next, we apply the substitution $y = \frac{\pi}{2} - u$ to the integral (16.45) to get

$$I = -\int_{\frac{\pi}{2}}^{-\frac{\pi}{2}} \left(\frac{\pi}{2} + y\right) \ln\left|\sin\left(\frac{\pi}{2} - y\right)\right| dy$$

$$= \int_{-\frac{\pi}{2}}^{\frac{\pi}{2}} \left(\frac{\pi}{2} + y\right) \ln|\cos y|\, dy$$

$$= \frac{\pi}{2} \int_{-\frac{\pi}{2}}^{\frac{\pi}{2}} \ln|\cos y|\, dy + \int_{-\frac{\pi}{2}}^{\frac{\pi}{2}} y \ln|\cos y|\, dy. \tag{16.46}$$

Since the first and the second integrands in the expression (16.46) are even and odd functions respectively, we are able to reduce it to

$$I = \pi \int_0^{\frac{\pi}{2}} \ln|\cos y|\, dy.$$

Notice that

$$\int_0^{\frac{\pi}{2}} \ln|\sin y|\, dy = \int_0^{\frac{\pi}{2}} \ln|\cos y|\, dy,$$

therefore we obtain

$$\frac{2}{\pi} I = 2 \int_0^{\frac{\pi}{2}} \ln|\cos y|\, dy$$

$$= \int_0^{\frac{\pi}{2}} \ln|\cos y|\, dy + \int_0^{\frac{\pi}{2}} \ln|\cos y|\, dy$$

$$= \int_0^{\frac{\pi}{2}} \ln|\cos y|\, dy + \int_0^{\frac{\pi}{2}} \ln|\sin y|\, dy$$

$$= \int_0^{\frac{\pi}{2}} \ln\left|\frac{2\sin y \cos y}{2}\right| dy$$

$$= \frac{1}{2} \int_0^{\frac{\pi}{2}} \ln|\sin 2y|\, d(2y) - \int_0^{\frac{\pi}{2}} \ln 2\, dy$$

$$= \frac{I}{\pi} - \frac{\pi}{2} \ln 2$$

which implies that

$$I = -\frac{\pi^2}{2} \ln 2.$$

We have completed the proof of the problem. ∎

Problem 16.25

(⋆) Prove that

$$\int_V (x^2 + y^2 + z^2)\, d(x, y, z) = \frac{4\pi}{5},$$

where $V = \{(x, y, z) \mid x^2 + y^2 + z^2 = 1\}$.

Proof. If $E = \{(r, \varphi, \theta) \,|\, 0 \le r \le 1,\, 0 \le \varphi \le \pi \text{ and } 0 \le \theta \le 2\pi\}$, then we have $\phi(E) = V$, where
$$\phi(r, \varphi, \theta) = (r \sin\varphi \cos\theta, r \sin\varphi \sin\theta, r \cos\theta).$$

Since $\det \mathbf{J}_\phi((r,\varphi,\theta)) = r^2 \sin\varphi$, we know from the spherical coordinates (with the aid of Remark 16.4(b)) that

$$\begin{aligned}
\int_V (x^2 + y^2 + z^2)\,\mathrm{d}(x,y,z) &= \int_{\phi(E)} (x^2 + y^2 + z^2)\,\mathrm{d}(x,y,z) \\
&= \int_E r^2 \cdot r^2 \sin\varphi\,\mathrm{d}r\,\mathrm{d}\varphi\,\mathrm{d}\theta \\
&= \int_0^1 \int_0^\pi \int_0^{2\pi} r^4 \sin\varphi\,\mathrm{d}r\,\mathrm{d}\varphi\,\mathrm{d}\theta.
\end{aligned} \qquad (16.47)$$

Applying Problem 16.11 to the integral (16.47), we establish immediately that

$$\int_V (x^2 + y^2 + z^2)\,\mathrm{d}(x,y,z) = \int_0^1 r^4\,\mathrm{d}r \int_0^\pi \sin\varphi\,\mathrm{d}\varphi \int_0^{2\pi} \mathrm{d}\theta = \frac{4\pi}{5},$$

completing the proof of the problem. ∎

Problem 16.26

(★)(★) Given that f is a differentiable function and
$$F(t) = \int_{D(t)} f(x^2 + y^2 + z^2)\,\mathrm{d}(x,y,z),$$
where $D(t) = \{(x,y,z)\,|\,x^2 + y^2 + z^2 \le t^2\}$. Prove that
$$F'(t) = 4\pi t^2 f(t^2).$$

Proof. Let $E(t) = \{(r,\varphi,\theta)\,|\,0 \le r \le t,\, 0 \le \varphi \le \pi \text{ and } 0 \le \theta \le 2\pi\}$. Then it is easy to see that
$$\phi(E(t)) = D(t),$$
where
$$\phi(r,\varphi,\theta) = (r \sin\varphi \cos\theta, r \sin\varphi \sin\theta, r \cos\theta).$$

Hence we follow from the formula (16.8) (with the aid of Remark 16.4(b)) and Problem 16.11 that

$$\begin{aligned}
F(t) &= \int_{\phi(E(t))} f(r^2) r^2\,\mathrm{d}r\,\mathrm{d}\varphi\,\mathrm{d}\theta \\
&= \int_0^t \int_0^\pi \int_0^{2\pi} r^2 f(r^2)\,\mathrm{d}r\,\mathrm{d}\varphi\,\mathrm{d}\theta \\
&= \int_0^{2\pi} \mathrm{d}\theta \int_0^\pi \sin\varphi\,\mathrm{d}\varphi \int_0^t r^2 f(r^2)\,\mathrm{d}r \\
&= 4\pi \int_0^t r^2 f(r^2)\,\mathrm{d}r.
\end{aligned}$$

16.5. Applications of the Change of Variables Theorem

Since f is differentiable, we deduce from the First Fundamental Theorem of Calculus that

$$F'(t) = 4\pi t^2 f(t^2),$$

completing the proof of the problem. ∎

APPENDIX A

Language of Mathematics

A.1 Fundamental Concepts

The goal of this appendix is to give a brief review to **mathematical logic** that we use frequently to write mathematical proofs logically and rigorously in this book. The main references we have used here are [1, Chap. 9] and [20, §1.1 - §1.3].

A.1.1 What is logic?

This may be the first question in your mind. In fact, the term "logic" came from the Greek word "logos" which can be translated as "sentence", "reason", "rule" and etc. Of course, these translations are not enough to explain the specialized meaning of "logic" when one uses it nowadays.

Roughly speaking, logic is the study of **principles of correct and incorrect reasoning**. It is a tool to establish reasonable conclusions based on a given set of assumptions. A "logical" person wants to figure out what makes good/bad reasoning good/bad. Understanding such principles can keep us away from making mistakes in our own reasoning and they allow us to judge others' reasoning.

A.1.2 What is mathematical logic?

I think this is the second question in your mind. Briefly speaking, mathematical logic is a subfield of mathematics and it is the application of the theory and techniques of formal logic to mathematics and mathematical reasoning. One of the main characteristic features of mathematical logic is the use of mathematical language of symbols and formulas.

In advanced mathematics, you will be studying a lot of mathematical concepts and well-known results you have been already familiar with. It may happen that your *computational skills* are excellent. However, this is not what you are going to learn or sharpen in your undergraduate mathematics courses. Instead, we will emphasize the backbone of the theory behind. For examples,

- What is a rational number/irrational number/real number? How do we "count" those numbers?

- What is a continuous/differentiable/integrable function?

- Why do the Intermediate Value Theorem, the Mean Value Theorem for Derivatives, the Integration by Parts and other theorems that you used in a calculus course work?

In real analysis, we will answer the above questions in a systematical way so that you know not only *how* to apply such mathematical theorems, but also understand *why* they are true.

A.2 Statements and Logical Connectives

A.2.1 Statements

A **mathematical statement** or simply **statement** is a sentence which is either true or false, but *not both*. Usually, we apply the lowercase letters (e.g. p, q and r) to represent statements. When a statement is true, we assign its **truth value** to be truth, denoted by **T**; When the statement is false, its truth value is false and it is denoted by **F**. A **compound statement** is a statement known as the composition of statements by applying **logical connectives** such as "and", "or", "not", "if ... then" and "if and only if".

A.2.2 Logical Connectives

Let p and q be statements. The truth value of a compound statement of p and q can be expressed in terms of a **truth table**.

- **Conjunction.** The conjunction of statements p and q, denoted by $p \wedge q$ and read as "p and q". It is defined as true **only** when both p and q are true.

p	q	$p \wedge q$
T	T	T
T	F	F
F	T	F
F	F	F

Table A.1: The truth table of $p \wedge q$.

- **Disjunction.** The disjunction of statements p and q, denoted by $p \vee q$ and read as "p or q". It is defined as false **only** when both p and q are false.

p	q	$p \vee q$
T	T	T
T	F	T
F	T	T
F	F	F

Table A.2: The truth table of $p \vee q$.

A.2. Statements and Logical Connectives

- **Negation.** The negation of a statement p, denoted by $\sim p$ and read as "the negation of p". It is defined as the **opposite value** of p.

p	$\sim p$
T	F
F	T

Table A.3: The truth table of $\sim p$.

- **Conditional statement.** A conditional statement, symbolized by $p \to q$, is an "if-then" statement and it is read as "if p, then q". Here p is called the **hypothesis** and q is the **conclusion**. It is defined as false **only** when the hypothesis p is true and the conclusion q is false.

p	q	$p \to q$
T	T	T
T	F	F
F	T	T
F	F	T

Table A.4: The truth table of $p \to q$.

We notice that the **inverse** and the **converse** of the conditional statement $p \to q$ are

$$\sim p \to \sim q \quad \text{and} \quad q \to p$$

respectively.[a]

- **Biconditional statement.** A biconditional statement, symbolized by $p \leftrightarrow q$, is an "if and only if" statement and it is read as "p if and only if q". It is defined as true when *both* p and q have the same truth value.

p	q	$p \leftrightarrow q$
T	T	T
T	F	F
F	T	F
F	F	T

Table A.5: The truth table of $p \to q$.

A.2.3 Equivalent statements, Tautologies and Contradictions

- **Equivalent statements.** If two statements p and q have the same truth table, then we say that they are **equivalent**. For example, the statement $p \to q$ is equivalent to the statement $\sim q \to \sim p$. Symbolically, we write

$$p \to q \equiv \sim q \to \sim p.$$

---

[a]It can be seen that the inverse and the converse are **equivalent** (see §A.2.3) and furthermore, the inverse is the **contrapositive** (see §A.5.2) of the converse.

p	q	$\sim q$	$\sim p$	$p \to q$	$\sim q \to \sim p$
T	T	F	F	T	T
T	F	T	F	F	F
F	T	F	T	T	T
F	F	T	T	T	T

Table A.6: The truth table of $p \to q$ and $\sim q \to \sim p$.

- **Tautologies.** If a compound statement always takes the truth value "**T**" no matter what the truth values of the variables are, then we call such a compound statement a **tautology**.

p	q	$\sim q$	$\sim p$	$p \vee q$	$p \wedge q$	$(p \vee q) \vee (\sim p)$
T	T	F	F	T	T	T
T	F	T	F	T	F	T
F	T	F	T	T	F	T
F	F	T	T	F	F	T

Table A.7: The tautology $(p \vee q) \vee (\sim p)$

Particularly, we write "$p \Rightarrow q$" if $p \to q$ is a tautology.[b] For example, we have
$$x > 2 \Rightarrow x^2 > 4.$$
Similarly, we write "$p \Leftrightarrow q$" if $p \leftrightarrow q$ is a tautology.

- **Contradictions.** If a compound statement always takes the truth value "**F**" no matter what the truth values of the variables are, then we call such a compound statement a **contradiction**.

p	q	$\sim q$	$\sim p$	$p \vee q$	$p \wedge q$	$(p \wedge q) \wedge (\sim p)$
T	T	F	F	T	T	F
T	F	T	F	T	F	F
F	T	F	T	T	F	F
F	F	T	T	F	F	F

Table A.8: The contradiction $(p \wedge q) \wedge (\sim p)$

A.3 Quantifiers and their Basic Properties

A.3.1 Existential quantifier and universal quantifier

There are two types of quantifiers: **existential quantifier** and **universal quantifier**.

- **Existential quantifier.** The expression "$\exists x \, P(x)$" means "there exists an x such that the property $P(x)$ holds" or "there is at least one x such that the property $P(x)$ holds". Here, the notation $\exists$ is called the **existential quantifier**, and $\exists x$ means that at least one element x.

---

[b]The notation "$\Rightarrow$" is read as "**implies**".

A.4. Necessity and Sufficiency

- **Universal quantifier.** The expression "$\forall x\ P(x)$" can be interpreted as "for each/for all/for every/for any x, the property $P(x)$ is true". Here, the notation $\forall$ is called the **universal quantifier** and $\forall x$ means all the elements x.

- **Order of quantifiers.** On the one hand, the positions of the **same type** of quantifiers can be interchanged without affecting the truth value. For examples,

$$\forall x\ \forall y\ P(x,y) \Leftrightarrow \forall y\ \forall x\ P(x,y) \quad \text{and} \quad \exists x\ \exists y\ P(x,y) \Leftrightarrow \exists y\ \exists x\ P(x,y).$$

On the other hand, we can't switch the positions of **different types** of quantifiers, i.e.,

$$\exists x\ \forall y\ P(x,y) \not\Leftrightarrow \forall y\ \exists x\ P(x,y).$$

A.3.2 Properties of quantifiers

In many theorems or proofs, you may see one of the following four statements

"$\forall x\ \forall y\ P(x,y)$", "$\forall x\ \exists y\ P(x,y)$", "$\exists x\ \forall y\ P(x,y)$" and "$\exists x\ \exists y\ P(x,y)$".

Understanding their interpretations can help you to *figure out* what the theorem is saying or what you are going to prove. Now their explanations are shown as follows:

- $\forall x\ \forall y\ P(x,y)$: For all x and for all y, the property $P(x,y)$ holds.

- $\forall x\ \exists y\ P(x,y)$: For all x, there exists y such that the property $P(x,y)$ holds.

- $\exists x\ \forall y\ P(x,y)$: There exists x such that for all y, the property $P(x,y)$ holds.

- $\exists x\ \exists y\ P(x,y)$: There exist x and y such that the property $P(x,y)$ holds.

Besides, we sometimes need to apply negation to quantifiers in writing a mathematical proof. Thus we have to understand what they are. In fact, the negation of the existential quantifier is the universal quantifier and vice versa. Symbolically, we have

$$\sim \forall x\ P(x) \equiv \exists x\ \sim P(x) \quad \text{and} \quad \sim \exists x\ P(x) \equiv \forall x\ \sim P(x).$$

> **Example A.1**
>
> The negation of the (true) statement "for all $n \in \mathbb{Z}$, we have $n^2 \geq 0$" is the (false) statement "there exists $n \in \mathbb{Z}$ such that $n^2 < 0$".

A.4 Necessity and Sufficiency

In mathematical logic, necessity and sufficiency are terms applied to describe an implicational relationship between statements. To say that p is a **necessary condition** for q means that it is impossible to have q without p. In other words, the nonexistence of p *guarantees* the nonexistence of q.

> **Example A.2**
>
> (a) The statement "$x^2 > 16$" is a **necessary condition** for the statement "$x > 5$". (It is impossible to have "$x > 5$" without having "$x^2 > 16$".)
>
> (b) Having four sides is a **necessary condition** for being a rectangle. (It is impossible to have a rectangle without having four sides.)
> $\underbrace{\text{Having four sides}}_{\text{Statement } p}$ $\underbrace{\text{being a rectangle}}_{\text{Statement } q}$

To say that the statement p is a **sufficient condition** for the statement q is to say that the existence of p *guarantees* the existence of q. In other words, it is impossible to have p without q: If p exists, then q must exist.

> **Example A.3**
>
> (a) The statement "$x > 5$" is a **sufficient condition** for the statement "$x^2 > 16$". (If "$x > 5$" is valid, then "$x^2 > 16$" is also valid.)
>
> (b) Being a rectangle is a **sufficient condition** for having four sides. (If "being a rectangle" is valid, then "having four sides" is also valid.)

For equivalent statements p and q, we say that p is a **necessary and sufficient condition** of q. For example, the statement "a is even" is a necessary and sufficient condition for the statement "$(a+1)^2 + 1$ is even".

A.5 Techniques of Proofs

Basically, there are three common ways of presenting a proof for a mathematical statement. They are

- **direct proof**,
- **proof by contrapositive** and
- **proof by contradiction**.

A.5.1 Direct Proof

To prove "$p \Rightarrow q$", we start with the hypothesis p and we proceed to show the truth of the conclusion q.

> **Example A.4**
>
> Prove that $\underbrace{\text{the sum of two even integers}}_{\text{hypothesis } p}$ is also an $\underbrace{\text{even integer}}_{\text{conlusion } q}$.

Proof. Let $2m$ and $2n$ be two even integers. Since we have

$$2m + 2n = 2(m+n),$$

A.5.2 Proof by Contrapositive

Sometimes, it is *hard* to give a direct proof. In this case, one may give an "indirect proof" called "proof by contrapositive". By definition, the contrapositive of the statement $p \to q$ is $\sim q \to \sim p$. We note that a statement and its contrapositive are actually equivalent, i.e.,

$$p \to q \equiv \sim q \to \sim p.$$

See Table A.6 for this. Thus, to prove "$p \Rightarrow q$", it is equivalent to suppose that q is false (i.e., $\sim q$) and then we prove that p is also false (i.e., $\sim p$).

Example A.5

Prove that if $\underbrace{x^2 \text{ is even}}_{\text{hypothesis } p}$, then $\underbrace{x \text{ is even}}_{\text{conlusion } q}$.

Proof. The contrapositive of the statement is "If x is not even, then x^2 is not even" and we prove this is true. Since x is not even, it is odd. Thus we have $x = 2n + 1$ for some $n \in \mathbb{Z}$. This fact implies that

$$x^2 = (2n+1)^2 = 4n^2 + 4n + 1 = 2(2n^2 + 2n) + 1$$

which is an odd integer. ∎

A.5.3 Proof by Contradiction

It is an indirect proof. The basic idea of "proof by contradiction" is to assume that the statement we want to prove is **false** and then we prove that this assumption leads to a **contradiction**: a statement p and its negation $\sim p$ **cannot** both be true. The usual way of presenting a proof by contradiction is given as follows:

- **Step 1:** Assume that the statement p was false, i.e., $\sim p$ was true.
- **Step 2:** $\sim p$ implies both q and $\sim q$ are true for some statement q.
- **Step 3:** Since q and $\sim q$ cannot be both true, the statement p is **true**.

Example A.6

Prove that $\underbrace{\text{for all } m, n \in \mathbb{Z}, \text{ we have } m^2 - 4n \neq 2}_{\text{statement } p}$.

Proof. The negation of p is that "there exists $m, n \in \mathbb{Z}$ such that $m^2 - 4n^2 = 2$". It is clear from $m^2 - 4n^2 = 2$ that

$$m^2 = 4n + 2 = 2(2n + 1)$$

so that m^2 is even. Since m^2 is even, m must also be even. Let $m = 2k$ for some $k \in \mathbb{Z}$. Put $m = 2k$ into the original equation $m^2 - 4n = 2$, we obtain

$$(2k)^2 - 4n = 2$$
$$2(k^2 - n) = 1. \qquad (A.1)$$

Since $k^2 - n \neq 0$, the equation (A.1) shows that $\underbrace{1 \text{ is even}}_{\text{statement } q}$. However, it is well-known that $\underbrace{1 \text{ is } \mathbf{not} \text{ even}}_{\text{negation } \sim q}$. This is obvious a contradiction. Hence the statement p must be true. ∎

Index

Symbols
F_σ set, 267
G_δ set, 249
σ-algebra, 245
n-dimensional interval in $\mathbb{R}^n$, 353
n-measure zero, 356
n-th partial sums, 75
nth Bernstein polynomial, 216
nth derivative of f, 131

A
A.M. $\geq$ G.M., 71, 93, 168, 202
Abel's Test, 78
Abel's Test for Improper Integrals, 234
Abel's Test for Uniform Convergence, 212
absolute convergence, 77
Absolute Convergence Test, 222
absolute maximum, 129
absolute minimum, 129
absolute value, 9
algebraic, 24
almost everywhere, 245
alternating series, 77
antiderivative, 161
Archimedean Property, 10
area of I, 353
Arzelà-Ascoli Theorem, 191
at most countable, 19

B
biconditional statement, 385
bijective, 2
Bolzano-Weierstrass Theorem, 68, 138, 190, 193
Bonnet's Theorem, 162
Borel function, 269
Borel measurable, 246, 268

Borel measurable function, 269
Borel set, 245
Borel-Cantelli Lemma, 254
boundary, 358
boundary point, 357
Bounded Convergence Theorem, 293
bounded sequence, 49
bounded set, 30, 100

C
Cantor set, 45, 256
Cantor's Intersection Theorem, 30
cardinality, 19
cartesian product, 2
Cauchy Criterion, 76
Cauchy Criterion for Improper Integrals, 233
Cauchy Criterion for Uniform Convergence, 189
Cauchy Mean Value Theorem, 129
Cauchy Principal Value, 220
Cauchy product, 78
Cauchy sequence, 52
Chain Rule, 128
Change of Variables Theorem, 160, 375
characteristic function, 270
Chebyshev's Inequality, 298
circle of convergence, 79
Clairaut's Theorem, 332
closed set, 27, 99
closure, 28
compact metric space, 30, 42, 52
compact set, 30, 100
comparability, 3
Comparison Test, 76, 222
complement, 2
complete metric space, 52, 70

Completeness Axiom, 10, 100
composition, 2
compound statement, 384
concave function, 132
conclusion, 385
conditional statement, 385
conjunction, 384
connected set, 31, 100
continuity of function, 98
continuous at p, 98
continuous on E, 98
continuously differentiable, 330
contradiction, 386
converge, 49
converges in measure, 284
converse, 385
convex function, 132
convexity, 132
countability, 19
countable, 19
countable additive, 244
countably additivity, 246
countably subadditivity, 244
cover, 29
cylindrical coordinates, 360

D

Darboux's Theorem, 129
definite integral, 157
derivative of f, 127
difference, 1
differentiable at x, 127
differentiable on E, 127
differential of $\mathbf{f}$ at $\mathbf{x}$, 328
differentiation, 127
Dini's Theorem, 193
direct proof, 388
directional derivative, 327
Dirichlet function, 104, 165, 371
Dirichlet product, 78
Dirichlet's Test, 78
Dirichlet's Test for Improper Integrals, 235
discontinuity of the first kind, 101
discontinuity of the second kind, 101
discontinuous at p, 101
disjunction, 384
distance, 9, 27
distance function, 27

diverge, 49
domain, 2
double integral, 355
dummy variable, 158

E

Egorov's Theorem, 271
empty set, 1
equicontinuity, 191
equivalence class, 3
equivalence relation, 3
equivalent, 385
equivalent statement, 385
existential quantifier, 386
Extended real number system, 51
Extreme Value Theorem, 100

F

Fatou's Lemma, 293
Fermat's Theorem, 129
finite intersection property, 43
First Fundamental Theorem of Calculus, 161
fixed point, 119
Froda's Theorem, 102
Frullani's Integral, 239
function, 2

G

Gamma function, 221
geometric series, 77
gradient of f at $\mathbf{x}$, 330
graph of f, 132
greatest integer function, 121
grid of I, 354

H

harmonic series, 81
Heaviside step function, 160
Heine-Borel Theorem, 30, 100
hypothesis, 385

I

implies, 386
Improper Integrals of the First Kind, 219
Improper Integrals of the Second Kind, 220
Improper Integrals of the Third Kind, 221
indeterminate form, 130
indirect proof, 389

Index

infinite series, 75
injective, 2
inner Jordan measure, 357
integer, 1
Integral Test for Convergence of Series, 223
Integration by Parts, 162
interior, 28
interior point, 28
Intermediate Value Theorem, 100, 129
Intermediate Value Theorem for Derivatives, 130
intersection, 1
inverse, 385
inverse function, 100
inverse map, 2
irrational number, 1

J
Jacobian matrix, 329
Jensen's inequality, 155
Jordan content, 356
Jordan measurable set, 356
Jordan measure, 356

L
L'Hôspital's Rule, 130
Lebesgue integrable function, 292
Lebesgue integral, 291
Lebesgue measurable function, 269
Lebesgue measurable set, 244
Lebesgue outer measure, 243
Lebesgue's Dominated Convergence Theorem, 294
Lebesgue's Integrability Condition, 159, 356
Lebesgue's Monotone Convergence Theorem, 294
left continuous at p, 120
left-hand derivative, 127
left-hand limit, 101
Leibniz's rule, 131
length of I, 353
Limit Comparison Test, 222
limit of function, 97
limit point, 28
limits at infinity, 102
limits of sequence, 49
Lipschitz condition, 142, 200
Lipschitz constant, 142, 200
Littlewood's Three Principles, 268, 270

local extreme of f, 128
local maximum at p, 117
local minimum at p, 117
logical connectives, 384
lower limit, 51
lower Riemann integral, 158, 355
lower Riemann sum, 355
Lusin's Theorem, 271

M
mapping, 2
mathematical statement, 384
Mean Value Theorem for Derivatives, 129, 162
measurable function, 269
measure zero, 159
metric, 27
metric space, 27
Monotonic Convergence Theorem, 50
monotonic function, 102
monotonic sequence, 50
monotonically decreasing, 50, 102, 129
monotonically increasing, 50, 102, 129

N
necessary and sufficient condition, 388
necessary condition, 387
necessity, 387
negation, 385
neighborhood, 27
nonreflexivity, 3

O
one-to-one, 2
onto, 2
open cover, 30
open set, 27, 99
order relation, 3
ordered pairs, 2
outer Jordan measure, 356

P
partial derivative, 327
partial summation formula, 78
partition, 3, 157
partition of I, 354
periodic, 116
point, 27
pointwise bounded, 190

polar coordinates, 360
popcorn function, 167
positive integer, 1
power series, 79
power set, 20
primitive function, 161
projectable region, 359
proof by contradiction, 388
proof by contrapositive, 388
proper subset, 1

R
radius, 27
radius of convergence, 79
range, 2
Ratio Test, 76
rational number, 1
real number, 1
real number line, 9
rearrangement of series, 79
refinement, 157
reflexivity, 3
relation, 3
Riemann function, 105, 167, 370
Riemann Integrability Condition, 159, 356
Riemann integrable on I, 355
Riemann integral, 158
Riemann sum, 355
Riemann-Stieltjes integral, 158
right continuous at p, 120
right-hand derivative, 127
right-hand limit, 101, 120
Root Test, 76
rule of assignment, 2
ruler function, 167

S
Sandwich Theorem, 50
Schwarz inequality, 90
Schwarz Inequality for Integral, 167
Second Fundamental Theorem of Calculus, 161
second-order partial derivatives of $\mathbf{f}$, 332
separation, 31
series, 75
set, 1
Sign-preserving Property, 193
sign-preserving Property, 112
simple discontinuity, 101

simple function, 270
spherical coordinates, 360
Squeeze Theorem for Convergent Sequences, 50, 98
statement, 384
step function, 271
strictly convex, 132
strictly decreasing, 102, 129
strictly increasing, 102, 129
subintervals in $\mathbb{R}^n$, 354
subsequence, 50
subsequential limit, 51
subset, 1
Substitution Theorem, 160
sufficiency, 387
sufficient condition, 388
supremum norm, 191
surjective, 2
symmetry, 3

T
tautologies, 386
Taylor's polynomial, 131
Taylor's Theorem, 131, 179
The Chain Rule, 329
The Change of Variables Theorem for Multiple Integrals, 360
The Implicit Function Theorem, 331
The Inverse Function Theorem, 331
The Mean Value Theorem for Differentiable Functions, 330
The Mean Value Theorem for Multiple Integrals, 359
The Simple Function Approximation Theorem, 270
Thomae's function, 167
total derivative of $\mathbf{f}$ at $\mathbf{x}$, 328
transitivity, 3
translation, 32
triple integral, 355
truth table, 384
truth value, 384

U
uncountable, 19
uniform closure, 192
uniform continuity, 99
uniformly bounded, 191
union, 1

Index

unit step function, 160
universal quantifier, 387
upper limit, 51
upper Riemann integral, 158, 355
upper Riemann sum, 355

V
value, 2

variable of integration, 158
vector space, 164
volume of I, 353

W
Weierstrass M-test, 189
Weierstrass Approximation Theorem, 192

Bibliography

[1] Z. Adamowicz and P. Zbierski, *Logic of Mathematics: A Modern Course of Classical Logic*, John Wiley & Sons Inc., 1997.

[2] T. M. Apostol, *Calculus Vol. 1*, 2nd ed., John Wiley & Sons, Inc., 1967.

[3] T. M. Apostol, *Mathematical Analysis*, 2nd ed., Addison-Wesley Publishing Company, 1974.

[4] A. G. Aksoy and M. A. Khamsi, *A Problem Book in Real Analysis*, Springer-Verlag, New York, 2009.

[5] R. G. Bartle and D. R. Sherbert, *Introduction to Real Analysis*, 4th ed., John Wiley & Sons, Inc., 2011.

[6] P. M. Fitzpatrick, *Advanced Calculus*, 2nd ed., Brooks/Cole, 2006.

[7] G. B. Folland, *Real Analysis: Modern Techniques and Their Applications*, 2nd ed., New York: Wiley, 1999.

[8] G. H. Hardy, *A Course of Pure Mathematics*, 10th ed., Cambridge University Press, 2002.

[9] M. Hata, *Problems and Solutions in Real Analysis*, Hackensack, N. J.: World Scientific, 2007.

[10] E. W. Hobson, On the Second Mean-Value Theorem of the Integral Calculus, *Proc. Lond. Math. Soc.* Ser. 2, Vol. 7, pp. 14 - 23, 1909.

[11] S. G. Krantz and H. R. Parks, *The Implicit Function Theorem: History, Theory, and Applications*, Boston: Birkhäuser, 2002.

[12] F. Jones, *Lebesgue Integration on Euclidean Space*, Rev. ed., Boston: Jones and Bartlett, 2001.

[13] D. C. Lay, *Linear Algebra and Its Applications*, 4th ed., Addison-Wesley Publishing Company, 2012.

[14] J. Lüroth, Bemerkung über gleichmässige Stetigkeit, *Math. Ann.*, Vol. 6, 1873, pp. 319, 320.

[15] P. A. Loeb and E. Talvila, Lusin's Theorem and Bochner Integration, *Scientiae Mathematicae Japonicae*, Vol. 10, pp. 55 - 62, 2004.

[16] J. R. Munkres, *Topology*, 2nd ed., Upper Saddle River, N.J.: Prentice-Hall, 2000.

[17] P. J. Olver and C. Shakiban, Applied Linear Algebra, 2nd ed., Cham: Springer International Publishing, 2018.

[18] O. de Oliveira, The Implicit and Inverse Function Theorems: Easy Proofs, *Real Anal. Exchange*, Vol. 39, No. 1, pp. 207 - 218, 2013.

[19] T.-L. T. Rădulescu, V. D. Rădulescu and T. Andreescu, *Problems in Real Analysis: Advanced Calculus on the Real Axis*, Springer-Verlag, New York, 2009.

[20] W. Rautenberg, *A Concise Introduction to Mathematical Logic*, 3rd ed., Springer-Verlag, New York, 2010.

[21] I. M. Roussos, *Improper Riemann Integrals*, Boca Raton: CRC Press, 2014.

[22] H. L. Royden and P. M. Fitzpatrick, *Real Analysis*, 4th ed., Boston: Prentice Hall, 2010.

[23] W. Rudin, *Principles of Mathematical Analysis*, 3rd ed., Mc-Graw Hill Inc., 1976.

[24] W. Rudin, *Real and Complex Analysis*, 3rd ed., Mc-Graw Hill Inc., 1987.

[25] R. Shakarchi, *Problems and Solutions for Undergraduate Analysis*, Springer-Verlag, New York, 1998.

[26] E. M. Stein and R. Shakarchi, *Real Analysis: Measure Theorey, Integration and Hilbert Spaces*, Princeton, N. J.: Princeton University Press, 2005.

[27] T. Tao, *Analysis I*, 3rd ed., Singapore: Springer, 2016.

[28] T. Tao, *An Introduction to Measure Theory*, Providence, R.I.: American Mathematical Society, 2011.

[29] W. R. Wade, *An Introduction to Analysis*, 3rd ed., Upper Saddle River: Pearson Prentice Hall, 2004.

[30] K. W. Yu, *Problems and Solutions for Undergraduate Real Analysis I*, Amazon.com, 2018.

[31] K. W. Yu, *Problems and Solutions for Undergraduate Real Analysis II*, Amazon.com, 2019.

[32] W. P. Ziemer, *Modern Real Analysis*, 2nd ed., Cham: Springer International Publishing, 2017.

[33] V. A. Zorich, *Mathematical Analysis I*, 2nd ed., Berlin: Springer, 2015.

[34] V. A. Zorich, *Mathematical Analysis II*, 2nd ed., Berlin: Springer, 2016.

Printed in Poland
by Amazon Fulfillment
Poland Sp. z o.o., Wrocław